Student Solutions Man

to accompany

FUNCTIONS MODELING CHANGE:

A Preparation for Calculus

Sixth Edition

Produced by the Calculus Consortium and initially funded by a National Science Foundation grant.

Eric Connally
Harvard University Extension

Deborah Hughes-Hallett
The University of Arizona

Andrew M. Gleason
Harvard University

Ann Davidian
Gen. Douglas MacArthur HS

Selin Kalaycıoğlu
New York University

Brigitte Lahme
Sonoma State University

Patti Frazer Lock
St. Lawrence University

Guadalupe I. Lozano
The University of Arizona

William G. McCallum
The University of Arizona

Jerry Morris
Sonoma State University

Cody Patterson
University of Texas at San Antonio

Karen Rhea
University of Michigan

Ayşe Şahin
Wright State University

Pat Shure
University of Michigan

Adam H. Spiegler
University of Colorado Denver

Carl Swenson
Seattle University

Aaron D. Wootton
University of Portland

with the assistance of

Philip Cheifetz
Nassau Community College

Douglas Quinney
University of Keele

Ellen Schmierer
Nassau Community College

Jeff Tecosky-Feldman
Haverford College

Enrique Acosta Jaramillo

Thomas W. Tucker
Colgate University

To order books or for customer service please, call 1-800-CALL WILEY (225-5945).

ISBN-13 978-1-119-56449-2

Printed in the United States of America

VC5BE4670-3F68-424F-8CC8-98A5D39B6E98_091119

CONTENTS

CHAPTER ONE

Solutions for Section 1.1

Skill Refresher

S1. Combining like terms, we see that $4x + 8x = 12x$.

S5. $2\pi r^2 + 2\pi r \cdot 2r = 2\pi r^2 + 4\pi r^2 = 6\pi r^2$.

S9. $\left(\frac{1}{2}\right) - 5(-5) = \frac{1}{2} + 25 = \frac{51}{2}$.

S13. The figure is a parallelogram, so $A = (-2, 8)$.

S17. The variable corresponding to 2 is on the horizontal axis and the variable corresponding to 6 is on the vertical axis, so the coordinates are $(2, 6)$.

EXERCISES

1. $m = f(v)$.

5. Appropriate axes are shown in Figure 1.1.

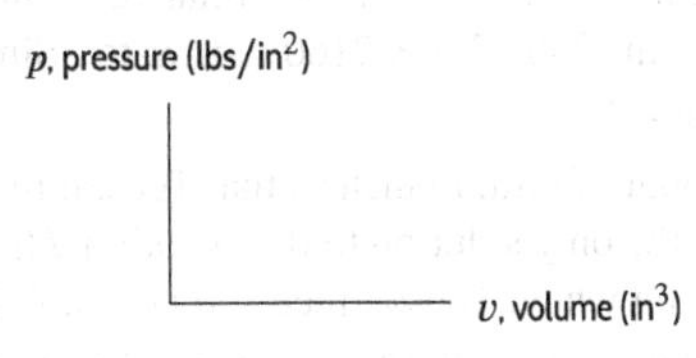

Figure 1.1

9. **(a)** Since the vertical intercept is $(0, 40)$, we have $f(0) = 40$.
(b) Since the horizontal intercept is $(2, 0)$, we have $f(2) = 0$.

13. Since $f(0) = f(4) = f(8) = 0$, the solutions are $x = 0, 4, 8$.

17. Note that the input variable is x in the expression $f(x)$, so we substitute the given input value for x, and evaluate.
(a) $f(0) = p(0) + q = q$.
(b) $f(1) = p(1) + q = p + q$.
(c) $f(2) = p(2) + q = 2p + q$.
(d) $f(-1) = p(-1) + q = -p + q$.

21. Here, m is a function of t. For any t, there is only one possible value of m. In addition, for any m, there is only one possible value of t, given by $t = m^2$. Thus, t is a function of m.

25. We apply the vertical-line test. As you can see in Figure 1.2, there is a vertical line meeting the graph in more than one point. Thus, this graph fails the vertical-line test and does not represent a function.

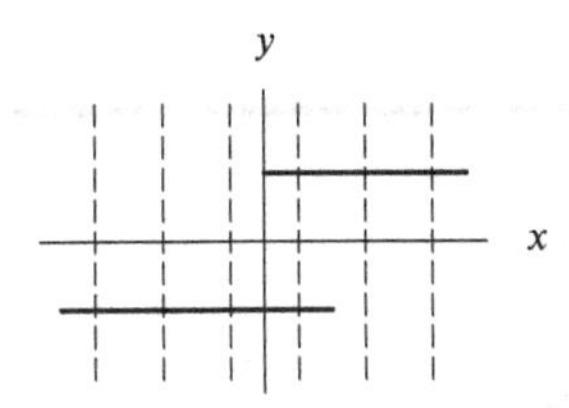

Figure 1.2

PROBLEMS

29. The value of N is not necessarily a function of G, since each value of G does not need to have a unique value of N associated to it. For example, suppose we choose the value of G to be a B. There may be more than one student who received a B, so there may be more than one ID number corresponding to B.

The value of G must be a function of N, because each ID number (each student) receives exactly one grade. Therefore each value of N has a unique value of G associated with it. Writing $G = f(N)$ indicates that the ID number is the input which uniquely determines the grade, the output.

33. **(a)** In 2010, or year $t = 0$, the ranking for Olivia was 4, making it most popular, and the ranking for Charlotte was 46, making it least popular.

(b) In 2015, or year $t = 5$, the ranking for Olivia was 2, making it still the most popular, and the ranking for Madison was 11, making it least popular.

37. **(a)** (i) The mass of water in 1 kg of air that has 100% relative humidity at 30° C is approximately 30 g.

(ii) The mass of water in 1 kg of air that has 50% relative humidity at 30° C is approximately 15 g. We can either read this from the graph of or use the answer to the previous part.

(iii) The mass of water in 1 kg of air that has 75% relative humidity at 30° C is approximately 22.5 g since it will be half way between the amount of water we found at 100% and 50% humidity.

(b) At 50% relative humidity, 1 kg of 20°air contains approximately 7.5 g of water. Therefore, 300 kg of air contains $7.5 \cdot 300 = 2250$ g of water. Since 1 kg = 1 liter of water, this is 2.25 liters.

(c) The answer will depend on the dimensions of the classroom. A classroom that is 8 meters by 10 meters with a height of 3 meters has volume $8 \cdot 10 \cdot 3 = 240\text{ m}^3$. Given that the density of air is approximately 1.2 kg/m^3, the mass of air in the classroom is $240\text{ m}^3 \cdot 1.2\text{ kg/m}^3 = 288$ kg. At 50% relative humidity and 20°C, 1 kg of air contains 7.5 g of water. Therefore the classroom contains $288 \cdot 7.5 = 2160$ g of water. Since 1000 g = 1 l of water, 2160 g = 2.16 l of water, which is a little over half a gallon.

41. **(a)** The number $f(5) = 0.35$ is the amount of rain, in inches, that Tucson received in the month of May.

(b) From the table, we see that $t = 1$ is the only solution to the equation $f(t) = 0$. This solution indicates that January is the only month among our data during which Tucson received no rainfall.

(c) From the table, we see that $t = 2$ and $t = 4$ are the only solutions to $f(t) = 0.1$. These solutions indicate that in February and April, Tucson received 0.1 inches of rain.

45. **(a)** No, in the year 1954 there were two world records; in the year 1981 there were three world records.

(b) Yes, each world record occurred in only one year.

(c) The world record of 3 minutes and 47.33 seconds was set in 1981.

(d) The statement $y(3{:}51.1) = 1967$ tells us that the world record of 3 minutes, 51.1 seconds was set in 1967.

49. **(a)** Adding the male total to the female total gives $x + y$, the total number of applicants.

(b) Of the men who apply, 15% are accepted. So $0.15x$ male applicants are accepted. Likewise, 18% of the women are accepted so we have $0.18y$ women accepted. Summing the two tells us that $0.15x + 0.18y$ applicants are accepted.

(c) The number accepted divided by the number who applied times 100 gives the percentage accepted. This expression is

$$\frac{(0.15)x + (0.18)y}{x + y}(100), \quad \text{or} \quad \frac{15x + 18y}{x + y}.$$

Solutions for Section 1.2

Skill Refresher

S1. $\frac{4-6}{3-2} = \frac{-2}{1} = -2.$

S5. $\frac{\frac{1}{2}-(-4)^2-\left(\frac{1}{2}-(5^2)\right)}{-4-5} = \frac{\frac{1}{2}-16-\frac{1}{2}+25}{-9} = \frac{9}{-9} = -1.$

S9. $\frac{x^2-\frac{3}{4}-\left(y^2-\frac{3}{4}\right)}{x-y} = \frac{x^2-\frac{3}{4}-y^2+\frac{3}{4}}{x-y} = \frac{x^2-y^2}{x-y} = \frac{(x-y)(x+y)}{x-y} = x + y.$

S13. **(a)** We see that $g(0) = 100$ and $g(5) = 25$ so $g(0)$ is larger.

(b) We have $g(5) - g(0) = 25 - 100 = -75.$

EXERCISES

1. The function is increasing for $x > 0$, since the graph rises there as we move to the right. The function is decreasing for $x < 0$, since the graph falls as we move to the right.

5. **(a)** Let $s = V(t)$ be the sales (in millions) of feature phones in year t. Then

$$\begin{aligned}\text{Average rate of change of } s \text{ from } t = 2012 \text{ to } t = 2014 &= \frac{\Delta s}{\Delta t} = \frac{V(2014) - V(2012)}{2014 - 2012}\\ &= \frac{634 - 1110}{2}\\ &= -238 \text{ million feature phones/year.}\end{aligned}$$

Let $q = D(t)$ be the sales (in millions) of smartphones in year t. Then

$$\begin{aligned}\text{Average rate of change of } q \text{ from } t = 2012 \text{ to } t = 2014 &= \frac{\Delta q}{\Delta t} = \frac{D(2014) - D(2012)}{2014 - 2012}\\ &= \frac{1245 - 681}{2}\\ &= 282 \text{ million smartphones/year.}\end{aligned}$$

(b) By the same argument

$$\begin{aligned}\text{Average rate of change of } s \text{ from } t = 2014 \text{ to } t = 2015 &= \frac{\Delta s}{\Delta t} = \frac{V(2015) - V(2014)}{2015 - 2014}\\ &= \frac{481 - 634}{1}\\ &= -153 \text{ million feature phones/year.}\end{aligned}$$

$$\begin{aligned}\text{Average rate of change of } q \text{ from } t = 2014 \text{ to } t = 2015 &= \frac{\Delta q}{\Delta t} = \frac{D(2015) - D(2014)}{2015 - 2014}\\ &= \frac{1462 - 1245}{1}\\ &= 217 \text{ million smartphones /year.}\end{aligned}$$

(c) The fact that $\Delta s/\Delta t = -238$ tells us that feature phone sales decreased at an average rate of 238 million feature phones/year between 2012 and 2014. The fact that the average rate of change is negative tells us that annual sales are decreasing.

The fact that $\Delta s/\Delta t = -153$ tells us that feature phone sales decreased at an average rate of 153 million feature phones/year between 2014 and 2015.

The fact that $\Delta q/\Delta t = 282$ means that smartphone sales increased at an average rate of 282 million players/year between 2012 and 2014. The fact that $\Delta q/\Delta t = 217$ means that smartphone sales increased at an average rate of 217 million smartphones/year between 2014 and 2015.

9. **(a)** (i) The average rate of change is

$$\frac{f(1) - f(0)}{1 - 0} = \frac{46.7 - 60}{1} = -13.3 \text{ microcoulombs per second.}$$

(ii) The average rate of change is

$$\frac{f(2) - f(1)}{2 - 1} = \frac{36.4 - 46.7}{1} = -10.3 \text{ microcoulombs per second.}$$

(iii) The average rate of change is

$$\frac{f(3) - f(2)}{3 - 2} = \frac{28.3 - 36.4}{1} = -8.1 \text{ microcoulombs per second.}$$

(iv) The average rate of change is

$$\frac{f(4)-f(3)}{4-3} = \frac{22.1-28.3}{1} = -6.2 \text{ microcoulombs per second.}$$

(b) As time goes on, the average rates of change in part (a) become less negative. This means that the rate of charge dissipation in the capacitor decreases as time goes on.

13. f appears to be increasing for x between 0 and 2.2 and for x between 4 and 6.1, so it is increasing on the intervals $(0, 2.2)$, and $(4, 6.1)$. f appears to be decreasing for x between 2.2 and 4 and for x between 6.1 and 8, so it is decreasing on the intervals $(2.2, 4)$, $(6.1, 8)$.

17. They are equal; both are given by

$$\frac{4.9-2.9}{6.1-2.2}.$$

PROBLEMS

21. **(a)** The function is increasing from, approximately, days 10 through 24, 39 through 54, and 69 through 84.
(b) The function is decreasing from, approximately, days 25 through 38, 55 through 68, and 85 through 98.

25. The average rate of change between $v = 30$ and $v = 35$ is

$$\frac{f(35)-f(30)}{35-30} = \frac{18(35)^2-18(30)^2}{5} = \frac{5850}{5} = 1170 \text{ kilojoules per meter per second.}$$

This means that, as the speed of the truck increases from 30 to 35 meters per second, the kinetic energy of the truck increases at an average rate of 1170 kilojoules for each meter per second by which the speed of the truck increases.

29.

$$\begin{aligned}\text{Average rate of change} &= \frac{\text{Temp at 7:32 am} - \text{Temp at 7:30 am}}{7{:}32 - 7{:}30}\\ &= \frac{45-(-4)}{2} = 24.5 \text{ degrees/minute.}\end{aligned}$$

33. According to the table in the text, the tree has $139\mu g$ of carbon-14 after 3000 years from death and $123\mu g$ of carbon-14 after 4000 years from death. Because the function $L = g(t)$ is decreasing, the tree must have died between 3000 and 4000 years ago.

37. **(a)** Since Δt refers to the change in the numbers of years, we calculate

$$\Delta t = 1970 - 1960 = 10, \quad \Delta t = 1980 - 1970 = 10, \quad \text{and so on until 2000.}$$

So from 1960 to 2000, $\Delta t = 10$ for all consecutive entries. From 2000 to 2010, $\Delta t = 5$.

(b) Since ΔG is the change in the amount of garbage produced per year, for the period 1960-1970 we have

$$\Delta G = 121.1 - 88.1 = 33.$$

Continuing in this way gives the Table 1.1:

Table 1.1

Time period	1960–70	1970-80	1980–90	1990–2000	2000–2005	2005–2010
ΔG	33	30.5	56.7	34.2	10.2	−2.8

(c) Not all of the ΔG values are the same. We also know that not all the values of Δt are the same. Computing $\Delta G/\Delta t$, we see the average rate of change in the amount of garbage produced each year, is not constant. This tells us that the amount of garbage being produced each year is changing, but not at a constant rate.

(d) In 2007 the United States embarked on a recycling and composting program.

Solutions for Section 1.3

Skill Refresher

S1. We have $f(0) = \frac{2}{3}(0) + 5 = 5$ and $f(3) = \frac{2}{3}(3) + 5 = 2 + 5 = 7$.

S5. To find the y-intercept, we let $x = 0$,

$$\begin{aligned} y &= -4(0) + 3 \\ &= 3. \end{aligned}$$

To find the x-intercept, we let $y = 0$,

$$\begin{aligned} 0 &= -4x + 3 \\ 4x &= 3 \\ x &= \frac{3}{4}. \end{aligned}$$

S9. The constant term is -4.18 and the coefficient of x is 1.35.

S13. Combining like terms we get

$$\frac{7}{2} - 2x.$$

Hence the constant term is $\frac{7}{2}$ and the coefficient of x is -2.

EXERCISES

1. (a) Since the slopes are 2 and 3, we see that $y = -2 + 3x$ has the greater slope.
   (b) Since the y-intercepts are -1 and -2, we see that $y = -1 + 2x$ has the greater y-intercept.

5. The values of x given in the table increases by $\Delta x = 10$ each time. But the corresponding values of $h(x)$ do not change by a constant amount each time. For example $h(x)$ increases by 20 when x change by 10 between 0 and 10, but $h(x)$ increases by 5 when x changes by 10 between 10 and 20. Since equal increments in x do not correspond to equal increments in $h(x)$ the function is not linear.

9. This table could represent a linear function because the rate of change of $p(\gamma)$ is constant. Between consecutive data points, $\Delta\gamma = -1$ and $\Delta p(\gamma) = 10$. Thus, the rate of change is $\Delta p(\gamma)/\Delta\gamma = -10$. Since this is constant, the function could be linear.

   As we just observed, the rate of change is

$$\frac{\Delta p(\gamma)}{\Delta \gamma} = \frac{10}{-1} = -10.$$

13. The vertical intercept is 29.99, which tells us that the company charges \$29.99 per month for the phone service, even if the person does not talk on the phone. The slope is 0.05. Since

$$\text{Slope} = \frac{\Delta\text{cost}}{\Delta\text{minutes}} = \frac{0.05}{1},$$

we see that, for each minute the phone is used, it costs an additional \$0.05.

PROBLEMS

17. Since the depreciation can be modeled linearly, we can write the formula for the value of the car, V, in terms of its age, t, in years, by the following formula:

$$V = b + mt.$$

Since the initial value of the car is \$26,500, we know that $b = 26{,}500$.

Hence,

$$V = 26{,}500 + mt.$$

To find m, we know that $V = 18{,}993$ when $t = 3$, so

$$\begin{aligned}19{,}999 &= 26{,}500 + m(3)\\ -6{,}501 &= 3m\\ \frac{-6{,}501}{3} &= m\\ -2167 &= m.\end{aligned}$$

So, $V = 26{,}500 - 2167t$.

21. **(a)** Any line with a slope of 2.1, using appropriate scales on the axes. The horizontal axis should be labeled "days" and the vertical axis should be labeled "inches." See Figure 1.3.

(b) Any line with a slope of -1.3, using appropriate scales on the axes. The horizontal axis should be labeled "miles" and the vertical axis should be labeled "gallons." See Figure 1.4.

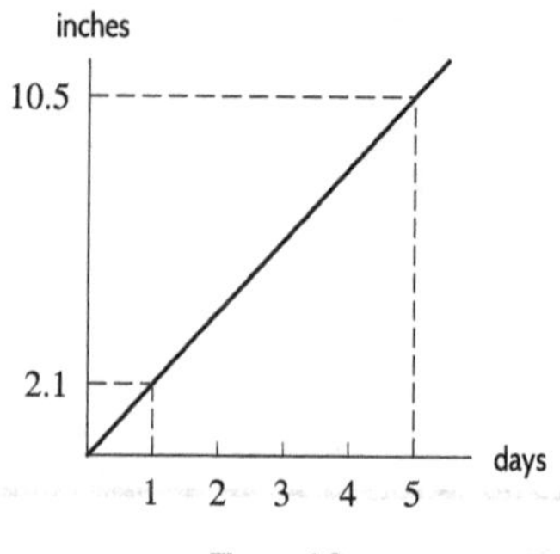

Figure 1.3

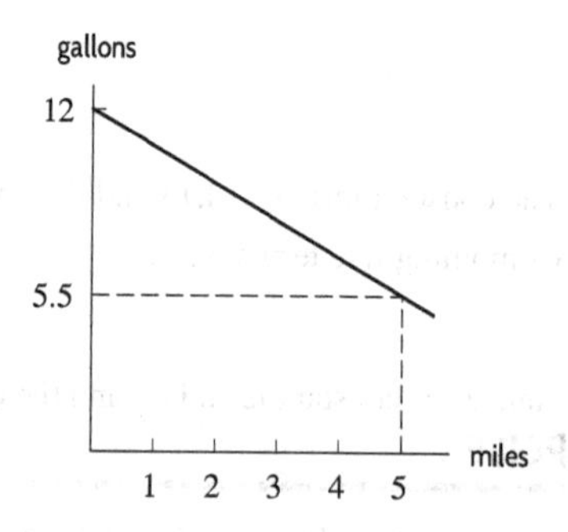

Figure 1.4

25. **(a)** We see that the population of Country B grows at the constant rate of roughly 2 million every ten years. Thus, Country B must be Australia. The population of country A did not change at a constant rate: In the ten years of 1970–1980 the population of Country A grew by 2.09 million while in the ten years of 1980–1990 its population dropped. Thus, Country A is Afghanistan.

(b) The rate of change of Country B is found by taking the population increase and dividing it by the corresponding time in which this increase occurred. Thus

$$\text{Rate of change of population} = \frac{22.16 - 10.29}{2010 - 1950} = \frac{11.87 \text{ million people}}{60 \text{ years}} = 0.20 \text{ million people/year.}$$

This rate of change tells us that on the average, the population of Australia increases by 0.20 million people every year.

(c) In 2010 the population of Australia was 22.16 million. If the population grows by 0.20 million every year, then in the eight years from 2010 to 2018

$$\text{Population increase} = 8 \cdot 0.20 \text{ million} = 1.60 \text{ million.}$$

Thus in 2018

$$\text{Population of Australia} = 22.16 + 1.60 \text{ million} \approx 23.76 \text{ million.}$$

29. **(a)** Since C is 8, we have $T = 300 + 200C = 300 + 200(8) = 1900$. Thus, taking 8 credits costs \$1900.

(b) Here, the value of T is 1700 and we solve for C.

$$\begin{aligned}T &= 300 + 200C\\ 1700 &= 300 + 200C\\ 7 &= C\end{aligned}$$

Thus, \$1, 700 is the cost of taking 7 credits.

(c) Table 1.2 is the table of costs.

Table 1.2

C	1	2	3	4	5	6	7	8	9	10	11	12
T	500	700	900	1100	1300	1500	1700	1900	2100	2300	2500	2700
$\frac{T}{C}$	500	350	300	275	260	250	243	238	233	230	227	225

(d) The largest value for C, that is, 12 credits, gives the smallest value of T/C. In general, the ratio of tuition cost to number of credits is getting smaller as C increases.

(e) The fixed cost of tuition (such as fees, registration, etc.), which is independent of the number of credits taken.

(f) The 200 represents the rate of change of cost with the number of credit hours. In other words, the cost of taking one additional credit hour.

33. As Figure 1.5 shows, the graph of $y = 2x + 400$ does not appear in the window $-10 \le x \le 10, -10 \le y \le 10$. This is because all the corresponding y-values are between 380 and 420, which are outside this window. The graph can be seen by using a different viewing window: for example, $380 \le y \le 420$.

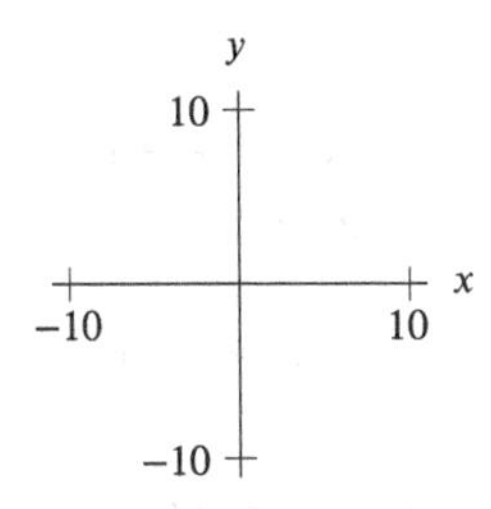

Figure 1.5

Solutions for Section 1.4

Skill Refresher

S1.

$$\begin{aligned} y - 5 &= 21 \\ y &= 26. \end{aligned}$$

S5. We first distribute $\frac{5}{3}(y+2)$ to obtain:

$$\begin{aligned} \frac{5}{3}(y+2) &= \frac{1}{2} - y \\ \frac{5}{3}y + \frac{10}{3} &= \frac{1}{2} - y \\ \frac{5}{3}y + y &= \frac{1}{2} - \frac{10}{3} \\ \frac{5}{3}y + \frac{3y}{3} &= \frac{3}{6} - \frac{20}{6} \\ \frac{8y}{3} &= -\frac{17}{6} \\ \left(\frac{3}{8}\right)\frac{8y}{3} &= \left(\frac{3}{8}\right)\left(-\frac{17}{6}\right) \\ y &= -\frac{17}{16}. \end{aligned}$$

S9. We collect all terms involving x and then divide by $2a$:

$$\begin{aligned} ab + ax &= c - ax \\ 2ax &= c - ab \\ x &= \frac{c - ab}{2a}. \end{aligned}$$

S13. Given that $f(x) = 0$, we have:

$$\begin{aligned} 7x + 1 &= 0 \\ 7x &= -1 \\ x &= -\frac{1}{7}. \end{aligned}$$

EXERCISES

1. We have the slope $m = -4$ so

$$y = b - 4x.$$

The line passes through $(7, 0)$ so

$$\begin{aligned} 0 &= b + (-4)(7) \\ 28 &= b \end{aligned}$$

and

$$y = 28 - 4x.$$

5. Since we know the x-intercept and y-intercepts are $(3, 0)$ and $(0, -5)$ respectively, we can find the slope:

$$\text{slope} = m = \frac{-5 - 0}{0 - 3} = \frac{-5}{-3} = \frac{5}{3}.$$

We can then put the slope and y-intercept into the general equation for a line.

$$y = -5 + \frac{5}{3}x.$$

9. **(a)** is (V), because slope is positive, vertical intercept is negative
(b) is (IV), because slope is negative, vertical intercept is positive
(c) is (I), because slope is 0, vertical intercept is positive
(d) is (VI), because slope and vertical intercept are both negative
(e) is (II), because slope and vertical intercept are both positive
(f) is (III), because slope is positive, vertical intercept is 0
(g) is (VII), because it is a vertical line with positive x-intercept.

13. Since the function is linear, we can use any two points to find its formula. We use the form

$$y = b + mx$$

to get temperature in °C, y, as a function of temperature in °F, x. We use the two points, $(32, 0)$ and $(41, 5)$. We begin by finding the slope, $\Delta y/\Delta x = (5 - 0)/(41 - 32) = 5/9$. Next, we substitute a point into our equation using our slope of $5/9$°C per °F and solve to find b, the y-intercept. We use the point $(32, 0)$:

$$\begin{aligned} 0 &= b + \frac{5}{9} \cdot 32 \\ -\frac{160}{9} &= b. \end{aligned}$$

Therefore,

$$y = -\frac{160}{9} + \frac{5}{9}x.$$

Traditionally, we give this formula as $y = (5/9)(x - 32)$, which is often easier to manipulate. You might want to check to see if the two are the same.

17. Since the function is linear, we can use any two points (from the graph) to find its formula. We use the form

$$s = b + mq$$

to get the number of hours of sleep obtained as a function of the quantity of tea drunk. We use the two points (4, 7) and (12, 3). We begin by finding the slope, $\Delta s/\Delta q = (3-7)/(12-4) = -0.5$. Next, we substitute a point into our equation using our slope of −0.5 hours of sleep per cup of tea and solve to find b, the s-intercept. We use the point (4, 7):

$$\begin{aligned} 7 &= b - 0.5 \cdot 4 \\ 9 &= b. \end{aligned}$$

Therefore,

$$s = 9 - 0.5q.$$

21. Rewriting in slope-intercept form:

$$\begin{aligned} 5x - 3y + 2 &= 0 \\ -3y &= -2 - 5x \\ y &= \frac{-2}{-3} - \frac{5}{-3}x \\ y &= \frac{2}{3} + \frac{5}{3}x \end{aligned}$$

25. Not possible, the slope is not defined (vertical line).

29. The function is not linear because the power of s is not 1.

33. These lines are parallel because they have the same slope, 5.

37. These lines are neither parallel nor perpendicular. They do not have the same slopes, nor are their slopes negative reciprocals (if they were, one of the slopes would be negative).

PROBLEMS

41. The starting value is $b = 12{,}000$, and the growth rate is $m = 225$, so $h(t) = 12{,}000 + 225t$.

45. $y = 5x - 3$. Since the slope of this line is 5, we want a line with slope $-\frac{1}{5}$ passing through the point (2, 1). The equation is $(y-1) = -\frac{1}{5}(x-2)$, or $y = -\frac{1}{5}x + \frac{7}{5}$.

49. Since P is the x-intercept, we know that point P has y-coordinate = 0, and if the x-coordinate is x_0, we can calculate the slope of line l using $P(x_0, 0)$ and the other given point $(0, -2)$.

$$m = \frac{-2-0}{0-x_0} = \frac{-2}{-x_0} = \frac{2}{x_0}.$$

We know this equals 2, since l is parallel to $y = 2x + 1$ and therefore must have the same slope. Thus we have

$$\frac{2}{x_0} = 2.$$

So $x_0 = 1$ and the coordinates of P are (1, 0).

53. (a) Since the starting temperature of the cobalt sample is 20° Celsius, we have $T = 20$ when $h = 0$. Since 10 calories of heat energy raise the temperature of the sample by 1° Celsius, 20 calories of heat energy raise the temperature by 2° Celsius, so $T = 22$ when $h = 20$. Similarly, $T = 24$ when $h = 40$ and $T = 26$ when $h = 60$. The result of these observations is shown in Table 1.3.

Table 1.3

h (calories)	0	20	40	60
T (°C)	20	22	24	26

(b) Since the values of h and T are both equally spaced in the table from part (a), we know that T is a linear function of h. In addition, since $T = 20$ when $h = 0$, we see that the vertical intercept is 20, so the linear equation has the form $T = 20 + mh$. The slope of the line is

$$m = \frac{\Delta T}{\Delta h} = \frac{22-20}{20-0} = \frac{1}{10},$$

so our formula is

$$T = 20 + \frac{1}{10}h.$$

(c) We want to find the value of h when $T = 95°$ Celsius. We therefore substitute $T = 95$ into our formula in part (b) and solve for h:

$$\begin{aligned} 20 + \frac{1}{10}h &= 95 \\ \frac{1}{10}h &= 75 \\ h &= 750. \end{aligned}$$

Therefore, 750 calories of heat energy must be applied to the cobalt sample to raise its temperature to 95° Celsius.

57. (a) A table of the allowable combinations of sesame and poppy-seed rolls is shown below.

Table 1.4

s, sesame seed rolls	0	1	2	3	4	5	6	7	8	9	10	11	12
p, poppy-seed rolls	12	11	10	9	8	7	6	5	4	3	2	1	0

(b) The sum of s and p is 12. So we can write $s + p = 12$, or $p = 12 - s$.

(c)

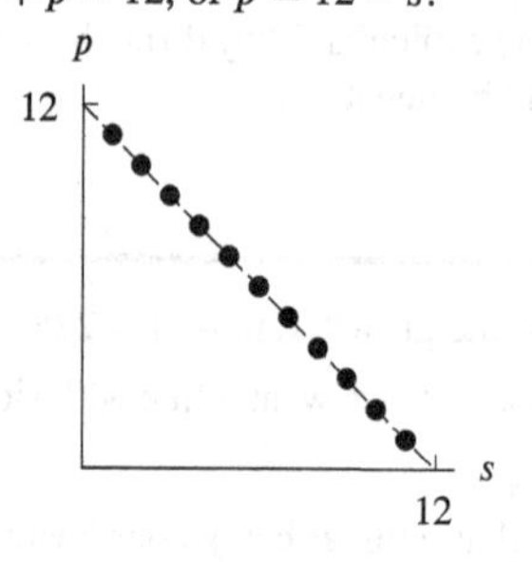

Figure 1.6

61. Using our formula for j, we have

$$\begin{aligned} h(-2) &= j(-2) = 30(0.2)^{-2} = 750 \\ h(1) &= j(1) = 30(0.2)^{1} = 6. \end{aligned}$$

This means that $h(t) = b + mt$ where

$$m = \frac{h(1) - h(-2)}{1-(-2)} = \frac{6-750}{3} = -248.$$

Solving for b, we have

$$\begin{aligned} h(-2) &= b - 248(-2) \\ b &= h(-2) + 248(-2) = 750 + 248(-2) = 254, \end{aligned}$$

so $h(t) = 254 - 248t$.

65. **(a)** Since i is linear, we can write

$$i(x) = b + mx.$$

Since $i(10) = 25$ and $i(20) = 50$, we have

$$m = \frac{50 - 25}{20 - 10} = 2.5.$$

So,

$$i(x) = b + 2.5x.$$

Using $i(10) = 25$, we can solve for b:

$$\begin{aligned} i(10) &= b + 2.5(10) \\ 25 &= b + 25 \\ b &= 0. \end{aligned}$$

Our formula then is

$$i(x) = 2.5x.$$

(b) The increase in risk associated with *not* smoking is $i(0)$. Since there is no increase in risk for a non-smoker, we have $i(0) = 0$.

(c) The slope of $i(x)$ tells us that the risk increases by a factor of 2.5 with each additional cigarette a person smokes per day.

69. Here, r is the independent variable, so we write

$$\begin{aligned} w(r) &= \pi x^2 - 3xr - 4rs - s\sqrt{x} \\ &= \pi x^2 - s\sqrt{x} - 3xr - 4rs \\ &= \underbrace{\pi x^2 - s\sqrt{x}}_{b} + \underbrace{(-3x - 4s)}_{m}\, r \end{aligned}$$

$$\text{so} \quad b = \pi x^2 - s\sqrt{x}$$
$$m = -3x - 4s.$$

Solutions for Section 1.5

Skill Refresher

S1. We set the equations $y = x$ and $y = 3 - x$ equal to one another.

$$\begin{aligned} x &= 3 - x \\ 2x &= 3 \\ x &= \frac{3}{2} \quad \text{and} \quad y = \frac{3}{2}. \end{aligned}$$

So the point of intersection is $x = 3/2$, $y = 3/2$.

S5. Multiplying the first equation by -2 and adding the two equations we get

$$0 = 3.$$

So this system has no solution.

S9. Subtract $2x$, then subtract 1, and then divide by 3.

$$\begin{aligned} 5x + 1 &> 2x + 10 \\ 3x + 1 &> 10 \\ 3x &> 9 \\ \frac{3x}{3} &> \frac{9}{3} \\ x &> 3. \end{aligned}$$

S13. Subtract $7x$, then subtract 5, and then divide by -5 with a direction change.

$$\begin{aligned} 2x + 5 &< 7x - 3 \\ -5x + 5 &< -3 \\ -5x &< -8 \\ \frac{-5x}{-5} &> \frac{-8}{-5} \\ x &> \frac{8}{5} \end{aligned}$$

EXERCISES

1. The functions f and g have the same y-intercept, $b = 20$. u and v both have y-intercept $b = 60$. f and g are increasing functions, with slopes $m = 2$ and $m = 4$, respectively. u and v are decreasing functions, with slopes $m = -1$ and $m = -2$, respectively.

The figure shows that graphs A and B describe increasing functions with the same y-intercept. The functions f and g are good candidates since they are both linear functions with positive slope and their y-intercepts coincide. Since graph A is steeper than graph B, the slope of A is greater than the slope of B. The slope of g is larger than the slope of f, so graph A corresponds to g and graph B corresponds to f.

Graphs D and E describe decreasing functions with the same y-intercept. u and v are good candidates since they both have negative slope and their y-intercepts coincide. Graph E is steeper than graph D. Thus, graph D corresponds to u, and graph E to v. Note that graphs D and E start at a higher point on the y-axis than A and B do. This corresponds to the fact that the y-intercept $b = 60$ of u and v is above the y-intercept $b = 20$ of f and g.

This leaves graph C and the function h. The y-intercept of h is -30, corresponding to the fact that graph C starts below the x-axis. The slope of h is 2, the same slope as f. Since graph C appears to climb at the same rate as graph B, it seems reasonable that f and h should have the same slope.

5. We have $C = f(d)$ so the slope of the linear function is the change in C over the change in d. The units are C-units over d-units, so the answer is dollars per day.

9. At the point of intersection, the two y-values must be the same, so we set the two expressions for y equal to each other, and solve for x:

$$\begin{aligned} 2x + 10 &= -3x - 7 \\ 5x &= -17 \\ x &= -3.4 \end{aligned}$$

We see that the lines intersect at $x = -3.4$. We can use either of the linear equations to find the corresponding y-value. Using the top equation, we have $y = 2(-3.4) + 10 = 3.2$. (We can check our work by noting that we get the same value for y using the bottom equation: $y = -3(-3.4) - 7 = 3.2$.) The coordinates of the point of intersection are $(-3.4, 3.2)$.

13. Since the number of gallons is changing at a constant rate, we use a linear function to model this relationship. The value at time 0 is 4, so the vertical intercept is 4. The constant rate of change is 10, so the slope is 10. We have

$$V = 4 + 10t.$$

PROBLEMS

17. **(a)** After one year, the value of the Frigbox refrigerator is $\$950 - \$50 = \$900$; after two years, its value is $\$950 - 2 \cdot \$50 = \$850$; after t years, the value, V, of the Frigbox is given by

$$V = 950 - t \cdot 50 \quad \text{or} \quad V = 950 - 50t.$$

Similarly, after t years, the value of the ArcticAir refrigerator is

$$V = 1200 - 100t.$$

The two refrigerators have equal value when

$$\begin{aligned} 950 - 50t &= 1200 - 100t \\ -250 &= -50t \\ 5 &= t. \end{aligned}$$

In five years the two refrigerators have equal value.

(b) According to the formula, in 20 years time, the value of the Frigbox refrigerator will be

$$\begin{aligned} V &= 950 - 50(20) \\ &= 950 - 1000 = -50 \end{aligned}$$

This negative value is not realistic, so after some time, the linear model is no longer appropriate. Similarly, the value of the ArcticAir refrigerator is predicted to be $V = 1200 - 100(20) = 1200 - 2000 = -800$, which is also not realistic.

21. Let t be the number of hours he waits. Since the BAC decreases linearly by 0.015% every hour, we see that

$$\underbrace{\text{BAC}}_{B} = \underbrace{\text{Starting amount}}_{0.13\%} - \underbrace{\text{Amount per hour}}_{0.015\%} \times \underbrace{\text{Number of hours}}_{t}$$

$$\text{so} \quad B = 0.13\% - (0.015\%)t.$$

To fall below the legal limit, we require:

$$0.13\% - (0.015\%)t < 0.08\%.$$

Solving for t gives:

$$\begin{aligned} 0.13\% - (0.015\%)t &< 0.08\% \\ 0.05\% - (0.015\%)t &< 0 && \text{subtract 0.08\% from both sides} \\ 0.05\% &< (0.015\%)t && \text{add } 0.015\%t \text{ to both sides} \\ 3.333 &< t. && \text{divide both sides by 0.015\%} \end{aligned}$$

Thus, we must wait for more than 3.333 hours, or 3 hours 20 minutes, in order to fall below the legal limit.

25. **(a)** $P = (a, 0)$

(b) $A = (0, b)$, $B = (-c, 0)$
$C = (a + c, b)$, $D = (a, 0)$

29.

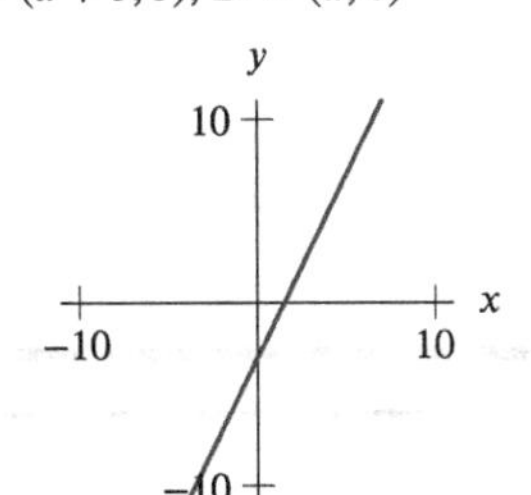

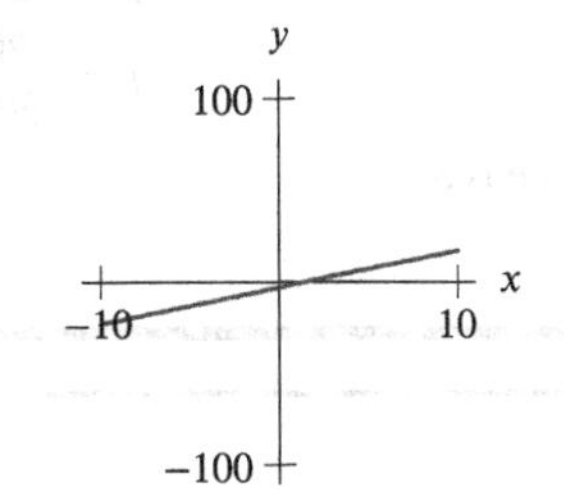

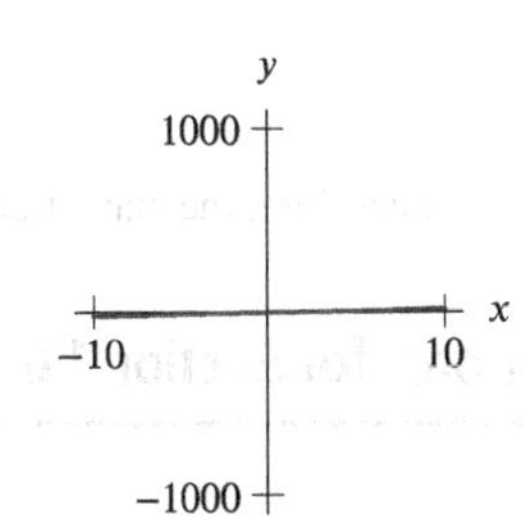

Figure 1.7

(d) If the width of the window remains constant and the height of the window increases, then the graph will appear less steep.

33. Writing this equation as

$$y = \underbrace{\frac{1}{\beta - 3}}_{m} \cdot x + \underbrace{\frac{1}{6 - \beta}}_{b},$$

we see that $m = 1/(\beta - 3)$ and $b = 1/(6 - \beta)$. For the slope to be positive, we require the denominator of m to be positive, so $\beta > 3$. For the y-intercept to be positive, we require the denominator of b to be positive, so $\beta < 6$. Putting together these two requirements gives $3 < \beta < 6$.

37. **(a)** Since the relationship is linear, the general formula for S in terms of p is

$$S = b + mp.$$

Since we know that the quantity supplied rises by 50 units when the rise in the price is \$0.50, we canwrite $\Delta S = 50$ units, when $\Delta p =$\$0.50. The slope is then:

$$m = \frac{\Delta S}{\Delta p} = \frac{50\text{ units}}{\$0.50} = 100\text{ units/dollar}.$$

Put this value of the slope into the formula for S and solve for b using $p = 2$ and $S = 100$:

$$\begin{aligned} S &= b + mp \\ 100 &= b + (100)(2) \\ 100 &= b + 200 \\ b &= -100. \end{aligned}$$

We now have the slope m and the S-intercept b. So, we know that

$$S = -100 + 100p.$$

(b) The slope in this problem is 100 units/dollar, which means that for every increase of \$1 in price, suppliers are willing to supply another 100 units.

(c) The price below which suppliers will not supply the good is represented by the point at which $S = 0$. Putting $S = 0$ into the equation found in (b) we get:

$$\begin{aligned} 0 &= -100 + 100p \\ 100 &= 100p \\ p &= 1. \end{aligned}$$

So when the price is \$1, or less, the suppliers will not want to produce anything.

(d) From Problem 36 we know that

$$D = 1100 - 200p.$$

To find when supply equals demand set the formulas for S and D equal and solve for p:

$$\begin{aligned} S &= D \\ -100 + 100p &= 1100 - 200p \\ 100p + 200p &= 1100 + 100 \\ 300p &= 1200 \\ p &= \frac{1200}{300} = 4. \end{aligned}$$

Therefore, the market clearing price is \$4.

Solutions for Section 1.6

Skill Refresher

S1. The wages you earn is a function of the number of hours worked, so the number hours is the independent variable and goes on the horizontal axis. See Figure 1.8.

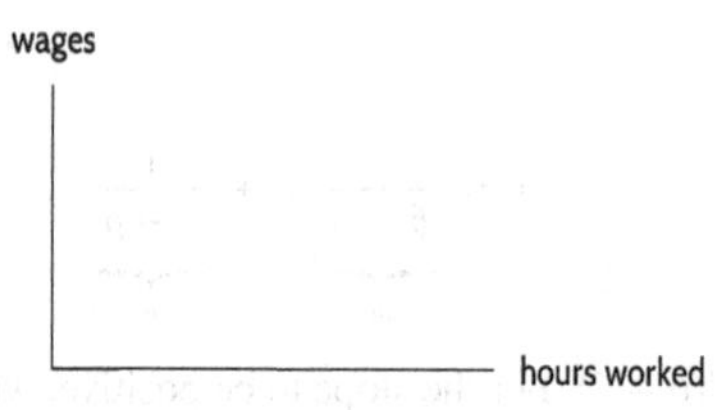

Figure 1.8

S5. **(a)** We see that y is the value of the laptop, and x is its age.
(b) We have $a = 750$. The units of a are dollars, the same units as the value of the laptop y.
(c) We have $b = -150$. Since x is age in years, and y is value, in dollars, the units of b must be dollars per year.

EXERCISES

1. These points are very close to a line with negative slope, so r is negative and $r = 1$ is not reasonable. (In fact, $r = -0.998$.)

5. A scatter plot of the data is shown in Figure 1.9. The value $r = 0.9$ is not reasonable. These points are very close to a line with negative slope, so r is negative. (In fact, $r = -0.99$.)

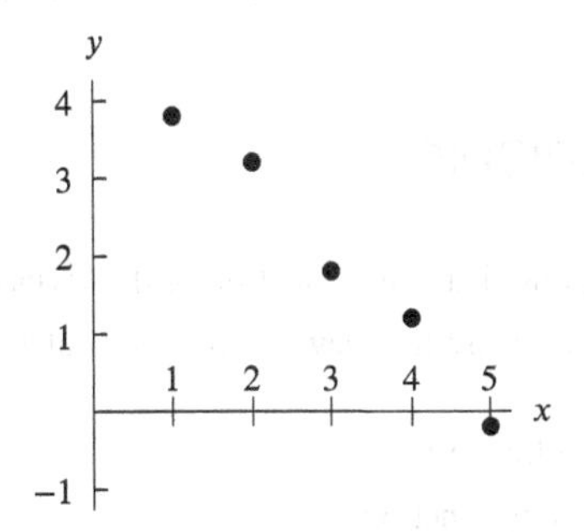

Figure 1.9

PROBLEMS

9. **(a)** When $n = 700$, we have $p = 4(700) - 2200 = 600$. When 700 people visit the park in a week, the profit is predicted to be \$600.
(b) The slope of 4 means that for every weekly visitor, the profit increases by \$4. The units are dollars per person.
(c) The vertical intercept of -2200 means that when nobody visits the park, that is, the number of weekly visitors is zero, the park loses \$2200.
(d) We solve $4n - 2200 = 0$ to find when the park has a profit of \$0. This happens when $n = 550$. With 550 weekly visitors, the park neither makes nor loses money. If more visitors come, it makes money.

13. **(a)** See Figure 1.10.

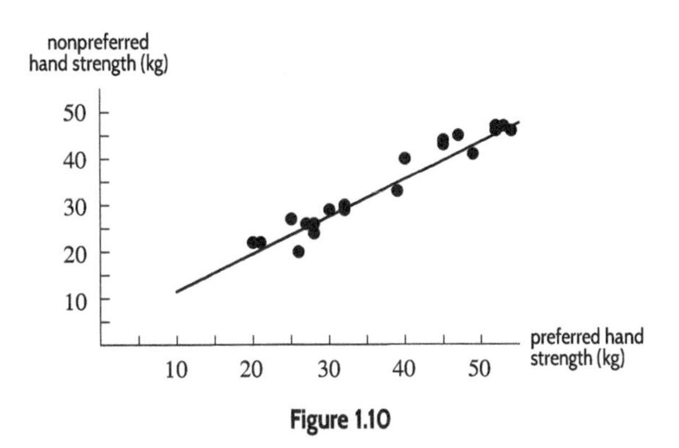

Figure 1.10

(b) Answers vary, but should be close to $y = 3.6 + 0.8x$.
(c) Answers may vary slightly. A possible equation is: $y = 3.623 + 0.825x$.
(d) The preferred hand strength is the independent quantity, so it is represented by x. Substituting $x = 37$ gives

$$y = 3.623 + 0.825(37) \approx 34.$$

So, the nonpreferred hand strength is about 34 kg.

(e) If we predict strength of the nonpreferred hand based on the strength of the preferred hand for values within the observed values of the preferred hand (such as 37), then we are interpolating. However, if we chose a value such as 10, which is below all the actual measurements, and use this to predict the nonpreferred hand strength, then we are extrapolating. Predicting from a value of 100 would be another example of extrapolation. In this case of hand strength, it seems safe to extrapolate; in other situations, extrapolation can be inaccurate.

(f) The correlation coefficient is positive because both hand strengths increase together, so the line has a positive slope. The value of r is close to 1 because the hand strengths lie close to a line of positive slope.

(g) The two clusters suggest that there are two distinct groups of students. These might be men and women, or perhaps students who are involved in college athletics (and therefore in excellent physical shape) and those who are not involved.

STRENGTHEN YOUR UNDERSTANDING

1. False. $f(t)$ is functional notation, meaning that f is a function of the variable t.

5. True. The number of people who enter a store in a day and the total sales for the day are related, but neither quantity is uniquely determined by the other.

9. True. A circle does not pass the vertical line test.

13. True. This is the definition of an increasing function.

17. False. Parentheses must be inserted. The correct ratio is $\dfrac{(10-2^2)-(10-1^2)}{2-1}=-3$.

21. False. Writing the equation as $y=(-3/2)x+7/2$ shows that the slope is $-3/2$.

25. True. A constant function has slope zero. Its graph is a horizontal line.

29. True. At $y=0$, we have $4x=52$, so $x=13$. The x-intercept is $(13,0)$.

33. False. Substitute the point's coordinates in the equation: $-3-4\neq-2(4+3)$.

37. False. The first line does but the second, in slope-intercept form, is $y=(1/8)x+(1/2)$, so it crosses the y-axis at $y=1/2$.

41. True. The point $(1,3)$ is on both lines because $3=-2\cdot1+5$ and $3=6\cdot1-3$.

45. True. The slope, $\Delta y/\Delta x$ is undefined because Δx is zero for any two points on a vertical line.

49. False. For example, in children there is a high correlation between height and reading ability, but it is clear that neither causes the other.

53. True. There is a perfect fit of the line to the data.

CHAPTER TWO

Solutions for Section 2.1

Skill Refresher

S1. $5(x-3) = 5x - 15$.

S5. $x(5x+1) - 3(5x+1) = 5x^2 + x - 15x - 3 = 5x^2 - 14x - 3$.

S9. $3 + 2\left(\frac{1}{x}\right)^2 - x = 3 + \frac{2}{x^2} - x = \frac{3x^2 + 2 - x^3}{x^2}$.

S13. We have

$$\begin{aligned} 8\sqrt{x} - 2 &= 10 \\ 8\sqrt{x} &= 12 \\ \sqrt{x} &= 1.5 \\ x &= (1.5)^2 = 2.25. \end{aligned}$$

S17. We have

$$\begin{aligned} \frac{3w+12}{5w-1} &= 0 \\ 3w + 12 &= 0 \\ 3w &= -12 \\ w &= -4 \end{aligned}$$

EXERCISES

1. Substituting -27 for x gives

$$g(-27) = -\frac{1}{2}(-27)^{1/3} = -\frac{1}{2}(-3) = \frac{3}{2}.$$

5. **(a)** Substituting $t = 0$ gives

$$g(0) = \frac{1}{0+2} - 1 = \frac{1}{2} - 1 = -\frac{1}{2}.$$

(b) Setting $g(t) = 0$ and solving gives

$$\begin{aligned} \frac{1}{t+2} - 1 &= 0 \\ \frac{1}{t+2} &= 1 \\ 1 &= t + 2 \\ t &= -1. \end{aligned}$$

9.

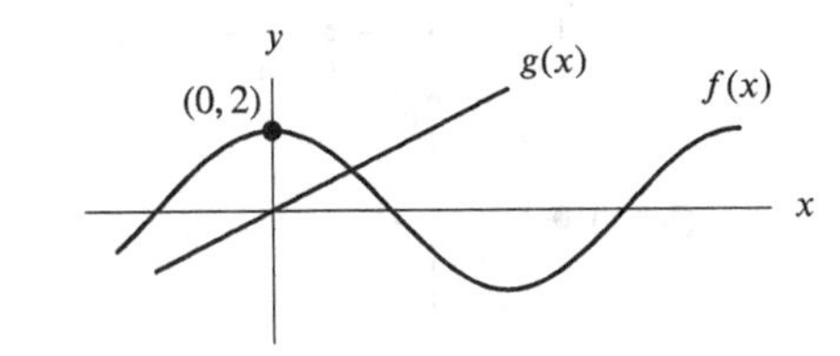

13. The solutions to $g(x) = 2$ are all x values that satisfy $2x + 2 = 2$. Subtracting 2 from both sides, we get $2x = 0$, and dividing by 2, we get $x = 0$. Thus there is only one x value, $x = 0$, with $g(x) = 2$.

PROBLEMS

17. The input, t, is the number of months since January 1, and the output, F, is the number of foxes. The expression $g(9)$ represents the number of foxes in the park on October 1. Table 1.3 on page 4 of the text gives $F = 100$ when $t = 9$. Thus, $g(9) = 100$. On October 1, there were 100 foxes in the park.

21. **(a)** From the graph, we estimate that $f(4) = 36$. This means that after 4 seconds, the electrical charge remaining in the circuit is 36 microcoulombs.

(b) From the graph, we estimate that $f(1.4) = 60$. This means that the remaining charge in the capacitor is 60 microcoulombs after 1.4 seconds.

25. **(a)** $g(x) = x^2 + x$
$g(-3x) = (-3x)^2 + (-3x)$
$g(-3x) = 9x^2 - 3x$

(b) $g(1-x) = (1-x)^2 + (1-x) = (1 - 2x + x^2) + (1-x) = x^2 - 3x + 2$

(c) $g(x+\pi) = (x+\pi)^2 + (x+\pi) = (x^2 + 2\pi x + \pi^2) + (x+\pi) = x^2 + (2\pi + 1)x + \pi^2 + \pi$

(d) $g(\sqrt{x}) = (\sqrt{x})^2 + \sqrt{x} = x + \sqrt{x}$

(e) $g\left(\frac{1}{x+1}\right) = \left(\frac{1}{(x+1)^2}\right) + \frac{1}{x+1} = \frac{1}{(x+1)^2} + \frac{x+1}{(x+1)^2} = \frac{x+2}{(x+1)^2}$

(f) $g(x^2) = (x^2)^2 + x^2 = x^4 + x^2$

29. **(a)**

x	−2	−1	0	1	2	3
$h(x)$	0	9	8	3	0	6

(b) From the table, we see that $h(3) = 6$, while $h(1) = 3$. Thus, $h(3) - h(1) = 6 - 3 = 3$.

(c) From the table, we see that $h(2) = 0$, and $h(0) = 8$. Thus, $h(2) - h(0) = 0 - 8 = -8$.

(d) From the table, we see that $h(0) = 8$. Thus, $2h(0) = 2(8) = 16$.

(e) From the table, we see that $h(1) = 3$. Thus, $h(1) + 3 = 3 + 3 = 6$.

33. **(a)** The car's position after 2 hours is denoted by the expression $s(2)$. The position after 2 hours is

$$s(2) = 11(2)^2 + 2 + 100 = 44 + 2 + 100 = 146.$$

(b) This is the same as asking the following question: "For what t is $v(t) = 65$?"

(c) To find out when the car is going 67 mph, we set $v(t) = 67$. We have

$$\begin{aligned} 22t + 1 &= 67 \\ 22t &= 66 \\ t &= 3. \end{aligned}$$

The car is going 67 mph at $t = 3$, that is, 3 hours after starting. Thus, when $t = 3$, $s(3) = 11(3^2) + 3 + 100 = 202$, so the car's position when it is going 67 mph is 202 miles.

37. **(a)** From Figure 2.1, we see that $P = (b, a)$ and $Q = (d, e)$.

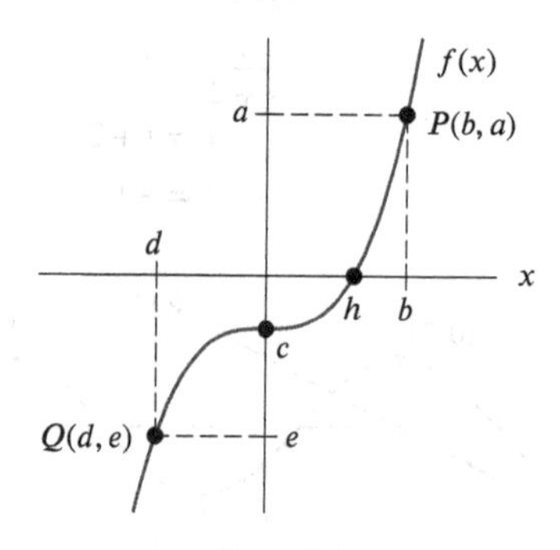

Figure 2.1

(b) To evaluate $f(b)$, we want to find the y-value when the x-value is b. Since (b, a) lies on this graph, we know that the y-value is a, so $f(b) = a$.

(c) To solve $f(x) = e$, we want to find the x-value for a y-value of e. Since (d, e) lies on this curve, $x = d$ is our solution.

(d) To solve $z = f(x)$, we need first to find a value for z; in other words, we need first to solve for $f(z) = c$. Since $(0, c)$ lies on this graph, we know that $z = 0$. Now we need to solve $0 = f(x)$ by finding the point whose y-value is 0. That point is $(h, 0)$, so $x = h$ is our solution.

(e) We know that $f(b) = a$ and $f(d) = e$. Thus, if $f(b) = -f(d)$, we know that $a = -e$.

41. **(a)** To evaluate $f(1)$, we need to find the value of f which corresponds to $x = 1$. Looking in the table, we see that that value is 2. So we can say $f(1) = 2$. Similarly, to find $g(3)$, we see in the table that the value of g which corresponds to $x = 3$ is 4. Thus, we know that $g(3) = 4$.

(b) The values of $f(x)$ increase by 3 as x increases by 1. For $x > 1$, the values of $g(x)$ are consecutive perfect squares. The entries for $g(x)$ are symmetric about $x = 1$. In other words, when $x < 1$ the values of $g(x)$ are the same as the values when $x > 1$, but the order is reversed.

(c) Since the values of $f(x)$ increase by 3 as x increases by 1 and $f(4) = 11$, we know that $f(5) = 11+3 = 14$. Similarly, $f(x)$ decreases by three as x goes down by one. Since $f(-1) = -4$, we conclude that $f(-2) = -4 - 3 = -7$.

The values of $g(x)$ are consecutive perfect squares. Since $g(4) = 9$, then $g(5)$ must be the next perfect square which is 16, so $g(5) = 16$. Since the values of $g(x)$ are symmetric about $x = 1$, the value of $g(-2)$ will equal $g(5)$ (since -2 and 4 are both a distance of 3 units from 1). Thus, $g(-2) = g(4) = 9$.

(d) To find a formula for $f(x)$, we begin by observing that $f(0) = -1$, so the value of $f(x)$ that corresponds to $x = 0$ is -1. We know that the value of $f(x)$ increases by 3 as x increases by 1, so

$$\begin{aligned} f(1) &= f(0) + 3 = -1 + 3 \\ f(2) &= f(1) + 3 = (-1 + 3) + 3 = -1 + 2 \cdot 3 \\ f(3) &= f(2) + 3 = (-1 + 2 \cdot 3) + 3 = -1 + 3 \cdot 3 \\ f(4) &= f(3) + 3 = (-1 + 3 \cdot 3) + 3 = -1 + 4 \cdot 3. \end{aligned}$$

The pattern is

$$f(x) = -1 + x \cdot 3 = -1 + 3x.$$

We can check this formula by choosing a value for x, such as $x = 4$, and use the formula to evaluate $f(4)$. We find that $f(4) = -1 + 3(4) = 11$, the same value we see in the table.

Since the values of $g(x)$ are all perfect squares, we expect the formula for $g(x)$ to have a square in it. We see that x^2 is not quite right since the table for such a function would look like Table 2.1.

Table 2.1

x	-1	0	1	2	3	4
x^2	1	0	1	4	9	16

But this table is very similar to the one that defines g. In order to make Table 2.1 look identical to the one given in the problem, we need to subtract 1 from each value of x so that $g(x) = (x - 1)^2$. We can check our formula by choosing a value for x, such as $x = 2$. Using our formula to evaluate $g(2)$, we have $g(2) = (2 - 1)^2 = 1^2 = 1$. This result agrees with the value given in the problem.

45. $f(a) = \dfrac{a \cdot a}{a + a} = \dfrac{a^2}{2a} = \dfrac{a}{2}.$

49. **(a)** To evaluate $f(2)$, we determine which value of I corresponds to $w = 2$. Looking at the graph, we see that $I \approx 7$ when $w = 2$. This means that ≈ 7000 people were infected two weeks after the epidemic began.

(b) The height of the epidemic occurred when the largest number of people were infected. To find this, we look on the graph to find the largest value of I, which seems to be approximately 8.5, or 8500 people. This seems to have occurred when $w = 4$, or four weeks after the epidemic began. We can say that at the height of the epidemic, at $w = 4$, $f(4) = 8.5$.

(c) To solve $f(x) = 4.5$, we must find the value of w for which $I = 4.5$, or 4500 people were infected. We see from the graph that there are actually two values of w at which $I = 4.5$, namely $w \approx 1$ and $w \approx 10$. This means that 4500 people were infected after the first week when the epidemic was on the rise, and that after the tenth week, when the epidemic was slowing, 4500 people remained infected.

(d) We are looking for all the values of w for which $f(w) \geq 6$. Looking at the graph, this seems to happen for all values of $w \geq 1.5$ and $w \leq 8$. This means that more than 6000 people were infected starting in the middle of the second week and lasting until the end of the eighth week, after which time the number of infected people fell below 6000.

Solutions for Section 2.2

Skill Refresher

S1. The function is undefined when the denominator is zero. Therefore, $x - 3 = 0$ tells us the function is undefined for $x = 3$.

S5. We solve for x as a function of y:

$$\begin{aligned} y &= \frac{x+3}{x-4} \\ y(x-4) &= x+3 \\ xy - 4y &= x+3 \\ xy - x &= 4y+3 \\ x(y-1) &= 4y+3 \\ x &= \frac{4y+3}{y-1} \end{aligned}$$

S9. Adding 8 to both sides of the inequality we get $x > 8$.

S13. $x^2 - 25 > 0$ is true when $x > 5$ or $x < -5$.

EXERCISES

1. The domain is $1 \leq x \leq 7$. The range is $2 \leq f(x) \leq 18$.

5. The graph of $f(x) = 1/x$ for $-2 \leq x \leq 2$ is shown in Figure 2.2. From the graph, we see that $f(x) = -(1/2)$ at $x = -2$. As we approach zero from the left, $f(x)$ gets more and more negative. On the other side of the y-axis, $f(x) = (1/2)$ at $x = 2$. As x approaches zero from the right, $f(x)$ grows larger and larger. Thus, on the domain $-2 \leq x \leq 2$, the range is $f(x) \leq -(1/2)$ or $f(x) \geq (1/2)$.

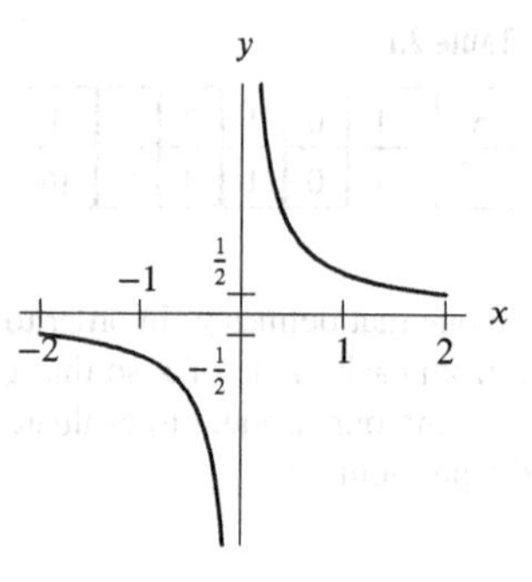

Figure 2.2

9. The domain is all real numbers except those which do not yield an output value. The expression $1/(x+3)$ is defined for any real number x except -3, since for $x = -3$ the denominator of $f(x)$, $x + 3$, is 0 and division by 0 is undefined. Therefore, the domain of $f(x)$ is all real numbers $\neq -3$.

13. To evaluate $f(x)$, we must have $9 + x > 0$. Thus

$$\text{Domain: } x > -9.$$

17. Any number can be squared, so the domain is all real numbers.

21. To evaluate $f(x)$, we must have $x - 4 > 0$. Thus

$$\text{Domain: } x > 4.$$

To find the range, we want to know all possible output values. We solve the equation $y = f(x)$ for x in terms of y. Since

$$y = \frac{1}{\sqrt{x-4}},$$

squaring gives

$$y^2 = \frac{1}{x-4},$$

and multiplying by $x - 4$ gives

$$\begin{aligned} y^2(x-4) &= 1 \\ y^2x - 4y^2 &= 1 \\ y^2x &= 1 + 4y^2 \\ x &= \frac{1+4y^2}{y^2}. \end{aligned}$$

This formula tells us how to find the x-value which corresponds to a given y-value. The formula works for any y except $y = 0$ (which puts a 0 in the denominator). We know that y must be positive, since $\sqrt{x-4}$ is positive, so we have

$$\text{Range: } y > 0.$$

PROBLEMS

25. The function $n(q)$ can be evaluated only when $q^2 + a \geq 0$. Since, when q is 0, any $a < 0$ gives $q^2 + a < 0$, we know that a must be greater than or equal to zero. Furthermore, for all $a \geq 0$, we have $q^2 + a \geq 0$. Hence the domain of $n(q)$ is all real numbers for any $a \geq 0$. So any nonnegative real number, such as 0, $\frac{1}{2}$, $\sqrt{2}$, or 77 is a possible value of a giving $n(q)$ a domain of all real numbers.

29. Since the restaurant opens at 2 pm, $t = 0$, and closes at 2 am, $t = 12$, a reasonable domain is $0 \leq t \leq 12$.

Since there cannot be fewer than 0 clients in the restaurant and 200 can fit inside, the range is the integers between 0 and 200.

33. We can put in any number for x except zero, which makes $1/x$ undefined. We note that as x approaches infinity or negative infinity, $1/x$ approaches zero, though it never arrives there, and that as x approaches zero, $1/x$ goes to negative or positive infinity. Thus, the range is all real numbers except a.

37. **(a)** Substituting $t = 0$ into the formula for $p(t)$ shows that $p(0) = 50$, meaning that there were 50 rabbits initially. Using a calculator, we see that $p(10) \approx 131$, which tells us there were about 131 rabbits after 10 months. Similarly, $p(50) \approx 911$ means there were about 911 rabbits after 50 months.

(b) The graph in Figure 2.3 tells us that the rabbit population grew quickly at first but then leveled off at about 1000 rabbits after around 75 months or so. It appears that the rabbit population increased until it reached the island's capacity.

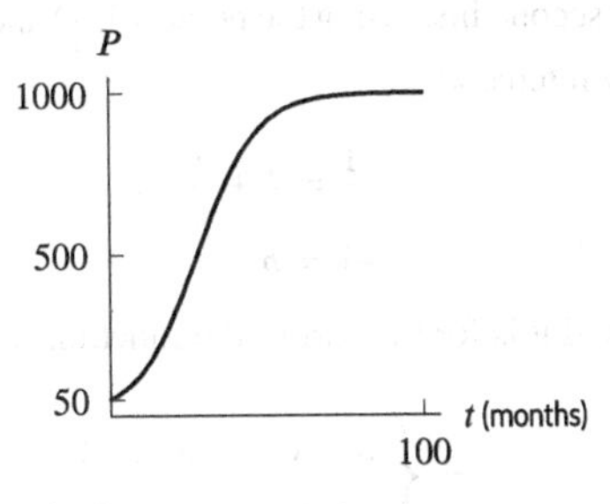

Figure 2.3

(c) From the graph in Figure 2.3, we see that the range is $50 \le p(t) \le 1000$. This tells us that (for $t \ge 0$) the number of rabbits is no less than 50 and no more than 1000.

(d) The smallest population occurred when $t = 0$. At that time, there were 50 rabbits. As t gets larger and larger, $(0.9)^t$ gets closer and closer to 0. Thus, as t increases, the denominator of

$$p(t) = \frac{1000}{1 + 19(0.9)^t}$$

decreases. As t increases, the denominator $1 + 19(0.9)^t$ gets close to 1 (try $t = 100$, for example). As the denominator gets closer to 1, the fraction gets closer to 1000. Thus, as t gets larger and larger, the population gets closer and closer to 1000. Thus, the range is $50 \le p(t) < 1000$.

Solutions for Section 2.3

Skill Refresher

S1. Since the point zero is not included, this graph represents $x > 0$.

S5. Since both end points of the interval are solid dots, this graph represents $x \le -1$ or $x \ge 2$.

S9. Since the point $(1, 1)$ lies on the graph of f, we have $f(1) = 1$.

S13. Domain: $-2 \le x \le 3$ and range: $-2 \le y \le 3$.

EXERCISES

1. $f(x) = \begin{cases} -1, & -1 \le x < 0 \\ 0, & 0 \le x < 1 \\ 1, & 1 \le x < 2 \end{cases}$ is shown in Figure 2.4.

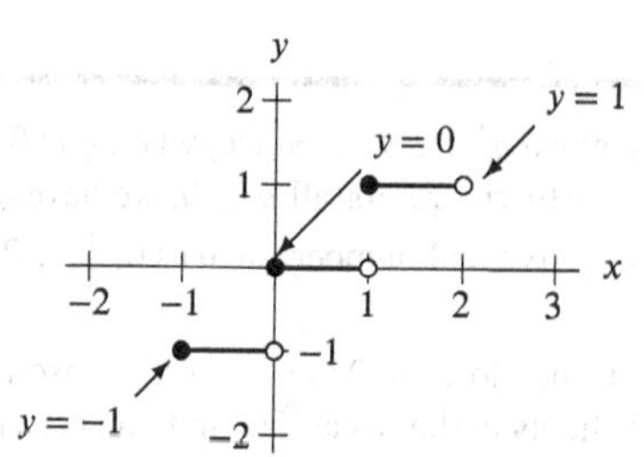

Figure 2.4

5. We find the formulas for each of the lines. For the first, we use the two points we have, $(1, 4)$ and $(3, 2)$. We find the slope: $(2 - 4)/(3 - 1) = -1$. Using the slope of -1, we solve for the y-intercept:

$$\begin{aligned} 4 &= b - 1 \cdot 1 \\ 5 &= b. \end{aligned}$$

Thus, the first line is $y = 5 - x$, and it is for the part of the function where $x < 3$. Notice that we do not use this formula for the value $x = 3$.

We follow the same method for the second line, using the points $(3, \frac{1}{2})$ and $(5, \frac{3}{2})$. We find the slope: $(\frac{3}{2} - \frac{1}{2})/(5-3) = \frac{1}{2}$. Using the slope of $\frac{1}{2}$, we solve for the y-intercept:

$$\begin{aligned} \frac{1}{2} &= b + \frac{1}{2} \cdot 3 \\ -1 &= b. \end{aligned}$$

Thus, the second line is $y = -1 + \frac{1}{2}x$, and it is for the part of the function where $x \ge 3$.

Therefore, the function is:

$$y = \begin{cases} 5 - x & \text{for } x < 3 \\ -1 + \frac{1}{2}x & \text{for } x \ge 3. \end{cases}$$

9. Since $G(x)$ is defined for all x, the domain is all real numbers. For $x < -1$ the values of the function are all negative numbers. For $-1 \geq x \geq 0$ the function's values are $4 \geq G(x) \geq 3$, while for $x > 0$ we see that $G(x) \geq 3$ and the values increase to infinity. The range is $G(x) < 0$ and $G(x) \geq 3$.

13. We want to find all numbers x such that $|x-3| = 7$. That is, we want the distance between x and 3 to be 7. Thus, x must be seven units to the left or seven units to the right of 3; that is, $x = -4$ and $x = 10$.

PROBLEMS

17. **(a)** Yes, because every value of x is associated with exactly one value of y.
(b) No, because some values of y are associated with more than one value of x.
(c) Only part (a) leads to a function. Its range is $y = 1, 2, 3, 4$.

21. **(a)** Since zero lies in the interval $-1 \leq x \leq 1$, we find the function value from the formula $f(x) = 3x$. This gives $f(0) = 3 \cdot 0 = 0$. To find $f(3)$, we first note that $x = 3$ lies in the interval $1 < x \leq 5$, so we find the function value from the formula $f(x) = -x + 4$. The result is $f(3) = -3 + 4 = 1$.
(b) By graphing f we can see in Figure 2.5 that the combined domain is $-1 \leq x \leq 5$ and the range is $-3 \leq f(x) \leq 3$.

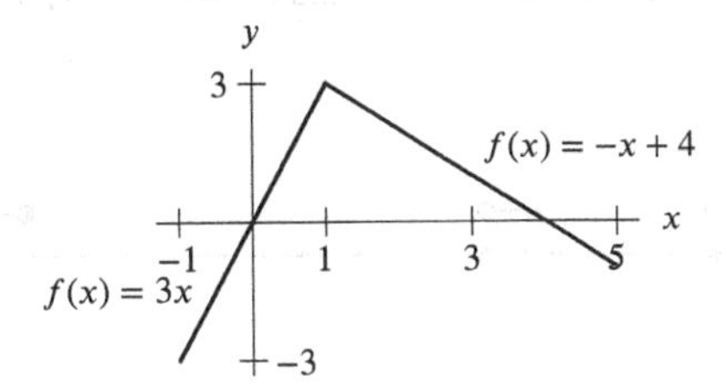

Figure 2.5

25. **(a)** Upon entry, the cost is \$2.50. The tax surcharge of \$0.50 is added to the fare. So, the initial cost will be \$3.00. The cost for the first $1/5$ mile adds \$0.50, giving a fare of \$3.50. For a journey of $2/5$ mile, another \$0.50 is added for a fare of \$4.00. Each additional $1/5$ mile gives another increment of \$0.50. See Table 2.2.

Table 2.2

Miles	0	0.2	0.4	0.6	0.8	1	1.2	1.4	1.6	1.8	2
Cost	3.00	3.50	4.00	4.50	5.00	5.50	6.00	6.50	7.00	7.50	8.00

(b) The table shows that the cost for a 1.2 mile trip is \$6.00.
(c) From the table, the maximum distance one can travel for \$7.00 is 1.6 miles.
(d) See Figure 2.6.

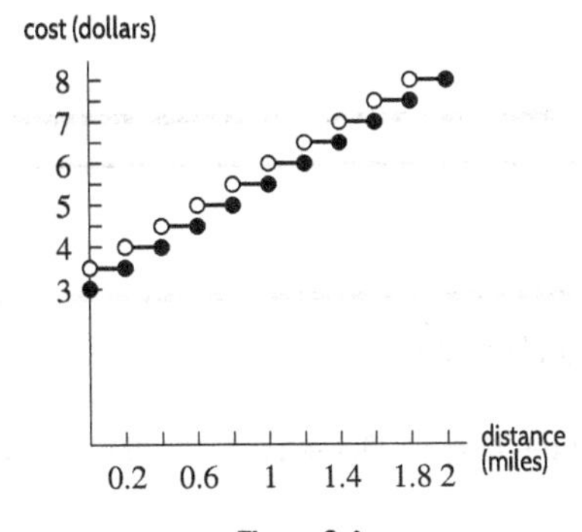

Figure 2.6

29. **(a)** Figure 2.7 shows the rates for off peak periods of the year. Figure 2.8 shows the rates for holiday periods (Dec 12–Jan 1, Jan 14–16, Feb 17–26).

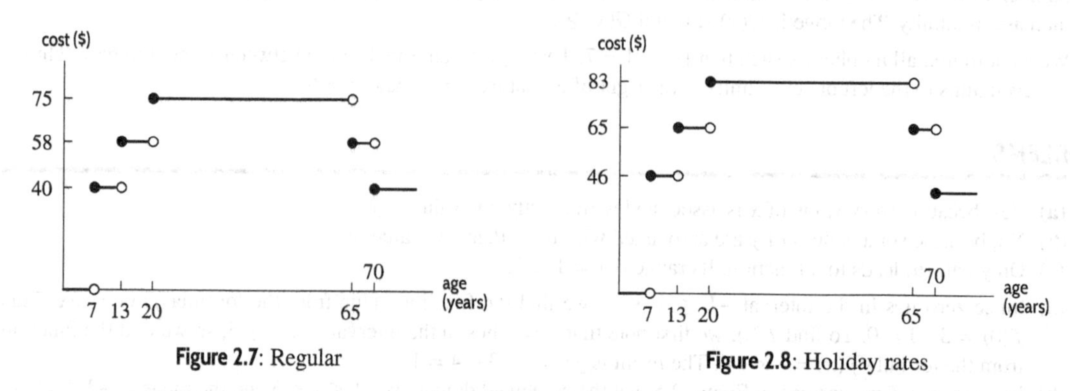

Figure 2.7: Regular   **Figure 2.8**: Holiday rates

(b) The holiday rate for juniors increases by \$6, for Teens and Seniors by \$7, but for Adults the increase is \$8.

(c) See Figure 2.9.

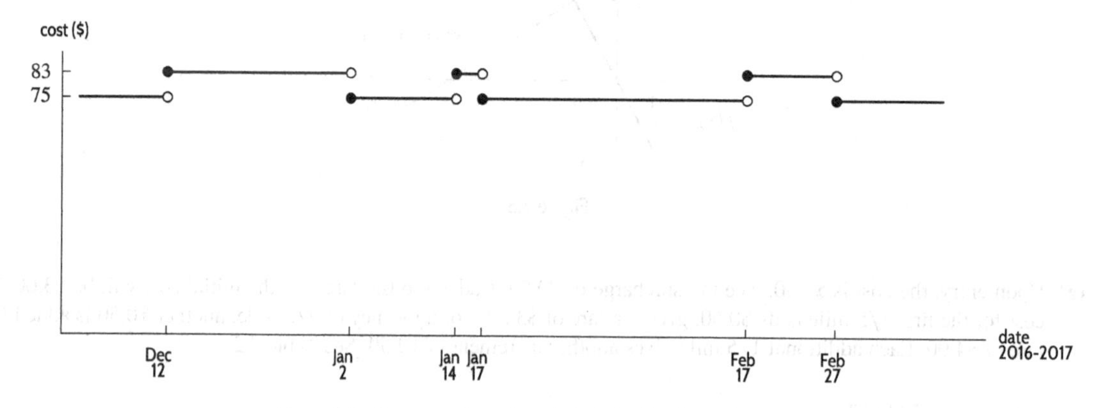

Figure 2.9: Cost as a function of 2016/2017 dates for Adults

(d) Rates through 11 December and after 27 February represent early-season and late-season rates, respectively; these are off-peak rates, since it makes economic sense to cut rates when there are fewer skiers. Holiday rates took effect from 12 December through 1 January because of the Christmas/New Year's holiday; they took effect from 14 January through 16 January for Martin Luther King's holiday; they took effect from 17 February through 26 February for the Presidents' Week holiday; it makes economic sense to charge peak rates during the holidays, as more skiers are available to use the facility. Other times represent rates during the heart of the winter skiing season; these are the regular rates.

Solutions for Section 2.4

Skill Refresher

S1. Substituting $x = 4$ into $f(x)$, we have $f(4) = \sqrt{4} = 2$.

S5. We have

$$f(x+1) = 1 + 3(x+1) = 1 + 3x + 3 = 4 + 3x,$$

and

$$f(x) + 1 = (1 + 3x) + 1 = 2 + 3x.$$

S9. The solutions are

$$x = -2 \quad \text{and} \quad x = 2.$$

S13. This appears to be f shifted up 2 units, so this is a vertical shift.

S17. This appears to be f shifted to the right one unit, so this is a horizontal shift of f.

EXERCISES

1. (a)

x	−1	0	1	2	3
$g(x)$	−3	0	2	1	−1

The graph of $g(x)$ is shifted one unit to the right of $f(x)$.

(b)

x	−3	−2	−1	0	1
$h(x)$	−3	0	2	1	−1

The graph of $h(x)$ is shifted one unit to the left of $f(x)$.

(c)

x	−2	−1	0	1	2
$k(x)$	0	3	5	4	2

The graph $k(x)$ is shifted up three units from $f(x)$.

(d)

x	−1	0	1	2	3
$m(x)$	0	3	5	4	2

The graph $m(x)$ is shifted one unit to the right and three units up from $f(x)$.

5. See Figure 2.10.

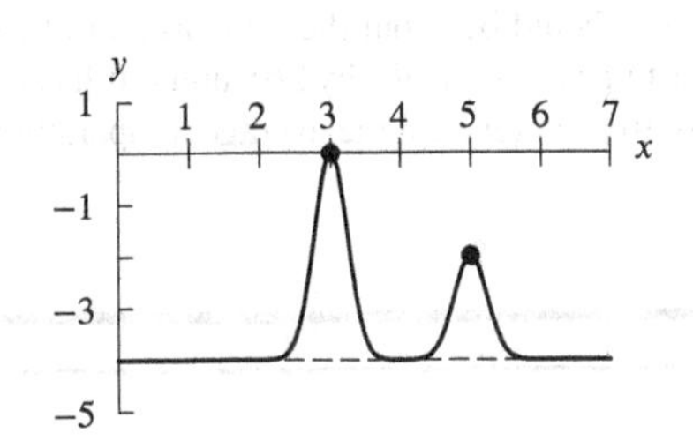

Figure 2.10

9. The range shifts the graph down 150 units, so the new range is $-50 \leq R(s) - 150 \leq 50$.

PROBLEMS

13. (a) The translation should leave the x-coordinate unchanged, and shift the y-coordinate up 3; so $y = g(x) + 3$.
(b) The translation should leave the y-coordinate unchanged, and shift the x-coordinate right by 2; so $y = g(x - 2)$.

17. Since $W = s(t + 4)$, at age $t = 3$ months Ben's weight is given by

$$W = s(3 + 4) = s(7).$$

We defined $s(7)$ to be the average weight of a 7-month old baby. At age 3 months, Ben's weight is the same as the average weight of 7-month old babies. Since, on average, a baby's weight increases as the baby grows, this means that Ben is heavier than the average for a 3-month old. Similarly, at age $t = 6$, Ben's weight is given by

$$W = s(6 + 4) = s(10).$$

Thus, at 6 months, Ben's weight is the same as the average weight of 10-month old babies. In both cases, we see that Ben is above average in weight.

21. Since this is an inside change, the graph is four units to the left of $q(z)$. That is, for any given z value, the value of $q(z+4)$ is the same as the value of the function q evaluated four units to the right of z (at $z+4$).

25. **(a)** There are many possible graphs, but all should show seasonally related cycles of temperature increases and decreases, as in Figure 2.11.

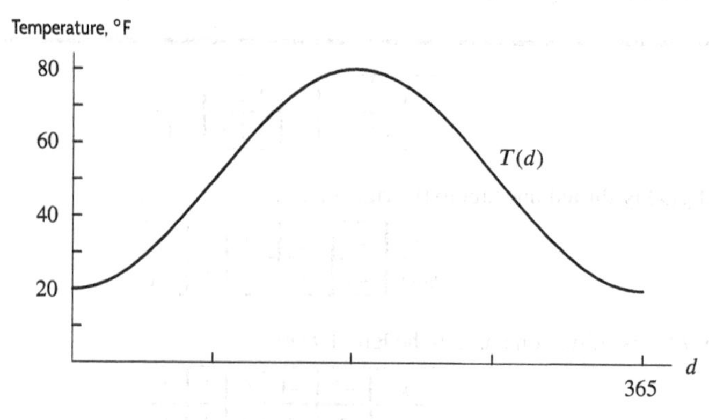

Figure 2.11

(b) While there are a wide variety of correct answers, the value of $T(6)$ is a temperature for a day in early January, $T(100)$ for a day in mid-April, and $T(215)$ for a day in early August. The value for $T(371) = T(365+6)$ should be close to that of $T(6)$.

(c) Since there are usually 365 days in a year, $T(d)$ and $T(d+365)$ represent average temperatures on days which are a year apart.

(d) $T(d+365)$ is the average temperature on the same day of the year a year earlier. They should be about the same value. Therefore, the graph of $T(d+365)$ should be about the same as that of $T(d)$.

(e) The graph of $T(d)+365$ is a shift upward of $T(d)$, by 365 units. It has no significance in practical terms, other than to represent a temperature that is 365° hotter than the average temperature on day d.

Solutions for Section 2.5

Skill Refresher

S1.

$$\begin{aligned} 5\left(\frac{1}{5}x-1\right)+5 &= x-5+5 \\ &= x. \end{aligned}$$

S5. Adding 4 to both sides and dividing by 3, we get $y = \frac{x+4}{3}$.

S9. Adding 2 to both sides, we get $\sqrt{y} = x+2$. Squaring both sides, we get

$$y = (x+2)^2 = x^2+4x+4.$$

EXERCISES

1. We use $5x+3$ as the input in the f function:

$$f(5x+3) = 2(5x+3)-1 = 10x+6-1 = 10x+5.$$

5. $f(g(0)) = f(1 - 0^2) = f(1) = 3 \cdot 1 - 1 = 2.$

9. $f(g(x)) = f(1 - x^2) = 3(1 - x^2) - 1 = 2 - 3x^2.$

13. Since $g(0) = 1$, we have $f(g(0)) = f(1) = -4$.

17. We begin by evaluating the inside function and see from the graph that $g(2) = 0$. We then substitute 0 for $g(2)$ in the expression, giving

$$f(g(2)) = f(0) = 1.$$

21. $A(f(t))$ is the area, in square centimeters, of the circle at time t minutes.

25. The inverse function, $f^{-1}(T)$, gives the temperature in °F needed if the cake is to bake in T minutes. Units of $f^{-1}(T)$ are °F.

29. The domain of $f^{-1}(y)$ is the range of $f(x)$, that is, all numbers between -3 and 2. Thus:

$$\text{Domain of } f^{-1}(y):\ -3 \leq \text{ all real numbers } \leq 2.$$

The range of $f^{-1}(y)$ is the domain of $f(x)$, that is all numbers between -2 and 3. Thus:

$$\text{Range of } f^{-1}(y):\ -2 \leq \text{ all real numbers } \leq 3.$$

33. Since $y = 2t + 3$, solving for t gives

$$\begin{aligned} 2t + 3 &= y \\ t &= \frac{y-3}{2} \\ f^{-1}(y) &= \frac{y-3}{2}. \end{aligned}$$

37. Since $A = \pi r^2$, solving for r gives

$$\begin{aligned} \frac{A}{\pi} &= r^2 \\ \sqrt{\frac{A}{\pi}} &= r \\ r = f^{-1}(A) &= \sqrt{\frac{A}{\pi}}. \end{aligned}$$

41. We have

$$k(3) = 4(3) - 10 = 12 - 10 = 2.$$

To find $k^{-1}(14)$, we find the value of x so that $k(x) = 14$. If $14 = 4x - 10$, then $24 = 4x$ so $x = 6$. Thus, $k^{-1}(14) = 6$.

45. **(a)** Since the vertical intercept of the graph of f is $(0, b)$, we have $f(0) = b$.
(b) Since the horizontal intercept of the graph of f is $(a, 0)$, we have $f(a) = 0$.
(c) The function f^{-1} goes from y-values to x-values, so to evaluate $f^{-1}(0)$, we want the x-value corresponding to $y = 0$. This is $x = a$, so $f^{-1}(0) = a$.
(d) Solving $f^{-1}(?) = 0$ means finding the y-value corresponding to $x = 0$. This is $y = b$, so $f^{-1}(b) = 0$.

PROBLEMS

49. **(a)** Since $k(2) = 1$, we have $h(k(2)) = h(1) = 3$.
(b) Since $h(2) = -1$, we have $k(h(2)) = k(-1) = 2$. Notice that $h(k(2)) \neq k(h(2))$.
(c) Since $k(0) = 5$, we have $h(k(0)) = h(5)$. We cannot find $h(k(0))$ since Table 2.24 does not give the value of $h(5)$. Notice that this does not mean $h(k(0))$ does not exist, just that we do not have enough information to find it.

53. **(a)** Since m is the number of daily miles driven, the domain of $f(m)$ is all numbers greater than or equal to 0 and up to 500, the maximum number of miles we can drive in a day. This is also the range of of $f^{-1}(C)$. Thus:

$$\text{Range of } f^{-1}(C)\text{: } 0 \le \text{ all integer numbers } \le 500.$$

The domain of $f^{-1}(C)$ is the range of $f(m)$, or the cost of driving 0 or more daily miles after renting the car. The base rental cost is 32 dollars regardless of miles driven. We also pay, at most, $32 + 0.19(500) = 127$ dollars (the cost of driving 500 miles). Thus, the domain of $f^{-1}(C)$ is:

$$\text{Domain of } f^{-1}(C)\text{: } 32 \le \text{ all real numbers } \le 127.$$

(b) We solve the equation $C = f(m) = 32 + 0.19m$ for d. Subtract 32 from both sides and divide both sides by 0.19 to get

$$m = \frac{1}{0.19}(C - 32).$$

So

$$f^{-1}(C) = \frac{100}{19}(C - 32).$$

57. We begin by observing that $C = f(T)$. When $C = 25$, we see in the table that $T = 18°$F. Since $T = g(t)$, we look for an entry in the other table where $T = 18$, and we see that $t = 12$. So $25 = f(g(12))$, and it cost \$25 to heat the house on January12.

61. From the first table, we see $f(7) = 9$, and so from the second table, we have $g(f(7)) = g(9) = 4000$. This tells us after 7 days, when the ice is 9 inches thick, the bearing strength is 4000 lbs.

65. At a temperature of $H = 250°$ F, we have $g(250) = \frac{3969000}{250+460} = 5590.1$. This tells us that if the temperature of the hot air inside the balloon is $H = 250°$ F, the air's weight is $W = 5590.1$ lbs. We can now use $f(W)$ to find the corresponding lift. Specifically, when the weight is $W = 5590.1$ lbs, the balloon's lift is $f(5590.1) = 7489 - 5590.1 = 1898.9$. Thus the lift generated at the maximum operating temperature is $L = 1898.9$ lbs.

69. Since we can cube any number, the domain of $n(r)$ is all real numbers. To find the range, we find the inverse function. Let $y = n(r)$. Solving for r, we get

$$\begin{aligned} y &= r^3 + 2 \\ y - 2 &= r^3 \\ r &= (y-2)^{1/3} \\ n^{-1}(y) &= (y-2)^{1/3}. \end{aligned}$$

Since the domain of $n^{-1}(y)$ is all real numbers, the range of $n(r)$ is all real numbers.

73. **(a)** The cost of producing 5000 loaves is \$653.
(b) $C^{-1}(80)$ is the number of loaves of bread that can be made for \$80, namely 0.62 thousand or 620.
(c) The solution is $q = 6.3$ thousand. It costs \$790 to make 6300 loaves.
(d) The solution is $x = 150$ dollars, so 1.2 thousand, or 1200, loaves can be made for \$150.

77. **(a)** We substitute zero into the function, giving:

$$H = f(0) = \frac{5}{9}(0 - 32) = -\frac{160}{9} = -17.778.$$

This means that zero degrees Fahrenheit is about -18 degrees Celsius.

(b) In Exercise 76, we found the inverse function. Using it with $H = 0$, we have:

$$t = f^{-1}(0) = \frac{9}{5}0 + 32 = 32.$$

This means that zero degrees Celsius is equivalent to 32 degrees Fahrenheit (the temperature at which water freezes).

(c) We substitute 100 into the function, giving:

$$H = f(100) = \frac{5}{9}(100 - 32) = \frac{340}{9} = 37.778.$$

This means that 100 degrees Fahrenheit is about 38 degrees Celsius.

(d) In Exercise 76, we found the inverse function. Using it with $H = 100$, we have:

$$t = f^{-1}(100) = \frac{9}{5}100 + 32 = 212.$$

This means that 100 degrees Celsius is equivalent to 212 degrees Fahrenheit (the temperature at which water boils).

81. Since $V = \frac{4}{3}\pi r^3$ and $r = 50 - 2.5t$, substituting r into V gives

$$V = f(t) = \frac{4}{3}\pi(50 - 2.5t)^3.$$

Solutions for Section 2.6

Skill Refresher

S1. Since $1 < 5$, $5 < 7$ and $7 < 9$, these numbers are increasing.

S5. The average rate of change is

$$\begin{aligned}\frac{f(3)-f(1)}{3-1} &= \frac{(1+5\cdot 3)-(1+5\cdot 1)}{2} \\ &= \frac{16-6}{2} = 5.\end{aligned}$$

S9. The average rate of change is

$$\begin{aligned}\frac{f(3)-f(1)}{3-1} &= \frac{3(3-2)^2 - 3(1-2)^2}{2} \\ &= \frac{3(1)-3(1)}{2} = 0.\end{aligned}$$

S13. The average rate of change is

$$\begin{aligned}\frac{f(3)-f(1)}{3-1} &= \frac{3-(-1)}{2} \\ &= \frac{4}{2} = 2.\end{aligned}$$

EXERCISES

1. To determine concavity, we calculate the rate of change:

$$\frac{\Delta f(x)}{\Delta x} = \frac{1.3-1.0}{1-0} = 0.3$$

$$\frac{\Delta f(x)}{\Delta x} = \frac{1.7-1.3}{3-1} = 0.2$$

$$\frac{\Delta f(x)}{\Delta x} = \frac{2.2-1.7}{6-3} \approx 0.167.$$

The rates of change are decreasing, so we expect the graph of $f(x)$ to be concave down.

5. The slope of $y = x^2$ is always increasing, so its graph is concave up. See Figure 2.12.

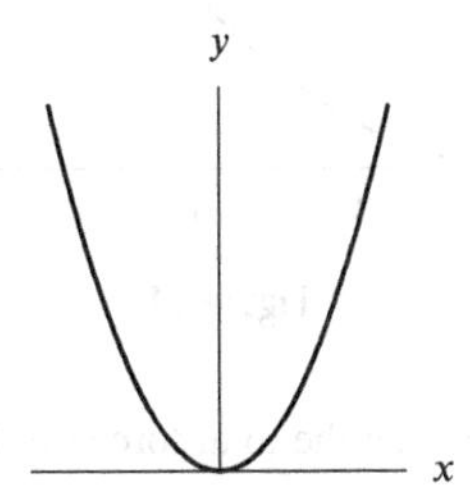

Figure 2.12

9. The rate of change between $x = 12$ and $x = 15$ is

$$\frac{\Delta H(x)}{\Delta x} = \frac{21.53 - 21.40}{15 - 12} \approx 0.043.$$

Similarly, we have

$$\frac{\Delta H(x)}{\Delta x} = \frac{21.75 - 21.53}{18 - 15} \approx 0.073$$
$$\frac{\Delta H(x)}{\Delta x} = \frac{22.02 - 21.75}{21 - 18} \approx 0.090.$$

The rate of change is increasing, so we expect the graph of $H(x)$ to be concave up.

PROBLEMS

13. **(a)** On the interval $0 \leq h \leq 0.5$, we have

$$\frac{\Delta v}{\Delta h} = \frac{3.13 - 0}{0.5 - 0} = 6.26 \text{ m/sec per meter.}$$

Continuing in the same way gives the result in Table 2.3.

Table 2.3

Interval	$0 \leq h \leq 0.5$	$0.5 \leq h \leq 1$	$1 \leq h \leq 1.5$
Rate of change $\Delta v/\Delta h$	6.26	2.6	1.98

(b) As h increases, the values of $h = f(v)$ increase, so f is an increasing function. However, as h increases, the average rates of change of f are decreasing. Therefore, f is concave down. This means that, as the height of the water in the tank increases, the speed of the water flowing through the spout increases, but the rate of increase is slowing down.

17. The graph is increasing for all x. It is concave down for $x < o$ and concave up for $x > 0$. Thus we have

(a) Increasing and concave down $x < 0$
(b) Decreasing and concave down nowhere
(c) Increasing and concave up $x > 0$
(d) Decreasing and concave down nowhere

21. Since the graph of f is bending upward on the interval $0 \leq t < 3$ and bending downward on the interval $3 < t \leq 6$, we estimate that the graph of f is concave up when $0 \leq t < 3$ and concave down when $3 < t \leq 6$. This means that between $t = 0$ and $t = 3$ years, the rate of growth of the fish population is speeding up, but between $t = 3$ and $t = 6$ years, the rate of growth is slowing down.

25. **(a)** The graph of $F = g(h)$ is given in Figure 2.13.

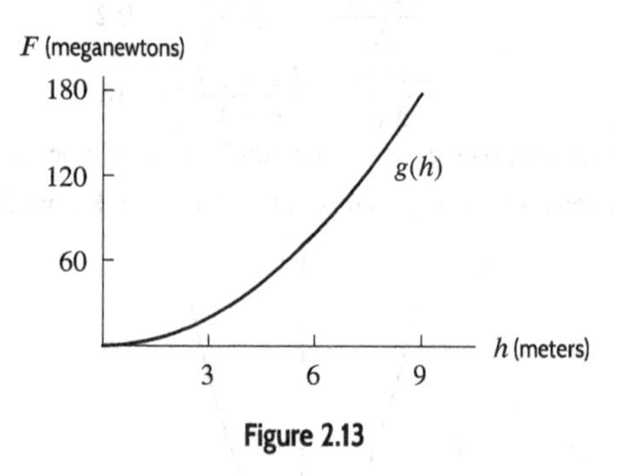

Figure 2.13

(b) The function g is increasing. This means that the total force on the dam increases as the height of the water it holds back increases.

(c) Since the graph of g is bending upward, the function g is concave up. This means, as the height of the water held back by the dam increases, the rate at which the force on the dam increases speeds up.

29. This function is increasing throughout and the rate of increase is increasing, so the graph is concave up.

33. Many answers are possible. See Figure 2.14.

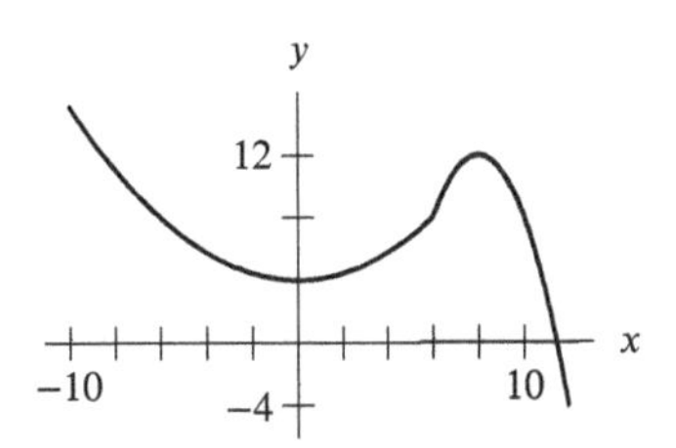

Figure 2.14

37. **(a)** This describes a situation in which y is increasing rapidly at first, then very slowly at the end. In Table (E), y increases dramatically at first (from 20 to 275) but is hardly growing at all by the end. In Graph (I), y is increasing at a constant rate, while in Graph (II), it is increasing faster at the end. Graph (III) increases rapidly at first, then slowly at the end. Thus, scenario (a) matches with Table (E) and Graph (III).

(b) Here, y is growing at a constant rate. In Table (G), y increases by 75 units for every 5-unit increase in x. A constant increase in y relative to x means a straight line, that is, a line with a constant slope. This is found in Graph (I).

(c) In this scenario, y is growing at a faster and faster rate as x gets larger. In Table (F), y starts out by growing by 16 units, then 30, then 54, and so on, so Table (F) refers to this case. In Graph (II), y is increasing faster and faster as x gets larger.

STRENGTHEN YOUR UNDERSTANDING

1. False. $f(2) = 3 \cdot 2^2 - 4 = 8$.

5. False. $W = (8 + 4)/(8 - 4) = 3$.

9. True. A fraction can only be zero if the numerator is zero.

13. False. The domain consists of all real numbers x, $x \neq 3$

17. True. Since f is an increasing function, the domain endpoints determine the range endpoints. We have $f(15) = 12$ and $f(20) = 14$.

21. True. $|x| = |-x|$ for all x.

25. True. If $x < 0$, then $f(x) = x < 0$, so $f(x) \neq 4$. If $x > 4$, then $f(x) = -x < 0$, so $f(x) \neq 4$. If $0 \leq x \leq 4$, then $f(x) = x^2 = 4$ only for $x = 2$. The only solution for the equation $f(x) = 4$ is $x = 2$.

29. True. This looks like the absolute value function shifted right 1 unit and down 2 units.

33. True. To find $f^{-1}(R)$, we solve $R = \frac{2}{3}S + 8$ for S by subtracting 8 from both sides and then multiplying both sides by $(3/2)$.

37. False. Since

$$f(g(x)) = 2\left(\frac{1}{2}x - 1\right) + 1 = x - 1 \neq x,$$

the functions do not undo each other.

41. True. Since the function is concave up, the average rate of change increases as we move right.

45. True. For $x > 0$, the function $f(x) = -x^2$ is both decreasing and concave down.

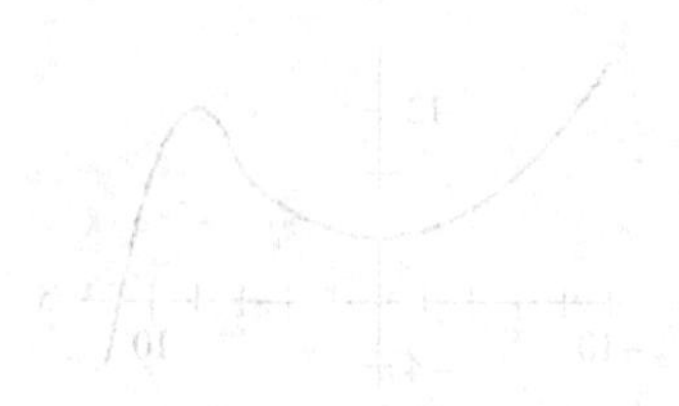

CHAPTER THREE

Solutions for Section 3.1

Skill Refresher

S1. In this example, we distribute the factors $50t$ and $2t$ across the two binomials $t^2 + 1$ and $25t^2 + 125$, respectively. Thus,

$$\begin{aligned}(t^2+1)(50t) - (25t^2+125)(2t) &= 50t^3 + 50t - (50t^3 + 250t)\\ &= 50t^3 + 50t - 50t^3 - 250t = -200t.\end{aligned}$$

S5. We expand and gather like terms.

$$\begin{aligned}3x(2x+5) + 4(x^2-1) &= (3x)(2x) + (3x)(5) + 4(x^2) + (4)(-1)\\ &= 6x^2 + 15x + 4x^2 - 4\\ &= 10x^2 + 15x - 4.\end{aligned}$$

S9. $3x^2 - x - 4 = (3x-4)(x+1)$

S13. There is a common factor of x in every term, so we can factor out the x:
$xy - x^2 = x(y - x)$.

S17.

$$\begin{aligned}x^2 + 7x + 6 &= 0\\ (x+6)(x+1) &= 0\\ x+6 = 0 \quad &\text{or} \quad x+1 = 0\\ x = -6 \quad &\text{or} \quad x = -1\end{aligned}$$

S21. We set the expression equal to zero, then factor and set each factor equal to zero:

$$\begin{aligned}x^2 + x &= 6\\ x^2 + x - 6 &= 0\\ (x+3)(x-2) &= 0\\ x+3 = 0 \ &\text{or} \ x-2 = 0\\ x = -3 \ &\text{or} \ x = 2.\end{aligned}$$

EXERCISES

1. We have:

$$g(t) = 3(t-2)^2 + 7 = 3(t^2 - 4t + 4) + 7 = 3t^2 - 12t + 19,$$

so $a = 3, b = -12, c = 19$.

5. No. We rewrite the function, giving

$$\begin{aligned}R(q) &= \frac{1}{q^2}(q^2+1)^2\\ &= \frac{1}{q^2}(q^4 + 2q^2 + 1)\\ &= q^2 + 2 + \frac{1}{q^2}\\ &= q^2 + 2 + q^{-2}.\end{aligned}$$

So $R(q)$ is not quadratic since it contains a term with q to a negative power.

9. We solve for r in the equation by factoring

$$\begin{aligned} 2r^2 - 6r - 36 &= 0 \\ 2(r^2 - 3r - 18) &= 0 \\ 2(r-6)(r+3) &= 0. \end{aligned}$$

The solutions are $r = 6$ and $r = -3$.

13. To find the zeros, we solve the equation

$$0 = 9x^2 + 6x + 1.$$

We see that this is factorable, as follows:

$$\begin{aligned} y &= (3x+1)(3x+1) \\ y &= (3x+1)^2. \end{aligned}$$

Therefore, there is only one zero at $x = -\frac{1}{3}$.

17. We solve for r in the equation $Q(r) = 2r^2 - 6r - 36 = 0$ using the quadratic formula with $a = 2$, $b = -6$ and $c = -36$.

$$\begin{aligned} r &= \frac{-(-6) \pm \sqrt{(-6)^2 - 4(2)(-36)}}{2(2)} \\ r &= \frac{6 \pm \sqrt{36 + 288}}{4} \\ r &= \frac{6 \pm \sqrt{324}}{4} \\ r &= \frac{6 \pm 18}{4}. \end{aligned}$$

Therefore $r = (6+18)/4 = 6$ and $r = (6-18)/4 = -3$. The zeros of $Q(r)$ are $r = 6$ and $r = -3$.

21. Without fully multiplying out, we can see that the coefficient of x^2 is 5, so this function has a graph which is concave up.

PROBLEMS

25. There is one zero at $x = -2$, so by symmetry the vertex is $(-2, 0)$. We have $y = a(x+2)^2$. Solving for a, we have

$$\begin{aligned} a(0+2)^2 &= 7 \\ 4a &= 7 \\ a &= \frac{7}{4}, \end{aligned}$$

so $y = (7/4)(x+2)^2$.

29. This is impossible. If the graph is concave down, it opens downward. Then the graph is above the x-axis between the two zeros so could not have a y-intercept of -6.

33. Factoring gives $y = -4cx + x^2 + 4c^2 = x^2 - 4ck + 4c^2 = (x - 2c)^2$. Since $c > 0$, this is the graph of $y = x^2$ shifted to the right $2c$ units. See Figure 3.1.

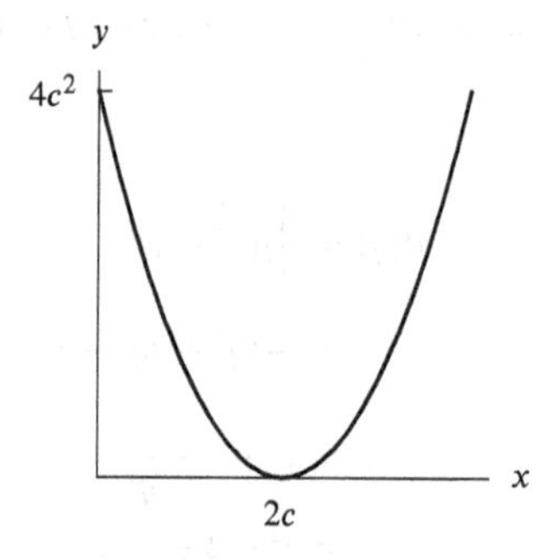

Figure 3.1: $y = -4cx + x^2 + 4c^2$ for $c > 0$

37. Since the parabola has x-intercepts at $x = -1$ and $x = 5$, its formula is:

$$y = a(x+1)(x-5).$$

The coordinates $(-2, 6)$ must satisfy the equation, so

$$6 = a(-2+1)(-2-5).$$

Solving for a gives $a = \frac{6}{7}$. The formula is:

$$y = \frac{6}{7}(x+1)(x-5).$$

41. **(a)** At $t = 0$ the snowboarder is 5 meters below the edge of the half-pipe.

(b) We find the zeros of $y = -4.9t^2 + 14t - 5$ using the quadratic formula:

$$t = \frac{-14 \pm \sqrt{14^2 - 4(-4.9)(-5)}}{2(-4.9)}$$
$$t = 0.4184 \text{ or } 2.4387.$$

Thus snowboarder leaves the pipe and flies into the air before she returns to the pipe, so we choose the lower zero. She reaches the air after 0.4184 seconds.

She comes back to the pipe at the second zero, after 2.4387 seconds.

(c) She is in the air from the time she leaves the pipe until the time she returns, from 0.4184 seconds to 2.4387 seconds. Thus, she spends $2.4387 - 0.4184 = 2.0203$ seconds in the air.

45. **(a)** According to the figure in the text, the package was dropped from a height of 5 km.

(b) When the package hits the ground, $h = 0$ and $d = 4430$. So, the package has moved 4430 meters forward when it lands.

(c) Since the maximum is at $d = 0$, the formula is of the form $h = ad^2 + b$ where a is negative and b is positive. Since $h = 5$ at $d = 0$, $5 = a(0)^2 + b = b$, so $b = 5$. We now know that $h = ad^2 + 5$. Since $h = 0$ when $d = 4430$, we have $0 = a(4430)^2 + 5$, giving $a = \frac{-5}{(4430)^2} \approx -0.000000255$. So $h \approx -0.000000255d^2 + 5$.

Solutions for Section 3.2

Skill Refresher

S1. $y^2 - 12y = y^2 - 12y + 36 - 36 = (y-6)^2 - 36$

S5. Get the variables on the left side, the constants on the right side and complete the square using $(\frac{-6}{2})^2 = 9$.

$$r^2 - 6r = -8$$
$$r^2 - 6r + 9 = 9 - 8$$
$$(r-3)^2 = 1.$$

Take the square root of both sides and solve for r.

$$r - 3 = \pm 1$$
$$r = 3 \pm 1.$$

So, $r = 4$ or $r = 2$.

S9. Rewrite the equation to equal zero, and factor.

$$n^2 + 4n - 5 = 0$$
$$(n+5)(n-1) = 0.$$

So, $n + 5 = 0$ or $n - 1 = 0$, thus $n = -5$ or $n = 1$.

EXERCISES

1. By comparing $f(x)$ to the vertex form, $y = a(x-h)^2 + k$, we see the vertex is $(h, k) = (1, 2)$. The axis of symmetry is the vertical line through the vertex, so the equation is $x = 1$. The parabola opens upward because the value of a is positive 3.

5. **(a)** See Figure 3.2. For g, we have $a = 1$, $b = 0$, and $c = 3$. Its vertex is at $(0, 3)$, and its axis of symmetry is the y-axis, or the line $x = 0$. This function has no zeros.

   (b) See Figure 3.3. For f, we have $a = -2$, $b = 4$, and $c = 16$. The axis of symmetry is the line $x = 1$ and the vertex is at $(1, 18)$. The zeros, or x-intercepts, are at $x = -2$ and $x = 4$. The y-intercept is at $y = 16$.

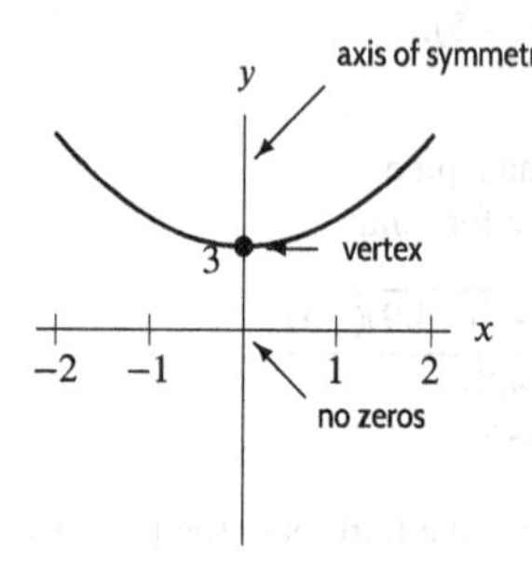

Figure 3.2: $g(x) = x^2 + 3$

Figure 3.3: $f(x) = -2x^2 + 4x + 16$

9. Since the vertex is $(6, 5)$, we use the form $y = a(x-h)^2 + k$, with $h = 6$ and $k = 5$. We solve for a, substituting in the second point, $(10, 8)$.

$$\begin{aligned} y &= a(x-6)^2 + 5 \\ 8 &= a(10-6)^2 + 5 \\ 3 &= 16a \\ \frac{3}{16} &= a. \end{aligned}$$

Thus, an equation for the parabola is

$$y = \frac{3}{16}(x-6)^2 + 5.$$

13. Letting $y = -3z^2 + 9z - 2$, we have:

$$\begin{aligned} y &= -3z^2 + 9z - 2 && \\ y + 2 &= -3z^2 + 9z && \text{add 2} \\ -\frac{1}{3}(y+2) &= z^2 - 3z && \text{multiply by } -1/3 \\ -\frac{1}{3}(y+2) + \left(\frac{3}{2}\right)^2 &= z^2 - 3x + \left(\frac{3}{2}\right)^2 && \text{complete the square} \\ -\frac{1}{3}(y+2) + \frac{9}{4} &= \left(z - \frac{3}{2}\right)^2 && \text{factor right-hand side} \\ -\frac{1}{3}(y+2) &= \left(z - \frac{3}{2}\right)^2 - \frac{9}{4} && \text{subtract } 9/4 \\ y + 2 &= -3\left(z - \frac{3}{2}\right)^2 + \frac{27}{4} && \text{multiply by } -3 \\ y &= -3\left(z - \frac{3}{2}\right)^2 + \frac{27}{4} - 2 && \text{subtract 2} \\ &= -3\left(z - \frac{3}{2}\right)^2 + \frac{19}{4} && \text{simplify,} \end{aligned}$$

so the vertex is $(3/2, 19/4)$ and the axis of symmetry is $z = 3/2$.

17. $m(n) + 1 = \frac{1}{2}n^2 + 1$

To graph this function, shift the graph of $m(n) = \frac{1}{2}n^2$ one unit up. See Figure 3.4.

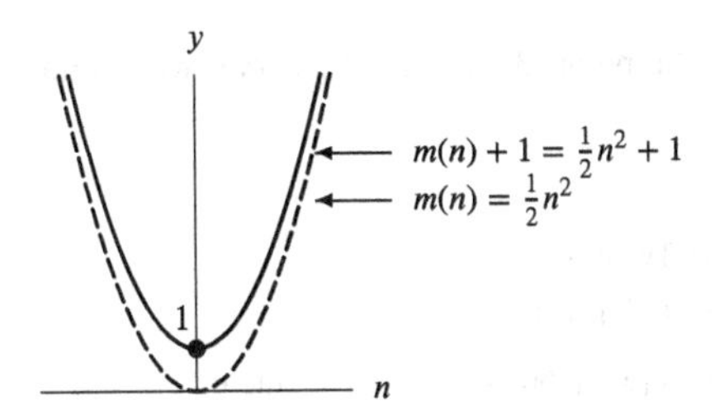

Figure 3.4

21. $m(n)+\sqrt{13}=\frac{1}{2}n^2+\sqrt{13}$

To sketch, shift the graph of $m(n)=\frac{1}{2}n^2$ up by $\sqrt{13}$ units, as in Figure 3.5.

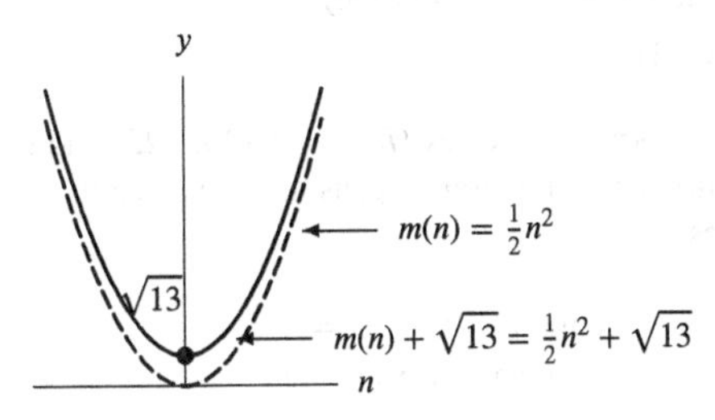

Figure 3.5

PROBLEMS

25. We complete the square to write the function in vertex form:

$$\begin{aligned}
y=f(t)&=-16\left(t^2-\frac{47}{16}t-\frac{3}{16}\right)\\
&=-16\left(t^2-\frac{47}{16}t+\left(-\frac{47}{32}\right)^2-\left(-\frac{47}{32}\right)^2-\frac{3}{16}\right)\\
&=-16\left(\left(t-\frac{47}{32}\right)^2-\frac{2209}{1024}-\frac{192}{1024}\right)\\
&=-16\left(\left(t-\frac{47}{32}\right)^2-\frac{2401}{1024}\right)\\
&=-16\left(t-\frac{47}{32}\right)^2+\frac{2401}{64}.
\end{aligned}$$

Thus, the vertex is at the point $(\frac{47}{32},\frac{2401}{64})$.

29. Using the vertex form $y=a(x-h)^2+k$, where $(h,k)=(2,5)$, we have

$$y=a(x-2)^2+5.$$

Since the parabola passes through $(1,2)$, these coordinates must satisfy the equation, so

$$2=a(1-2)^2+5.$$

Solving for a gives $a=-3$. The formula is:

$$y=-3(x-2)^2+5.$$

33. We have $y = a(x-7)^2 + 3$. Since the point $(3,7)$ is on the curve, we obtain $7 = a(-4)^2 + 3$, so $a = 1/4$. Therefore $(1/4)(x-7)^2 + 3$.

37. We have

$$
\begin{aligned}
y &= 0.03x^2 + 1.8x + 2 \\
y - 2 &= 0.03x^2 + 1.8x \\
&= 0.03\left(x^2 + 60x\right) && \text{factor} \\
\frac{y-2}{0.03} &= x^2 + 60x \\
\frac{y-2}{0.03} + (30)^2 &= x^2 + 60x + (30)^2 && \text{complete the square} \\
\frac{y-2}{0.03} + 900 &= (x+30)^2 && \text{factor} \\
\frac{y-2}{0.03} &= (x+30)^2 - 900 \\
y - 2 &= 0.03(x+30)^2 - 0.03(900) \\
y &= 0.03\,(x-(-30))^2 - 25.
\end{aligned}
$$

This is a quadratic function in vertex form with vertex $(h,k) = (-30,-25)$ and $a = 0.03$. From the original equation, we see that the y-intercept is $y = 2$. Since $a > 0$, the graph opens up, and since the vertex lies below the x-axis, there are two x-intercepts. Solving for $y = 0$ gives

$$
\begin{aligned}
0.03(x+30)^2 - 25 &= 0 \\
0.03(x+30)^2 &= 25 \\
(x+30)^2 &= \frac{25}{0.03} \\
&= \frac{2500}{3} \\
x + 30 &= \pm\sqrt{\frac{2500}{3}} \\
x &= -30 \pm \sqrt{\frac{2500}{3}},
\end{aligned}
$$

so the x-intercepts are $x = -58.868$ and $x = -1.133$. See Figure 3.6.

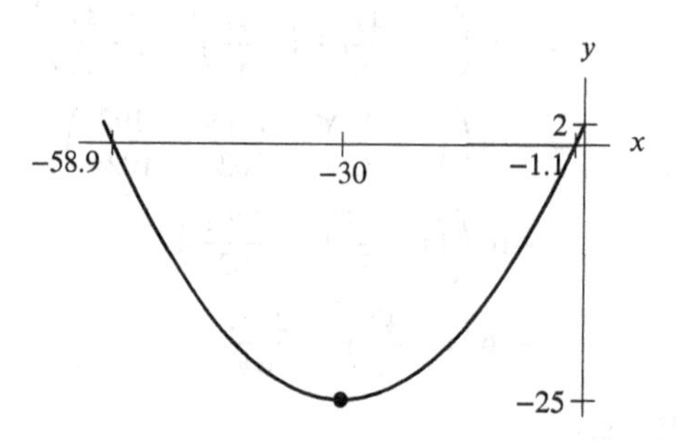

Figure 3.6

41. The graph of $y = x^2 - 10x + 25$ appears to be the graph of $y = x^2$ moved to the right by 5 units. See Figure 3.7. If this were so, then its formula would be $y = (x-5)^2$. Since $(x-5)^2 = x^2 - 10x + 25$, $y = x^2 - 10x + 25$ is, indeed, a horizontal shift of $y = x^2$.

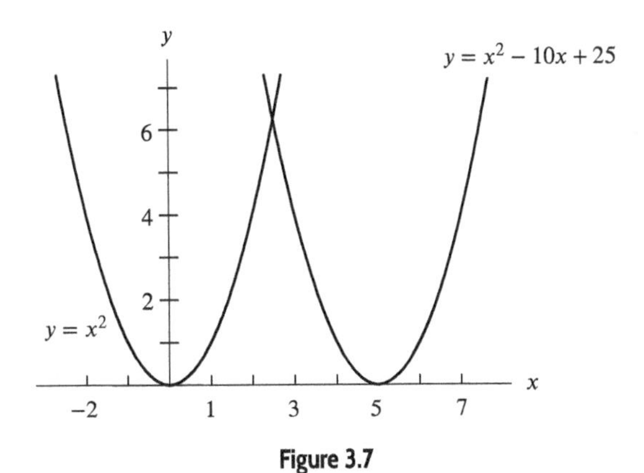

Figure 3.7

45. **(a)** See Figure 3.8.

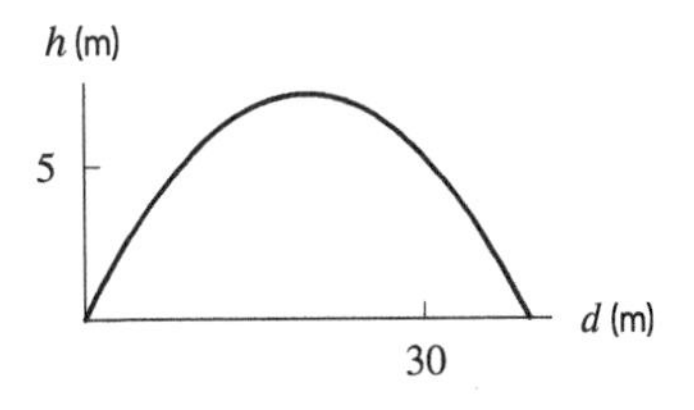

Figure 3.8

(b) When the ball hits the ground $h = 0$, so $h = 0.75d - 0.0192d^2 = d(0.75 - 0.0192d) = 0$ and we get $d = 0$ or $d \approx 39.063$ m. Since $d = 0$ is the position where the kicker is standing, the ball must hit the ground about 39.063 meters from the point where it is kicked.

(c) The path is parabolic and the maximum height occurs at the vertex, which lies on the axis of symmetry, midway between the zeros at $d \approx 19.531$ m. Since $h = 0.75(19.531) - 0.0192(19.531)^2 \approx 7.324$, we know that the ball reaches 7.324 meters above the ground before it begins to fall.

(d) From part (c), the horizontal distance traveled when the ball reaches its maximum height is ≈ 19.531 m.

STRENGTHEN YOUR UNDERSTANDING

1. True. It is of the form $f(x) = a(x - r)(x - s)$ where $a = 1$, $r = 0$ and $s = -2$.
5. False. The time when the object hits the ground is when the height is zero, $s(t) = 0$. The value $s(0)$ gives the height when the object is launched at $t = 0$.
9. False. A quadratic function may have two, one, or no zeros.
13. False. The vertex is located at the point (h, k).

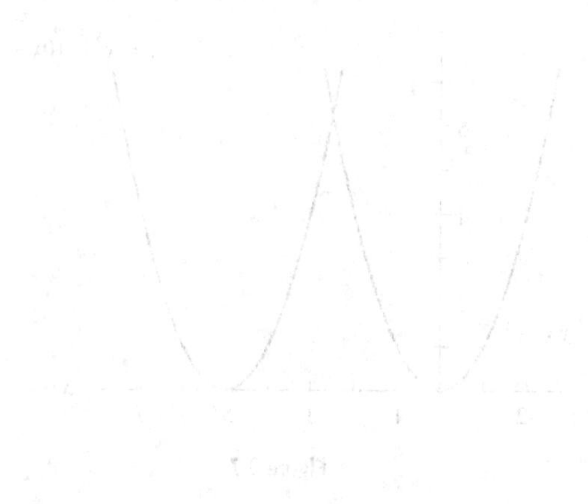

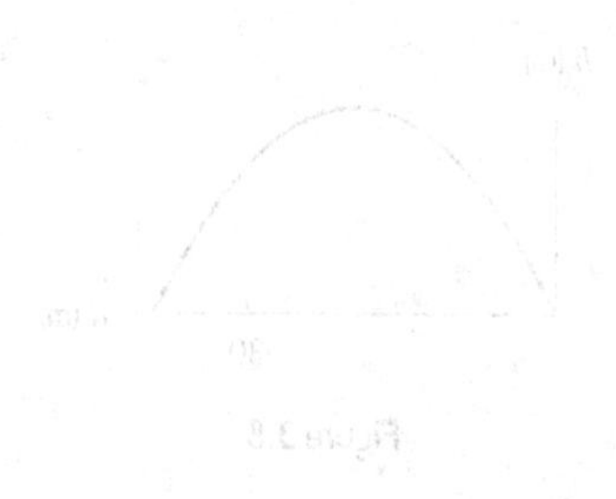

CHAPTER FOUR

Solutions for Section 4.1

Skill Refresher

S1. We have $6\% = 0.06$.

S5. $0.15 \cdot 50 = 7.5$

S9. Since $1.1 \cdot 40 = 44$, the new price is $44.

S13. Since $63/60 = 1.05$, there is a 5% increase.

S17. Factor out 25 to get $25(1 + 0.02) = (1.02)25$, so $P = 1.02$.

S21. We have $4 \cdot 5 \cdot 5 \cdot 5 \cdot 5 \cdot 5 \cdot 5 \cdot 5 \cdot 5 \cdot 5 = 4 \cdot 5^9$.

S25. We have

$$\begin{aligned}\frac{(2a^3b^2)^3}{(4ab^{-4})^2} &= \frac{2^3a^{3\cdot 3}b^{2\cdot 3)}}{16a^2b^{-4\cdot 2}}\\ &= \frac{8a^9b^6}{16a^2b^{-8}}\\ &= \frac{a^{9-2}b^{6-(-8)}}{2}\\ &= \frac{a^7b^{14}}{2}.\end{aligned}$$

EXERCISES

1. Yes. Writing the function as

$$g(w) = 2\left(2^{-w}\right) = 2\left(2^{-1}\right)^w = 2\left(\frac{1}{2}\right)^w,$$

we have $a = 2$ and $b = 1/2$.

5. No. The base must be a constant.

9. Yes. Writing the function as

$$K(x) = \frac{2^x}{3 \cdot 3^x} = \frac{1}{3}\left(\frac{2^x}{3^x}\right) = \frac{1}{3}\left(\frac{2}{3}\right)^x,$$

we have $a = 1/3$ and $b = 2/3$.

13. We can rewrite this as

$$\begin{aligned}Q &= 0.0022(2.31^{-3})^t\\ &= 0.0022(0.0811)^t,\end{aligned}$$

so $a = 0.0022$, $b = 0.0811$, and $r = b - 1 = -0.9189 = -91.89\%$.

17. The growth factor per century is 1+ the growth per century. Since the forest is shrinking, the growth is negative, so we subtract 0.80, giving 0.20.

21. **(a)** Since the initial amount is 112.8 and the quantity is decreasing at a rate of 23.4% per year, the formula is $Q = 112.8(1 - 0.234)^t = 112.8(0.766)^t$.

(b) At $t = 10$, we have $Q = 112.8(0.766)^{10} = 7.845$.

PROBLEMS

25. **(a)** The formula $f(t) = ab^t$ represents exponential growth if the base $b > 1$ and exponential decay if $0 < b < 1$. Towns (i), (ii), and (iv) are growing and towns (iii), (v), and (vi) are shrinking.

(b) Town (iv) is growing the fastest since its growth factor of 1.185 is the largest. Since $1.185 = 1 + 0.185$, it is growing at a rate of 18.5% per year.

(c) Town (v) is shrinking the fastest since its growth factor of 0.78 is the smallest. Since $0.78 = 1 - 0.22$, it is shrinking at a rate of 22% per year.

(d) In the exponential function $f(t) = ab^t$, the parameter a gives the value of the function when $t = 0$. We see that town (iii) has the largest initial population (2500) and town (ii) has the smallest initial population (600).

29. The population is growing at a rate of 1.9% per year. So, at the end of each year, the population is $100\% + 1.9\% = 101.9\%$ of what it had been the previous year. The growth factor is 1.019. If P is the population of this country, in millions, and t is the number of years since 2014, then, after one year,

$$\begin{aligned} & P = 70(1.019). \\ \text{After two years,}\quad & P = 70(1.019)(1.019) = 70(1.019)^2 \\ \text{After three years,}\quad & P = 70(1.019)(1.019)(1.019) = 70(1.019)^3 \\ \text{After } t \text{ years,}\quad & P = 70\underbrace{(1.019)(1.019)\ldots(1.019)}_{t \text{ times}} = 70(1.019)^t \end{aligned}$$

33. **(a)** We have $C = C_0(1-r)^t = 100(1-0.16)^t = 100(0.84)^t$, so

$$C = 100(0.84)^t.$$

(b) At $t = 5$, we have $C = 100(0.84)^5 = 41.821$ mg

37. Let $D(t)$ be the difference between the oven's temperature and the yam's temperature, which is given by an exponential function $D(t) = ab^t$. The initial temperature difference is $300°\text{F} - 0°\text{F} = 300°\text{F}$, so $a = 300$. The temperature difference decreases by 3% per minute, so $b = 1 - 0.03 = 0.97$. Thus,

$$D(t) = 300(0.97)^t.$$

If the yam's temperature is represented by $Y(t)$, then the temperature difference is given by

$$D(t) = 300 - Y(t),$$

so, solving for $Y(t)$, we have

$$Y(t) = 300 - D(t),$$

giving

$$Y(t) = 300 - 300(0.97)^t.$$

41. For $f(x) = 3^x$ on the interval $1 \le x \le 2$ we have

$$\text{Rate of change} = \frac{f(2) - f(1)}{2 - 1} = \frac{3^2 - 3^1}{1} = \frac{9 - 3}{1} = 6.$$

45. **(a)** (i) On the interval $0 \le x \le 1$ we have

$$\text{Rate of change} = \frac{f(1) - f(0)}{1 - 0} = \frac{2^1 - 2^0}{1} = \frac{2 - 1}{1} = 1.$$

(ii) On the interval $1 \le x \le 2$ we have

$$\text{Rate of change} = \frac{f(2) - f(1)}{2 - 1} = \frac{2^2 - 2^1}{1} = \frac{4 - 2}{1} = 2.$$

(iii) On the interval $2 \le x \le 3$ we have

$$\text{Rate of change} = \frac{f(3) - f(2)}{3 - 2} = \frac{2^3 - 2^2}{1} = \frac{8 - 4}{1} = 4.$$

(iv) On the interval $3 \le x \le 4$ we have

$$\text{Rate of change } = \frac{f(4)-f(3)}{4-3} = \frac{2^4-2^3}{1} = \frac{16-8}{1} = 8.$$

(b) The rate of change is increasing by a factor of 2 over consecutive intervals. This means for the interval $19 \le x \le 20$, we expect a rate of change of $2^{19} = 524{,}288$ since the interval $19 \le x \le 20$ is 19 intervals past $0 \le x \le 1$. Checking, we get, as predicted,

$$\text{Rate of change } = \frac{f(20)-f(19)}{20-19} = \frac{2^{20}-2^{19}}{1} = \frac{1{,}048{,}576-524{,}288}{1} = 524{,}288.$$

49. We have

$$\begin{aligned} y &= 5(0.5)^{t/3} \\ &= 5\left(2^{-1}\right)^{\frac{1}{3}\cdot t} \\ &= 5\cdot 2^{-\frac{1}{3}\cdot t} \\ &= 5\left(4^{1/2}\right)^{-\frac{1}{3}\cdot t} \\ &= 5\cdot 4^{-\frac{1}{6}\cdot t}, \end{aligned}$$

so $a = 5, k = -1/6$.

53. (a) Since $N(0)$ gives the number of teams remaining in the tournament after no rounds have been played, we have $N(0) = 64$. After 1 round, half of the original 64 teams remain in the competition, so

$$N(1) = 64\left(\frac{1}{2}\right).$$

After 2 rounds, half of these teams remain, so

$$N(2) = 64\left(\frac{1}{2}\right)\left(\frac{1}{2}\right).$$

And, after r rounds, the original pool of 64 teams has been halved r times, so that

$$N(r) = 64\underbrace{\left(\frac{1}{2}\right)\left(\frac{1}{2}\right)\cdots\left(\frac{1}{2}\right)}_{\text{Pool halved } r \text{ times}},$$

giving

$$N(r) = 64\left(\frac{1}{2}\right)^r.$$

The graph of $y = N(r)$ is given in Figure 4.1. The domain of N is $0 \le r \le 6$, for r an integer. A curve has been dashed in to help you see the overall shape of the function.

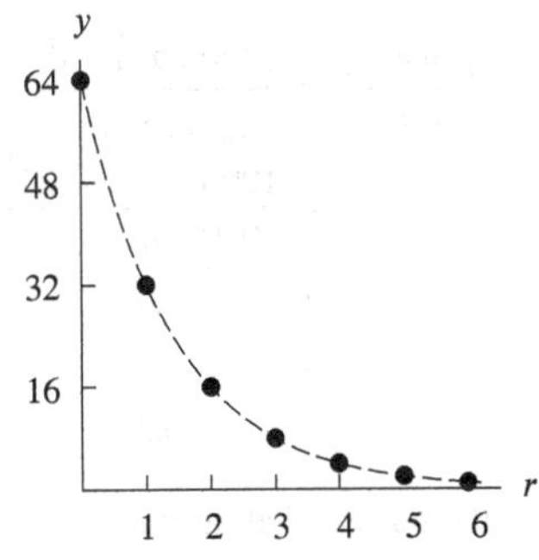

Figure 4.1: The graph of $y = N(r) = 64 \cdot \left(\frac{1}{2}\right)^r$

(b) There is a winner when there is only one team left, Connecticut. So, $N(r) = 1$.

$$\begin{aligned}64\left(\frac{1}{2}\right)^r &= 1\\ \left(\frac{1}{2}\right)^r &= \frac{1}{64}\\ \frac{1}{2^r} &= \frac{1}{64}\\ 2^r &= 64\\ r &= 6.\end{aligned}$$

You can solve $2^r = 64$ either by taking successive powers of 2 until you get to 64 or by substituting values for r until you get the one that works.

Connecticut won after winning 6 rounds.

57. **(a)** (i) The monthly payment on \$1000 each month at 4% for a loan period of 15 years is \$7.40. For \$60,000, the payment would be $\$7.40 \times 60 = \$444$ per month.

(ii) The monthly payment on \$1000 each month at 4% for a loan period of 30 years is \$4.77. For \$60,000, the payment would be $\$4.77 \times 60 = \$286.20$ per month.

(iii) The monthly payment on \$1000 each month at 6% for a loan period of 15 years is \$8.44. For \$60,000, the payment would be $\$8.44 \times 60 = \$506.40$ per month.

(b) As calculated in part (a)-(i), the monthly payment on a \$60,000 loan at 4% for 15 years would be \$444 per month. In part (a)-(iii) we showed that the monthly payment on a \$60,000 loan at 6% for 15 years would be \$506.40 per month. So taking the loan out at 4% rather that 6% would save the difference:

$$\text{Amount saved} = \$506.40 - \$444 = \$62.40 \text{ per month}$$

Since there are $15 \times 12 = 180$ months in 15 years,

$$\text{Total amount saved} = \$62.40 \text{ per month} \times 180 \text{ months} = \$11{,}232.$$

(c) In part (a)-(i) we found the monthly payment on a 4% mortgage of \$60,000 for 15 years to be \$444. The total amount paid over 15 years is then

$$\$444 \text{ per month} \times 180 \text{ months} = \$79{,}920.$$

In part (a)-(ii) we found the monthly payment on a 4% mortgage of \$60,000 for 30 years to be \$286.20. The total amount paid over 30 years is then

$$286.20 \text{ per month} \times 360 \text{ months} = \$103{,}032.$$

The amount saved by taking the mortgage over a shorter period of time is the difference:

$$\$103{,}032 - \$79{,}920 = \$23{,}112.$$

61. We have

$$\begin{aligned}\frac{f(n+2)}{f(n)} &= \frac{1000 \cdot 2^{-\frac{1}{4}-\frac{n+2}{2}}}{1000 \cdot 2^{-\frac{1}{4}-\frac{n}{2}}}\\ &= \frac{1000}{1000} \cdot \frac{2^{-\frac{1}{4}}}{2^{-\frac{1}{4}}} \cdot \frac{2^{-\frac{n+2}{2}}}{2^{-\frac{n}{2}}}\\ &= 2^{-\frac{n+2}{2}} \cdot 2^{\frac{n}{2}}\\ &= 2^{-\frac{n}{2}-\frac{2}{2}+\frac{n}{2}}\\ &= 2^{-1} = 0.5.\end{aligned}$$

This means a sheet two numbers higher in the series is half as wide. For instance, a sheet of $A3$ is half as wide as a sheet of $A1$.

65. The graph $a_0(b_0)^t$ climbs faster than that of $a_1(b_1)^t$, so $b_0 > b_1$.

Solutions for Section 4.2

Skill Refresher

S1. We have $f(0) = 5.6(1.043)^0 = 5.6$ and $f(3) = 5.6(1.043)^3 = 6.354$.

S5. We have

$$\begin{aligned} \frac{5}{x^2} &= 125 \\ \frac{1}{x^2} &= 25 \\ x^2 &= 1/25 \\ x &= \pm(1/25)^{1/2} = \pm 1/5 = \pm 0.2. \end{aligned}$$

EXERCISES

1. The formula $P_A = 200 + 1.3t$ for City A shows that its population is growing linearly. In year $t = 0$, the city has 200,000 people and the population grows by 1.3 thousand people, or 1,300 people, each year.

The formulas for cities B, C, and D show that these populations are changing exponentially. Since $P_B = 270(1.021)^t$, City B starts with 270,000 people and grows at an annual rate of 2.1%. Similarly, City C starts with 150,000 people and grows at 4.5% annually.

Since $P_D = 600(0.978)^t$, City D starts with 600,000 people, but its population decreases at a rate of 2.2% per year. We find the annual percent rate by taking $b = 0.978 = 1 + r$, which gives $r = -0.022 = -2.2\%$. So City D starts out with more people than the other three but is shrinking.

Figure 4.2 gives the graphs of the three exponential populations. Notice that the P-intercepts of the graphs correspond to the initial populations (when $t = 0$) of the towns. Although the graph of P_C starts below the graph of P_B, it eventually catches up and rises above the graph of P_B, because City C is growing faster than City B.

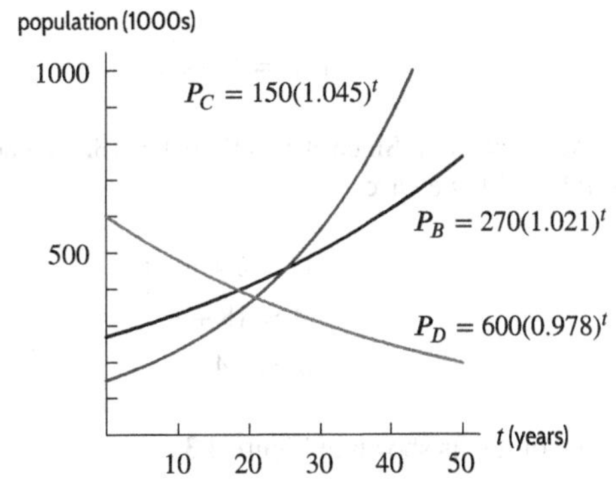

Figure 4.2: The graphs of the three exponentially changing populations

5. **(a)** See Table 4.1.

Table 4.1

t	0	1	2	3	4	5
$f(t)$	1000	1200	1440	1728	2073.6	2488.32

(b) We have

$$\frac{f(1)}{f(0)} = \frac{1200}{1000} = 1.2$$

$$\frac{f(2)}{f(1)} = \frac{1440}{1200} = 1.2$$
$$\frac{f(3)}{f(2)} = \frac{1728}{1440} = 1.2$$
$$\frac{f(4)}{f(3)} = \frac{2073.6}{1728} = 1.2$$
$$\frac{f(5)}{f(4)} = \frac{2488.32}{2073.6} = 1.2.$$

(c) All of the ratios of successive terms are 1.2. This makes sense because we have

$$f(0) = 1000$$
$$f(1) = 1000 \cdot 1.2$$
$$f(2) = 1000 \cdot 1.2 \cdot 1.2$$
$$f(3) = 1000 \cdot 1.2 \cdot 1.2 \cdot 1.2$$

and so on. Each term is the previous term multiplied by 1.2. It follows that the ratio of successive terms will always be the growth factor, which is 1.2 in this case.

9. (a) If a function is linear, then the differences in successive function values will be constant. If a function is exponential, the ratios of successive function values will remain constant. Now

$$i(1) - i(0) = 14 - 18 = -4$$

and

$$i(2) - i(1) = 10 - 14 = -4.$$

Checking the rest of the data, we see that the differences remain constant, so $i(x)$ is linear.

(b) We know that $i(x)$ is linear, so it must be of the form

$$i(x) = b + mx,$$

where m is the slope and b is the y-intercept. Since at $x = 0$, $i(0) = 18$, we know that the y-intercept is 18, so $b = 18$. Also, we know that at $x = 1$, $i(1) = 14$, we have

$$i(1) = b + m \cdot 1$$
$$14 = 18 + m$$
$$m = -4.$$

Thus, $i(x) = 18 - 4x$. The graph of $i(x)$ is shown in Figure 4.3.

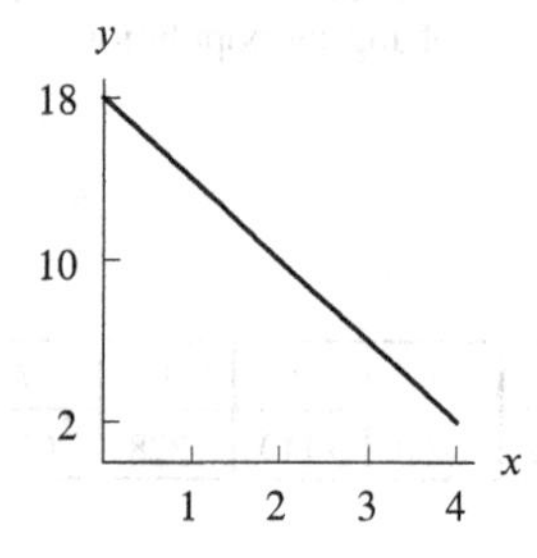

Figure 4.3

13. We know that $f(x) = ab^x$. Taking the ratio of $f(2)$ to $f(-1)$ we have

$$\begin{aligned}\frac{f(2)}{f(1)} &= \frac{1/27}{27} = \frac{ab^2}{ab^{-1}}\\ \frac{1}{(27)^2} &= b^3\\ b^3 &= \frac{1}{27^2}\\ b &= (\frac{1}{27^2})^{\frac{1}{3}}.\end{aligned}$$

Thus, $b = \frac{1}{9}$. Therefore, $f(x) = a(\frac{1}{9})^x$.

Using the fact that $f(-1) = 27$, we have

$$f(-1) = a\left(\frac{1}{9}\right)^{-1} = a \cdot 9 = 27,$$

which means $a = 3$. Thus,

$$f(x) = 3\left(\frac{1}{9}\right)^x.$$

17. Since the function is exponential, we know $y = ab^x$. We also know that $(0, 1/2)$ and $(3, 1/54)$ are on the graph of this function, so $1/2 = ab^0$ and $1/54 = ab^3$. The first equation implies that $a = 1/2$. Substituting this value in the second equation gives $1/54 = (1/2)b^3$ or $b^3 = 1/27$, or $b = 1/3$. Thus, $y = \frac{1}{2}\left(\frac{1}{3}\right)^x$.

21. Let f be the function whose graph is shown. Were f exponential, it would increase by equal factors on equal intervals. However, we see that

$$\begin{aligned}\frac{f(3)}{f(1)} &= \frac{11}{5} = 2.2\\ \frac{f(5)}{f(3)} &= \frac{30}{11} = 2.7.\end{aligned}$$

On these equal intervals, the value of f does not increase by equal factors, so f is not exponential.

25. **(a)** The function h is linear because equal spacing between successive input values ($\Delta x = 3$) results in equal spacing between successive output values ($\Delta h = 1.47$). On the other hand, the function f is exponential because ratios of successive rounded output values equal 1.10.

(b) Since f is exponential, we know that $f(x) = ab^x$, and we must figure out the values of the constants a and b. From our given information, we have $f(0) = 2.23$, which yields

$$a = ab^0 = 2.23.$$

Also, since $f(3) = 2.45$, we have

$$\begin{aligned}2.23b^3 &= 2.45\\ b &= \left(\frac{2.45}{2.23}\right)^{1/3}\\ &= 1.10^{1/3}\\ &= 1.03,\end{aligned}$$

so our final answer is $f(x) = 2.23((1.10)^{1/3})^x = 2.23(1.10)^{x/3} = 2.23(1.03)^x$.

PROBLEMS

29. Testing the rates of change for $R(t)$, we find that

$$\frac{2.61 - 2.32}{9 - 5} = 0.0725$$

and

$$\frac{3.12 - 2.61}{15 - 9} = 0.085,$$

so we know that $R(t)$ is not linear. If $R(t)$ is exponential, then $R(t) = ab^t$, and

$$R(5) = a(b)^5 = 2.32$$

and

$$R(9) = a(b)^9 = 2.61.$$

So

$$\begin{aligned}\frac{R(9)}{R(5)} &= \frac{ab^9}{ab^5} = \frac{2.61}{2.32}\\ \frac{b^9}{b^5} &= \frac{2.61}{2.32}\\ b^4 &= \frac{2.61}{2.32}\\ b &= (\frac{2.61}{2.32})^{\frac{1}{4}} \approx 1.030.\end{aligned}$$

Since

$$\begin{aligned}R(15) &= a(b)^{15} = 3.12\\ \frac{R(15)}{R(9)} &= \frac{ab^{15}}{ab^9} = \frac{3.12}{2.61}\\ b^6 &= \frac{3.12}{2.61}\\ b &= \left(\frac{3.12}{2.61}\right)^{\frac{1}{6}} \approx 1.030.\end{aligned}$$

Since the growth factor, b, is constant, we know that $R(t)$ could be an exponential function and that $R(t) = ab^t$. Taking the ratios of $R(5)$ and $R(9)$, we have

$$\begin{aligned}\frac{R(9)}{R(5)} &= \frac{ab^9}{ab^5} = \frac{2.61}{2.32}\\ b^4 &= 1.125\\ b &= 1.030.\end{aligned}$$

So $R(t) = a(1.030)^t$. We now solve for a by using $R(5) = 2.32$,

$$\begin{aligned}R(5) &= a(1.030)^5\\ 2.32 &= a(1.030)^5\\ a &= \frac{2.32}{1.030^5} \approx 2.001.\end{aligned}$$

Thus, $R(t) = 2.001(1.030)^t$.

33. One approach is to graph both functions and to see where the graph of $p(x)$ is below the graph of $q(x)$. From Figure 4.4, we see that $p(x)$ intersects $q(x)$ in two places: namely, at $x \approx -1.69$ and $x = 2$. We notice that $p(x)$ is above $q(x)$ between these two points and below $q(x)$ outside the segment defined by these two points. Hence $p(x) < q(x)$ for $x < -1.69$ and for $x > 2$.

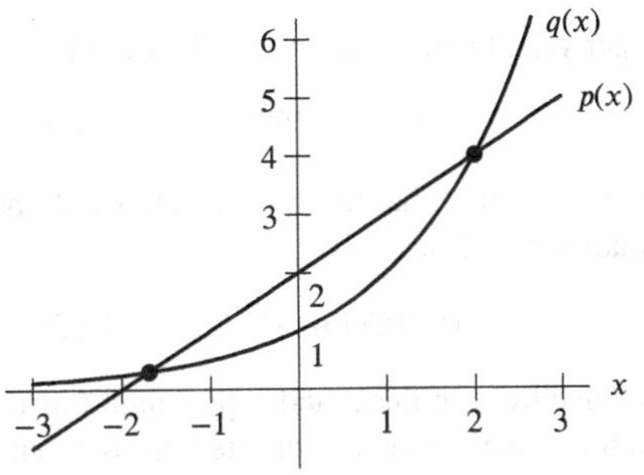

Figure 4.4

37. **(a)** Since, for $t < 0$, we know that the voltage is a constant 80 volts, $V(t) = 80$ on that interval.

For $t \geq 0$, we know that $v(t)$ is an exponential function, so $V(t) = ab^t$. According to this formula, $V(0) = ab^0 = a(1) = a$. According to the graph, $V(0) = 80$. From these two facts, we know that $a = 80$, so $V(t) = 80b^t$. If $V(10) = 80b^{10}$ and $V(10) = 15$ (from the graph), then

$$\begin{aligned} 80b^{10} &= 15 \\ b^{10} &= \frac{15}{80} \\ (b^{10})^{\frac{1}{10}} &= (\frac{15}{80})^{\frac{1}{10}} \\ b &\approx 0.8459 \end{aligned}$$

so that $V(t) = 80(0.8459)^t$ on this interval. Combining the two pieces, we have

$$V(t) = \begin{cases} 80 & \text{for } t < 0 \\ 80(0.8459)^t & \text{for } t \geq 0. \end{cases}$$

(b) Using a computer or graphing calculator, we can find the intersection of the line $y = 0.1$ with $y = 80(0.8459)^t$. We find $t \approx 39.933$ seconds.

41. We let W represent the winning time and t represent the number of years since 1994.

(a) To find the linear function, we first find the slope:

$$\text{Slope} = \frac{\Delta W}{\Delta t} = \frac{40.981 - 43.45}{16 - 0} = -0.154.$$

The vertical intercept is 43.45 so the linear function is $W = 43.45 - 0.154t$. The predicted winning time in 2018 is $W = 43.45 - 0.154(24) = 39.754$ seconds.

(b) The time at $t = 0$ is 43.45, so the exponential function is $W = 43.45(a)^t$. We use the fact that $W = 40.981$ when $t = 16$ to find a:

$$\begin{aligned} 40.981 &= 43.45(a)^{16} \\ 0.943176 &= a^{16} \\ a &= (0.943176)^{1/16} = 0.99635. \end{aligned}$$

The exponential function is $W = 43.45(0.99635)^t$. The predicted winning time in 2018 is $W = 43.45(0.99635)^{24} = 39.799$ seconds.

45. **(a)** Since the rate of change is constant, the increase is linear.

(b) Life expectancy is increasing at a constant rate of 3 months, or 0.25 years, each year. The slope is 0.25. When $t = 17$ we have $L = 79.1$. We use the point-slope form to find the linear function:

$$\begin{aligned} L - 79.1 &= 0.25(t - 17) \\ L - 79.1 &= 0.25t - 4.25 \\ L &= 0.25t + 74.85. \end{aligned}$$

(c) When $t = 50$, we have $L = 0.25(50) + 74.85 = 87.35$. If the rate of increase continues, babies born in 2050 will have a life expectancy of 87.35 years.

49. **(a)** Assuming linear growth at 250 per year, the population in 2013 would be

$$18{,}500 + 250 \cdot 10 = 21{,}000.$$

Using the population after one year, we find that the percent rate would be $250/18{,}500 \approx 0.013514 = 1.351\%$ per year, so after 10 years the population would be

$$18{,}500(1.013514)^{10} \approx 21{,}158.$$

The town's growth is poorly modeled by both linear and exponential functions.

(b) We do not have enough information to make even an educated guess about a formula.

Solutions for Section 4.3

Skill Refresher

S1. The intersection point is approximately $(1.65, 9)$. This means that $f(1.65) = g(1.65) = 9$. One solution to the equation $f(t) = g(t)$ is $t = 1.65$.

S5. **(a)** As x gets large and positive, y also gets large and positive.

(b) As x gets large and negative, y gets smaller and goes to 0.

EXERCISES

1. The function could not be exponential because it first decreases and then increases.

5. **(a)** See Table 4.2.

Table 4.2

x	−3	−2	−1	0	1	2	3
$f(x)$	1/8	1/4	1/2	1	2	4	8

(b) For large negative values of x, $f(x)$ is close to the x-axis. But for large positive values of x, $f(x)$ climbs rapidly away from the x-axis. As x gets larger, y grows more and more rapidly. See Figure 4.5.

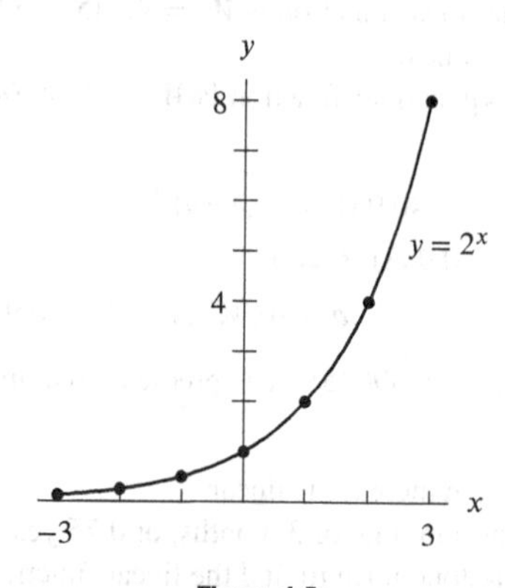

Figure 4.5

9. Yes, the graphs will cross. The graph of $g(x)$ has a smaller y-intercept but increases faster and will eventually overtake the graph of $f(x)$.

13. No, the graphs will not cross. Both functions are decreasing but the graph of $f(x)$ has a larger y-intercept and is decreasing at a slower rate than $g(x)$, so it will always be above the graph of $g(x)$.

17. The graph of $g(x)$ is shifted four units to the left of $f(x)$, and the graph of $h(x)$ is shifted two units to the right of $f(x)$.

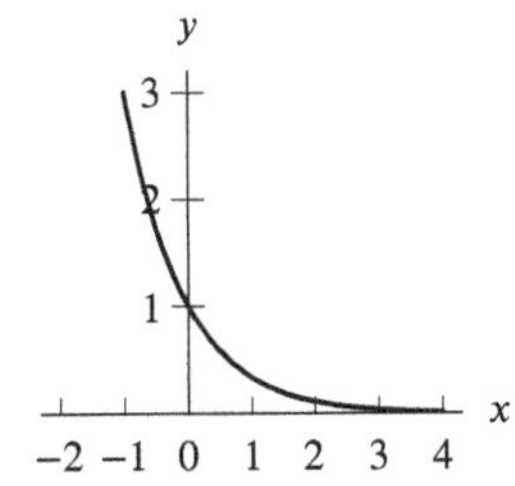

Figure 4.6: $f(x) = \left(\frac{1}{3}\right)^x$

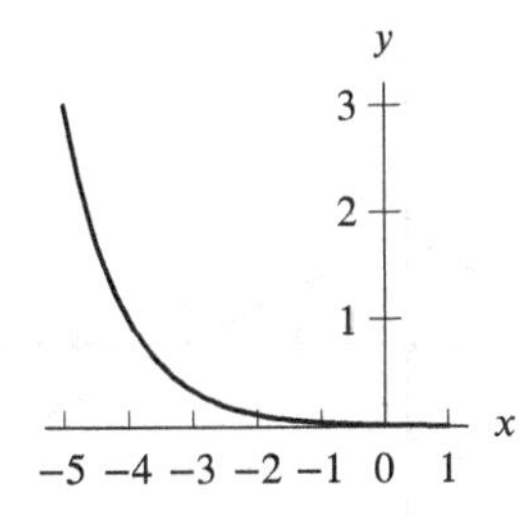

Figure 4.7: $g(x) = \left(\frac{1}{3}\right)^{x+4}$

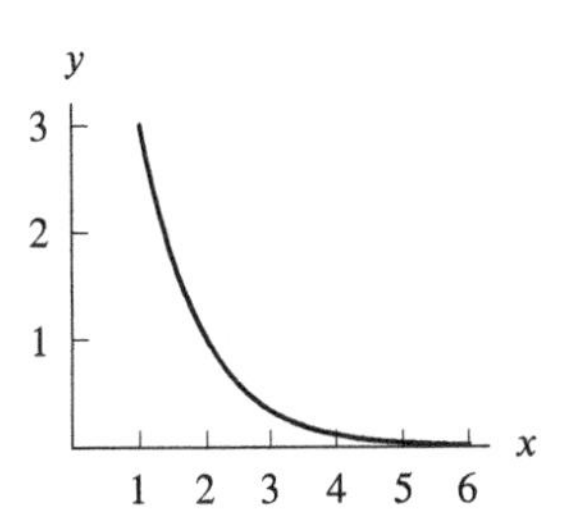

Figure 4.8: $h(x) = \left(\frac{1}{3}\right)^{x-2}$

21. Solve for P to obtain $P = 7(0.6)^t$. Graphing $P = 7(0.6)^t$ and tracing along the graph on a calculator gives us an answer of $t = 2.452$. See Figure 4.9.

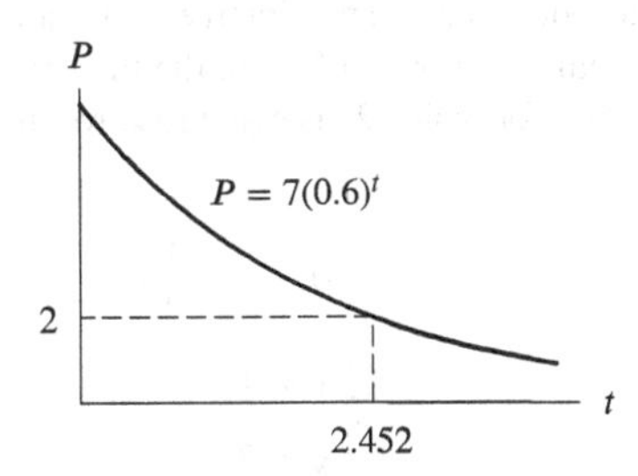

Figure 4.9

PROBLEMS

25. Since $y = a$ when $t = 0$ in $y = ab^t$, a is the y-intercept. Thus, the function with the greatest y-intercept, D, has the largest a.

29. As t approaches $-\infty$, the value of ab^t approaches zero for any a, so the horizontal asymptote is $y = 0$ (the t-axis).

33. A possible graph is shown in Figure 4.10. There are many possible answers.

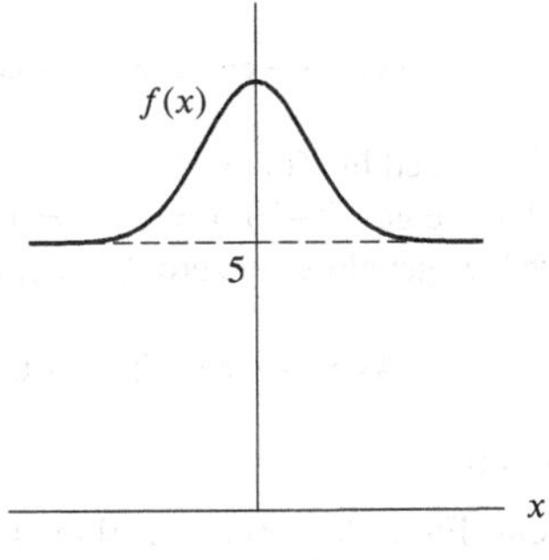

Figure 4.10

37. As r increases, the graph of $y = a(1+r)^t$ rises more steeply, so the point of intersection moves to the left and down. However, no matter how steep the graph becomes, the point of intersection remains above and to the right of the y-intercept of the second curve, or the point $(0, b)$. Thus, the value of y_0 decreases but does not reach b.

41. The function, when entered as $y = 1.04\char`^5x$, is interpreted as $y = (1.04^5)x = 1.217x$. This function's graph is a straight line in all windows. Parentheses must be used to ensure that x is in the exponent.

45. **(a)** See Figure 4.11.

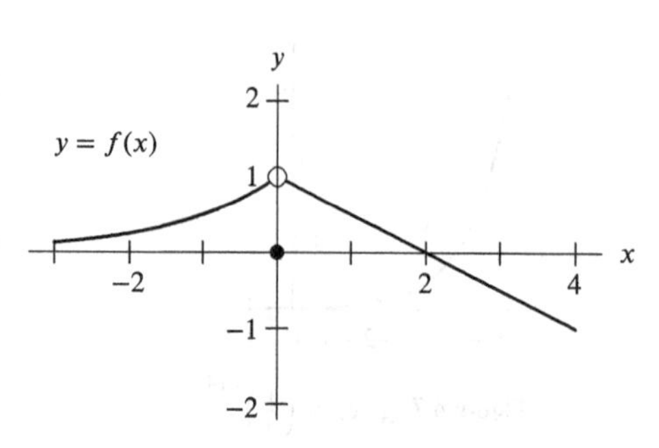

Figure 4.11

(b) The range of this function is all real numbers less than one — i.e. $f(x) < 1$.

(c) The y-intercept occurs at $(0, 0)$. This point is also an x-intercept. To solve for other x-intercepts we must attempt to solve $f(x) = 0$ for each of the two remaining parts of f. In the first case, we know that the function $f(x) = 2^x$ has no x-intercepts, as there is no value of x for which 2^x is equal to zero. In the last case, for $x > 0$, we set $f(x) = 0$ and solve for x:

$$\begin{aligned} 0 &= 1 - \frac{1}{2}x \\ \frac{1}{2}x &= 1 \\ x &= 2. \end{aligned}$$

Hence $x = 2$ is another x-intercept of f.

(d) As x gets large, the function is defined by $f(x) = 1 - 1/2x$. To determine what happens to f as $x \to +\infty$, find values of f for very large values of x. For example,

$$f(100) = 1 - \frac{1}{2}(100) = -49, \quad f(10000) = 1 - \frac{1}{2}(10000) = -4999$$

$$\text{and} \quad f(1{,}000{,}000) = 1 - \frac{1}{2}(1{,}000{,}000) = -499,999.$$

As x becomes larger, $f(x)$ becomes more and more negative. A way to write this is:

$$\text{As } x \to +\infty, \ f(x) \to -\infty.$$

As x gets very negative, the function is defined by $f(x) = 2^x$.

Choosing very negative values of x, we get $f(-100) = 2^{-100} = 1/2^{100}$, and $f(-1000) = 2^{-1000} = 1/2^{1000}$. As x becomes more negative the function values get closer to zero. We write

$$\text{As } x \to -\infty, \ f(x) \to 0.$$

(e) Increasing for $x < 0$, decreasing for $x > 0$.

49. **(a)** Figure 4.12 shows the three populations. From this graph, the three models seem to be in good agreement. Models 1 and 3 are indistinguishable; model 2 appears to rise a little faster. However, notice that we cannot see the behavior beyond 50 months because our function values go beyond the top of the viewing window.

(b) Figure 4.13 shows the population differences. The graph of $y = f_2(x) - f_1(x) = 3(1.21)^x - 3(1.2)^x$ grows very rapidly, especially after 40 months. The graph of $y = f_3(x) - f_1(x) = 3.01(1.2)^x - 3(1.2)^x$ is hardly visible on this scale.

(c) Models 1 and 3 are in good agreement, but model 2 predicts a much larger mussel population than does model 1 after only 50 months. We can come to at least two conclusions. First, even small differences in the base of an exponential function can be highly significant, while differences in initial values are not as significant. Second, although two exponential curves can look very similar, they can actually be making very different predictions as time increases.

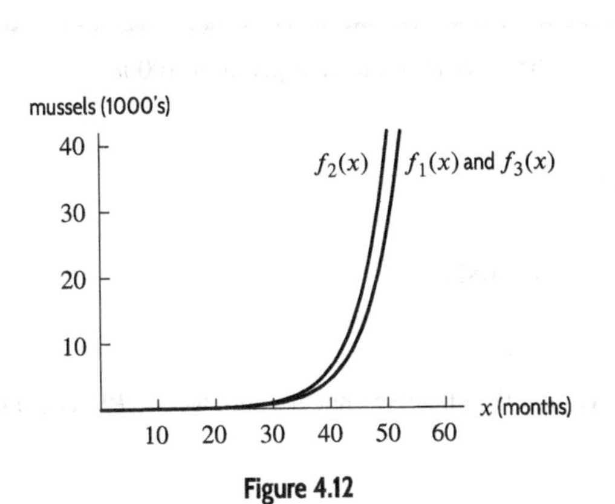

Figure 4.12

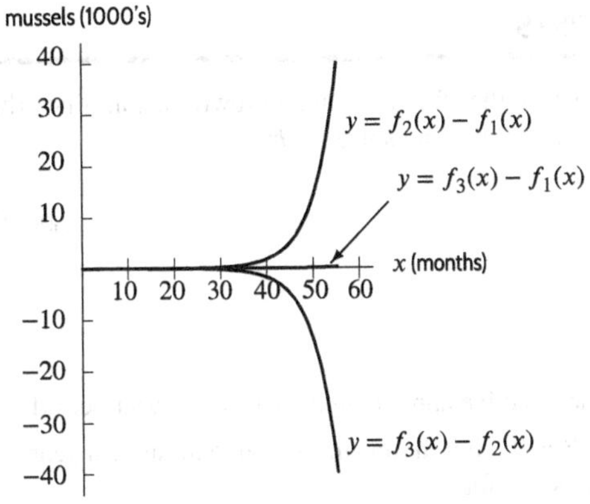

Figure 4.13

53. It appears in the graph that

(a) $\lim_{x\to-\infty} f(x) = 5$

(b) $\lim_{x\to\infty} f(x) = -3$.

Of course, we need to be sure that we are seeing all the important features of the graph in order to have confidence in these estimates.

Solutions for Section 4.4

Skill Refresher

S1. Expressing r in decimal form, we have $r = 0.05$. Therefore, We have

$$\left(1+\frac{0.05}{12}\right)^{12\cdot 5} = (1+0.0042)^{60} = (1.0042)^{60} = 1.286$$

S5. **(a)** (i) We have

$$2\left(1+\frac{0.5}{2}\right)^{2\cdot 3} = 2\,(1+0.25)^6 = 2\,(1.25)^6 = 2\cdot 3.8147 = 7.6294$$

.

(ii) We have

$$\left(2\left(1+\frac{0.5}{2}\right)\right)^{2\cdot 3} = (2\,(1+0.25))^6 = (2\,(1.25))^6 = (2.5)^6 = 244.1406$$

.

(iii) We have

$$2\left(1+\left(\frac{0.5}{2}\right)^{2\cdot 3}\right) = 2\,(1+(0.25)^6) = 2\,(1+0.0002) = 2\cdot 1.0002 = 2.0004.$$

(iv) We have

$$2\left(1+\frac{0.5^{2\cdot 3}}{2}\right) = 2\left(1+\frac{0.5^6}{2}\right) = 2\left(1+\frac{0.01563}{2}\right) = 2\,(1+0.0078)) = 2\cdot 1.0078 = 2.0156$$

.

(b) None are equal.

EXERCISES

1. Let b be the effective annual growth factor. Since the amount in the account at time t is given by $1000b^t$, we set $1000b^{15}$ equal to 3500 and solve for b:

$$\begin{aligned} 1000b^{15} &= 3500 \\ b^{15} &= 3.5 \\ b &= (3.5)^{1/15} = 1.0871. \end{aligned}$$

The effective annual yield over the 15-year period was 8.71% per year.

5. If P is the initial amount, the amount after 8 years is $0.5P$. To find the effective annual yield, we set Pb^8 equal to $0.5P$ and solve for b:

$$\begin{aligned} b^8 &= 0.5 \\ b &= (0.5)^{1/8} = 0.917. \end{aligned}$$

Since $0.917 = 1 - 0.083$, the investment has decreased by an effective annual rate of -8.3% per year.

9. **(a)** If the interest is compounded annually, there will be $\$500 \cdot 1.05 = \$525$ after one year.
(b) If the interest is compounded weekly, there will be $500 \cdot (1 + 0.05/52)^{52} = \$525.62$ after one year.
(c) If the interest is compounded every minute, there will be $500 \cdot (1 + 0.05/525{,}600)^{525600} = \$525.64$ after one year.

13. **(a)** The nominal rate is the stated annual interest without compounding, thus 3%.
The effective annual rate for an account paying 1% compounded annually is 3%.
(b) The nominal rate is the stated annual interest without compounding, thus 3%.
With quarterly compounding, there are four interest payments per year, each of which is $3/4 = 0.75\%$. Over the course of the year, this occurs four times, giving an effective annual rate of $1.0075^4 = 1.03034$, which is 3.034%.
(c) The nominal rate is the stated annual interest without compounding, thus 3%.
With daily compounding, there are 365 interest payments per year, each of which is $(3/365)\%$. Over the course of the year, this occurs 365 times, giving an effective annual rate of $(1 + 0.03/365)^{365} = 1.03045$, which is 3.045%.

PROBLEMS

17. If the investment is growing by 3% per year, we know that, at the end of one year, the investment will be worth 103% of what it had been the previous year. At the end of two years, it will be 103% of 103% $= (1.03)^2$ as large. At the end of 10 years, it will have grown by a factor of $(1.03)^{10}$, or 1.34392. The investment will be 134.392% of what it had been, so we know that it will have increased by 34.392%. Since $(1.03)^{10} \approx 1.34392$, it increases by 34.392%.

21. (i) Equation (b). Since the growth factor is 1.12, or 112%, the annual interest rate is 12%.
(ii) Equation (a). An account earning at least 1% monthly will have a monthly growth factor of at least 1.01, which means that the annual (12-month) growth factor will be at least

$$(1.01)^{12} = 1.1268.$$

Thus, an account earning at least 1% monthly will earn at least 12.68% yearly. The only account that earns this much interest is account (a).
(iii) Equation (c). An account earning 12% annually compounded semi-annually will earn 6% twice yearly. In t years, there are $2t$ half-years.
(iv) Equations (b), (c) and (d). An account that earns 3% each quarter ends up with a yearly growth factor of $(1.03)^4 = 1.1255$. This corresponds to an annual percentage rate of 12.55%. Accounts (b), (c) and (d) earn less than this. Check this by determining the growth factor in each case.
(v) Equations (a) and (e). An account that earns 6% every 6 months will have a growth factor, after 1 year, of $(1+0.06)^2 = 1.1236$, which is equivalent to a 12.36% annual interest rate, compounded annually. Account (a), earning 20% each year, earns more than 6% twice each year, or 12.36% annually. Account (e), which earns 3% each quarter, earns $(1.03)^2 = 1.0609$, or 6.09% every 6 months, which is greater than 6%.

Solutions for Section 4.5

Skill Refresher

S1. We have $e^{0.07} = 1.073$.

S5. We have $f(0) = 2.3e^{0.3(0)} = 2.3$ and $f(4) = 2.3e^{0.3(4)} = 7.636$.

S9. Writing the function as

$$f(t) = \left(3e^{0.04t}\right)^3 = 3^3 e^{0.04t\cdot 3} = 27e^{0.12t},$$

we have $a = 27$ and $k = 0.12$.

S13. Writing the function as

$$m(x) = \frac{7e^{0.2x}}{\sqrt{3e^x}} = \frac{7}{\sqrt{3}}e^{0.2x} \cdot e^{-0.5x} = \frac{7}{\sqrt{3}}e^{-0.3x},$$

we have $a = \frac{7}{\sqrt{3}}$ and $k = -0.3$.

EXERCISES

1. We know that $e \approx 2.71828$, so $2 < e < 3$. Since e lies between 2 and 3, the graph of $y = e^x$ lies between the graphs of $y = 2^x$ and $y = 3^x$. Since 3^x increases faster than 2^x, the correct matching is shown in Figure 4.14.

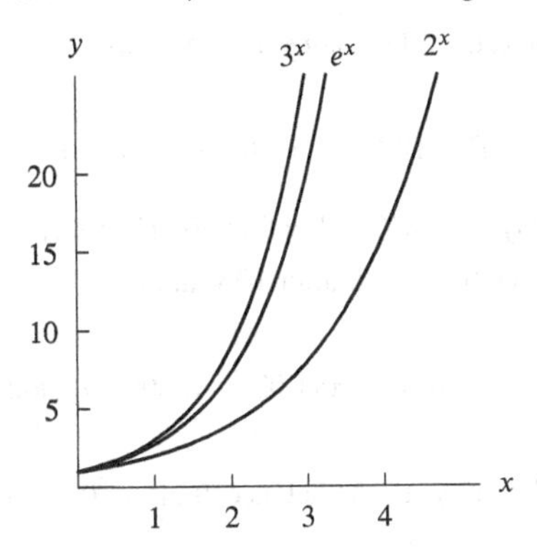

Figure 4.14

5. Calculating the equivalent continuous rates, we find $e^{0.45} = 1.568$, $e^{0.47} = 1.600$, $e^{0.5} = 1.649$. Thus the functions to be matched are

(a) $(1.5)^x$ **(b)** $(1.568)^x$ **(c)** $(1.6)^x$ **(d)** $(1.649)^x$

So (a) is (IV), (b) is (III), (c) is (II), (d) is (I).

9. **(a)** We see that $Q_0 = 2.7$
(b) Since the base is less than one, the quantity is decreasing.
(c) Since the base is $0.12 = 1 - 0.88$, the decay rate is 88% per unit time.
(d) No, the growth rate is not continuous.

13. **(a)** We see that $Q_0 = 1$
(b) Since the base is greater than one, the quantity is increasing.
(c) Since the base is $2 = 1 + 1$, the growth rate is 100% per unit time. The quantity is doubling every time unit.
(d) No, the growth rate is not continuous.

PROBLEMS

17. **(a)** Since the decay rate is not continuous, we have $Q = 500(0.93)^t$. At $t = 10$ we have $Q = 500(0.93)^{10} = 241.991$.
(b) Since the decay rate is continuous, we have $Q = 500e^{-0.07t}$. At $t = 10$ we have $Q = 500e^{-0.07(10)} = 248.293$. As we expect, the results are similar for continuous and not-continuous assumptions, but slightly larger if we assume a continuous decay rate.

21. **(a)** Using $P = P_0 e^{kt}$ where $P_0 = 25{,}000$ and $k = 7.5\%$, we have

$$P(t) = 25{,}000e^{0.075t}.$$

(b) We first need to find the growth factor so will rewrite

$$P = 25{,}000e^{0.075t} = 25{,}000(e^{0.075})^t \approx 25{,}000(1.07788)^t.$$

At the end of a year, the population is 107.788% of what it had been at the end of the previous year. This corresponds to an increase of approximately 7.788%. This is greater than 7.5% because the rate of 7.5% per year is being applied to larger and larger amounts. In one instant, the population is growing at a rate of 7.5% per year. In the next instant, it grows again at a rate of 7.5% a year, but 7.5% of a slightly larger number. The fact that the population is increasing in tiny increments continuously results in an actual increase greater than the 7.5% increase that would result from one single jump of 7.5% at the end of the year.

25. **(a)** First, we note that after t hours, the population, P, of the colony is given by $P = 100e^{0.25t}$. Substituting 4 for t in this equation, we obtain

$$P = 100e^{0.25(4)} = 271.828.$$

Therefore, there are about 272 bacteria in the colony after 4 hours.

(b) Since

$$P = 100e^{0.25t} = 100(e^{0.25})^t = 100(1.2840)^t,$$

the hourly growth rate of the colony is $1.2840 - 1 = 0.2840$. Therefore, the colony grows by 28.4% each hour.

29. **(a)** For an annual interest rate of 2%, the balance B after 10 years is

$$B = 5000(1.02)^{10} = 6094.97 \text{ dollars}.$$

(b) For a continuous interest rate of 2% per year, the balance B after 10 years is

$$B = 5000e^{0.02\cdot 10} = 6107.01 \text{ dollars}.$$

33. With continuous compounding, the interest earns interest during the year, so the balance grows faster with continuous compounding than with annual compounding. Curve A corresponds to continuous compounding and curve B corresponds to annual compounding. The initial amount in both cases is the vertical intercept, \$500.

37. Since $e^{0.053} = 1.0544$, the effective annual yield of the account paying 5.3% interest compounded continuously is 5.44%. Since this is less than the effective annual yield of 5.5% from the 5.5% compounded annually, we see that the account paying 5.5% interest compounded annually is slightly better.

41. The balance in the first bank is $10{,}000(1.05)^8 = \$14{,}774.55$. The balance in the second bank is $10{,}000e^{0.05(8)} = \$14{,}918.25$. The bank with continuously compounded interest has a balance \$143.70 higher.

45. **(a)** Since poultry production is increasing at a constant continuous percent rate, we use the exponential formula $P = ae^{kt}$. Since $P = 116$ when $t = 0$, we have $a = 116$. Since $k = 0.01$, we have

$$P = 116e^{0.01t}.$$

(b) When $t = 4$, we have $P = 116e^{0.01(4)} = 120.734$. In the year 2020, the formula predicts that world poultry production will be about 121 million tons.

(c) A graph of $P = 116e^{0.01t}$ is given in Figure 4.15. We see that when $P = 125$ we have $t = 7.5$. We expect production to be 125 million tons in 2023 or 2024.

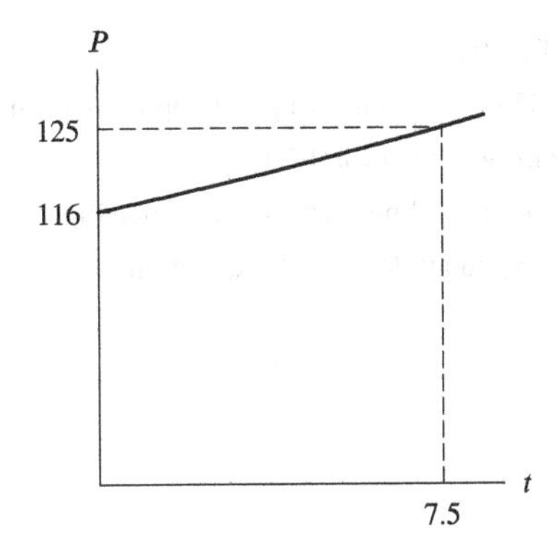

Figure 4.15

49. **(a)** Let p_0 be the price of an item at the beginning of 2000. At the beginning of 2001, its price will be 103.4% of that initial price or $1.034p_0$. At the beginning of 2002, its price will be 102.8% of the price from the year before, that is:

$$\text{Price beginning 2002} = (1.028)(1.034p_0).$$

By the beginning of 2003, the price will be 101.6% of its price the previous year.

$$\begin{aligned}\text{Price beginning 2003} &= 1.016(\text{price beginning 2002}) \\ &= 1.016(1.028)(1.034p_0).\end{aligned}$$

Continuing this process,

$$\begin{aligned}(\text{Price beginning 2005}) &= (1.027)(1.023)(1.016)(1.028)(1.034)p_0 \\ &\approx 1.135p_0.\end{aligned}$$

So, the cost at the beginning of 2005 is 113.5% of the cost at the beginning of 2000 and the total percent increase is 13.5%.

(b) If r is the average inflation rate for this time period, then $b = 1 + r$ is the factor by which the population on the average grows each year. Using this average growth factor, if the price of an item is initially p_0, at the end of a year its value would be $p_0 b$, at the end of two years it would be $(p_0 b)b = p_0 b^2$, and at the end of five years $p_0 b^5$. According to the answer in part (a), the price at the end of five years is $1.135p_0$. So

$$\begin{aligned}p_0 b^5 &= 1.135p_0 \\ b^5 &= 1.135 \\ b &= (1.135)^{1/5} \approx 1.026.\end{aligned}$$

If $b = 1.026$, then $r = 0.026$ or 2.6%, the average annual inflation rate.

(c) We assume that the average rate of 2.6% inflation for 2000 through 2004 holds through the beginning of 2010. So, on average, the price of the shower curtain is 102.6% of what it was the previous year for ten years. Then the price of the shower curtain would be $20(1.026)^{10} \approx \$25.85$.

53. $\lim_{x\to\infty} e^{-3x} = 0.$

57. The values of a and k are both positive.

STRENGTHEN YOUR UNDERSTANDING

1. True. If the constant rate is r then the formula is $f(t) = a \cdot (1 + r)^t$. The function decreases when $0 < 1 + r < 1$ and increases when $1 + r > 1$.

5. False. The annual growth factor would be 1.04, so $S = S_0(1.04)^t$.

9. True. The initial value means the value of Q when $t = 0$, so $Q = f(0) = a \cdot b^0 = a \cdot 1 = a$.

13. False. This is the formula of a linear function.

17. True. The irrational number $e = 2.71828\cdots$ has this as a good approximation.

21. True. The initial value is 200 and the growth factor is 1.04.

25. True. Since k is the continuous growth rate and negative, Q is decreasing.

29. True. The interest from any quarter is compounded in subsequent quarters.

CHAPTER FIVE

Solutions for Section 5.1

Skill Refresher

S1. Since $1000 = (10)(10)(10) = 10^3$, our answer is 10^3.

S5. Since $10^0 = 1$, our answer is 10^0.

S9. Since $0.1 = 10^{-1}$, we have $x = -1$.

S13. Since $10^0 = 1$, we have $x = 0$. Similarly, it follows for any constant b, if $b^x = 1$, then $x = 0$.

S17. Since $\sqrt{e^{9t}} = e^{\frac{9}{2}t}$, we have $e^{\frac{9}{2}t} = e^7$. Solving the equation $\frac{9}{2}t = 7$, we have $t = \frac{14}{9}$.

EXERCISES

1. The equation $\log 0.01 = -2$ is equivalent to $10^{-2} = 0.01$.

5. The statement is equivalent to $0.646 = e^{-0.437}$.

9. The equation $\ln 4 = x^2$ is equivalent to $e^{x^2} = 4$.

13. The statement is equivalent to $-4 = \ln(0.0183)$.

17. The equation $e^{2x} = 7$ is equivalent to $2x = \ln 7$.

21. We are looking for a power of 10, and $\log 0.1$ is asking for the power of 10 which gives 0.1. Since $0.1 = \frac{1}{10} = 10^{-1}$, we know that $\log 0.1 = \log 10^{-1} = -1$.

25. Using the identity $\log 10^N = N$, we have $\log 10^2 = 2$.

29. By definition, $10^{\log 7} = 7$.

33. Recall that $\sqrt{e} = e^{1/2}$. Using the identity $\ln e^N = N$, we get $\ln\sqrt{e} = \ln e^{1/2} = \frac{1}{2}$.

37. Since $\frac{1}{\sqrt{e}} = e^{-1/2}$, $\ln\frac{1}{\sqrt{e}} = \ln e^{-1/2} = -\frac{1}{2}$.

41. $\ln(\ln e) = \ln(1) = 0$.

45. By definition, $\log_5 125$ is the power of 5 which gives 125. Since $5^3 = 125$, we get $\log_5 125 = 3$.

49. We are solving for an exponent, so we use logarithms. We can use either the common logarithm or the natural logarithm. Using the log rules, we have

$$\begin{aligned}
1.45^x &= 25\\
\log(1.45^x) &= \log(25)\\
x\log(1.45) &= \log(25)\\
x &= \frac{\log(25)}{\log(1.45)} = 8.663.
\end{aligned}$$

If we had used the natural logarithm, we would have

$$x = \frac{\ln(25)}{\ln(1.45)} = 8.663.$$

53. We begin by dividing both sides by 17 to isolate the exponent:

$$\frac{48}{17} = (2.3)^w.$$

We then take the log of both sides and use the rules of logs to solve for w:

$$\log\frac{48}{17} = \log(2.3)^w$$
$$\log\frac{48}{17} = w\log(2.3)$$
$$w = \frac{\log\frac{48}{17}}{\log(2.3)} = 1.246.$$

PROBLEMS

57. **(a)** Using the identity $\ln e^N = N$, we get $\ln e^{2x} = 2x$.
(b) Using the identity $e^{\ln N} = N$, we get $e^{\ln(3x+2)} = 3x + 2$.
(c) Since $\frac{1}{e^{5x}} = e^{-5x}$, we get $\ln\left(\frac{1}{e^{5x}}\right) = \ln e^{-5x} = -5x$.
(d) Since $\sqrt{e^x} = (e^x)^{1/2} = e^{\frac{1}{2}x}$, we have $\ln\sqrt{e^x} = \ln e^{\frac{1}{2}x} = \frac{1}{2}x$.

61. False. $\frac{\log A}{\log B}$ cannot be rewritten.

65. False. $\sqrt{\log x} = (\log x)^{1/2}$.

69. Rewrite as $2\ln x^{-2} + \ln x^4 = 2(-2)\ln x + 4\ln x = 0$.

73. Rewrite as $\ln\frac{1}{e^x+1} = \ln 1 - \ln(e^x+1) = -\ln(e^x+1)$.

77. Taking natural logs of both sides, we get

$$\ln(e^{-5x}) = \ln 9.$$

This gives

$$-5x = \ln 9$$
$$x = -\frac{\ln 9}{5} \approx -0.439.$$

81. We have

$$e^{x+4} = 10$$
$$\ln e^{x+4} = \ln 10$$
$$x + 4 = \ln 10$$
$$x = \ln 10 - 4$$

85.

$$0.4\left(\frac{1}{3}\right)^{3x} = 7\cdot 2^{-x}$$
$$0.4\left(\frac{1}{3}\right)^{3x}\cdot 2^x = 7\cdot 2^{-x}\cdot 2^x = 7$$
$$0.4\left(\left(\frac{1}{3}\right)^3\right)^x\cdot 2^x = 7$$
$$\left(\left(\frac{1}{3}\right)^3\cdot 2\right)^x = \frac{7}{0.4} = 7\left(\frac{5}{2}\right) = \frac{35}{2}$$
$$\log\left(\frac{2}{27}\right)^x = \log\frac{35}{2}$$
$$x\log\left(\frac{2}{27}\right) = \log\frac{35}{2}$$

$$x = \frac{\log\left(\frac{35}{2}\right)}{\log\left(\frac{2}{27}\right)}.$$

89. Taking logs of both sides, we get

$$\log 19^{6x} = \log(77 \cdot 7^{4x}).$$

This gives

$$\begin{aligned}
6x \log 19 &= \log 77 + \log 7^{4x} \\
6x \log 19 &= \log 77 + 4x \log 7 \\
6x \log 19 - 4x \log 7 &= \log 77 \\
x(6 \log 19 - 4 \log 7) &= \log 77 \\
x &= \frac{\log 77}{6 \log 19 - 4 \log 7} \approx 0.440.
\end{aligned}$$

93. The properties of logarithms do not say anything about the log of a difference. So we move the second term to the right side of the equation before introducing logs:

$$P = P_0 e^{kx}.$$

Now take natural logs and use their properties:

$$\begin{aligned}
\ln P &= \ln(P_0 e^{kx}) \\
\ln P &= \ln P_0 + \ln(e^{kx}) \\
\ln P &= \ln P_0 + kx \\
kx &= \ln P - \ln P_0 \\
x &= \frac{\ln P - \ln P_0}{k}.
\end{aligned}$$

97.

$$\begin{aligned}
\frac{\log x^2 + \log x^3}{\log(100x)} &= 3 \\
\log x^2 + \log x^3 &= 3 \log(100x) \\
2 \log x + 3 \log x &= 3(\log 100 + \log x) \\
5 \log x &= 3(2 + \log x) \\
5 \log x &= 6 + 3 \log x \\
2 \log x &= 6 \\
\log x &= 3 \\
x &= 10^3 = 1000.
\end{aligned}$$

To check, we see that

$$\begin{aligned}
\frac{\log x^2 + \log x^3}{\log(100x)} &= \frac{\log(1000^2) + \log(1000^3)}{\log(100 \cdot 1000)} \\
&= \frac{\log(1{,}000{,}000) + \log(1{,}000{,}000{,}000)}{\log(100{,}000)} \\
&= \frac{6+9}{5} \\
&= 3,
\end{aligned}$$

as required.

101. We first rearrange the equation so that the natural log is alone on one side, and we then convert to exponential form:

$$\begin{aligned} 2\ln(6x-1)+5 &= 7 \\ 2\ln(6x-1) &= 2 \\ \ln(6x-1) &= 1 \\ e^{\ln(6x-1)} &= e^1 \\ 6x-1 &= e \\ 6x &= e+1 \\ x &= \frac{e+1}{6} \approx 0.620. \end{aligned}$$

105. Using properties of logs, we have

$$\begin{aligned} \log(MN^x) &= a \\ \log M + x\log N &= a \\ x\log N &= a - \log M \\ x &= \frac{a-\log M}{\log N}. \end{aligned}$$

109. **(a)** $\log 3 = \log\frac{15}{5} = \log 15 - \log 5$
(b) $\log 25 = \log 5^2 = 2\log 5$
(c) $\log 75 = \log(15\cdot 5) = \log 15 + \log 5$

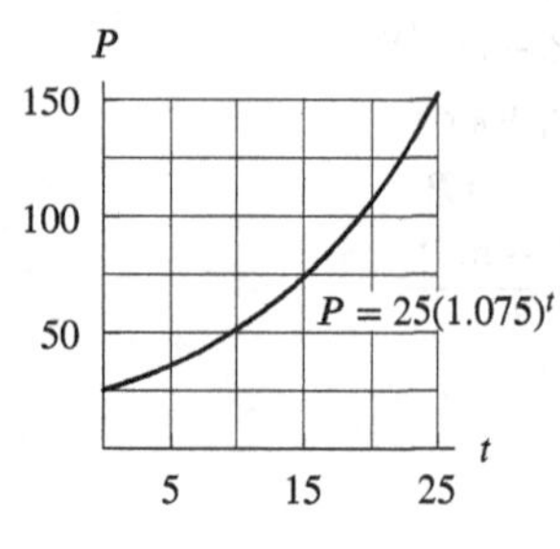

Figure 5.1

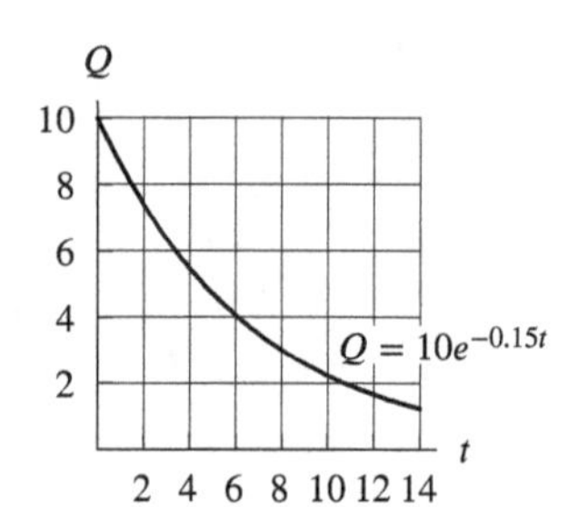

Figure 5.2

113. To find a formula for P, we find the points labeled (x_0, y_0) and (x_1, y_1) in Figure 5.3. We see that $x_1 = 8$ and that $y_1 = 50$. From the graph of Q, we see that

$$y_0 = Q(8) = 117.7181(0.7517)^8 = 12.$$

To find x_0 we use the fact that $Q(x_0) = 50$:

$$\begin{aligned} 117.7181(0.7517)^{x_0} &= 50 \\ (0.7517)^{x_0} &= \frac{50}{117.7181} \\ x_0 &= \frac{\log(50/117.7181)}{\log 0.7517} \\ &= 3. \end{aligned}$$

We have $P(3) = 12$ and $P(8) = 50$. Using the ratio method, we have

$$\begin{aligned} \frac{ab^8}{ab^3} &= \frac{P(8)}{P(3)} \\ b^5 &= \frac{50}{12} \\ b &= \left(\frac{50}{12}\right)^{1/5} \approx 1.3303. \end{aligned}$$

Now we can solve for a:

$$\begin{aligned} a(1.3303)^3 &= 12 \\ a &= \frac{12}{(1.3303)^3} \\ &\approx 5.0969. \end{aligned}$$

so $P(x) = 5.0969(1.3303)^x$.

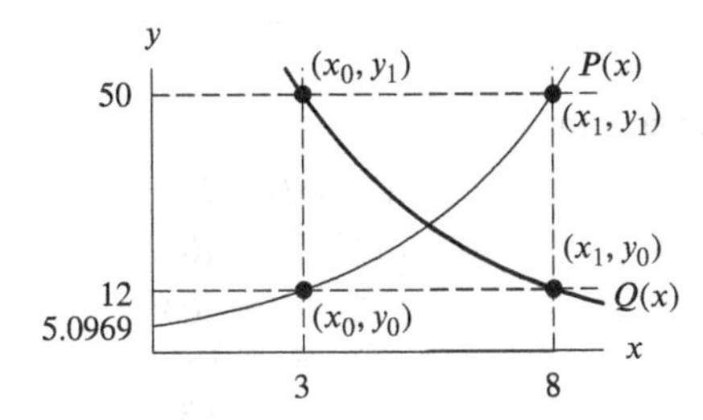

Figure 5.3

117. We have

$$\begin{aligned} \ln A &= \ln 2^{2^{283}} \\ &= 2^{283} \cdot \ln 2 \\ \text{so} \quad \ln(\ln A) &= \ln\left(2^{283} \cdot \ln 2\right) \\ &= \ln\left(2^{283}\right) + \ln(\ln 2) \\ &= 2^{83} \cdot \ln 2 + \ln(\ln 2) \\ &= 6.704 \times 10^{24} \qquad \text{with a calculator} \\ \text{and} \quad \ln B &= \ln 3^{3^{352}} \\ &= 3^{352} \cdot \ln 3 \\ \text{so} \quad \ln(\ln B) &= \ln\left(3^{352} \cdot \ln 3\right) \\ &= \ln\left(3^{352}\right) + \ln(\ln 3) \\ &= 3^{52} \cdot \ln 3 + \ln(\ln 3) \\ &= 7.098 \times 10^{24}. \qquad \text{with a calculator} \end{aligned}$$

Thus, since $\ln(\ln B)$ is larger than $\ln(\ln A)$, we infer that $B > A$.

Solutions for Section 5.2

Skill Refresher

S1. Rewrite as $10^{-\log 5x} = 10^{\log(5x)^{-1}} = (5x)^{-1}$.

S5. Taking logs of both sides, we get

$$\log(4^x) = \log 9.$$

This gives

$$x \log 4 = \log 9$$

or in other words

$$x = \frac{\log 9}{\log 4} = 1.585.$$

S9. Dividing both sides by 16 gives

$$e^{-2x} = \frac{4}{16} = 0.25.$$

Taking natural logs of both sides and then solving for x, we get

$$\begin{aligned} \ln(e^{-2x}) &= \ln(0.25) \\ -2x &= \ln(0.25) \\ x &= \frac{\ln(0.25)}{-2} = 0.693. \end{aligned}$$

We see that $x = 0.693$.

S13. We begin by converting to exponential form:

$$\begin{aligned} \log(2x+7) &= 2 \\ 10^{\log(2x+7)} &= 10^2 \\ 2x+7 &= 100 \\ 2x &= 93 \\ x &= \frac{93}{2}. \end{aligned}$$

EXERCISES

1. The continuous percent growth rate is the value of k in the equation $Q = ae^{kt}$, which is 7.

To convert to the form $Q = ab^t$, we first say that the right sides of the two equations equal each other (since each equals Q), and then we solve for a and b. Thus, we have $ab^t = 4e^{7t}$. At $t = 0$, we can solve for a:

$$\begin{aligned} ab^0 &= 4e^{7\cdot 0} \\ a \cdot 1 &= 4 \cdot 1 \\ a &= 4. \end{aligned}$$

Thus, we have $4b^t = 4e^{7t}$, and we solve for b:

$$\begin{aligned} 4b^t &= 4e^{7t} \\ b^t &= e^{7t} \\ b^t &= \left(e^7\right)^t \\ b &= e^7 \approx 1096.633. \end{aligned}$$

Therefore, the equation is $Q = 4 \cdot 1096.633^t$.

5. We want $25e^{0.053t} = 25(e^{0.053})^t = ab^t$, so we choose $a = 25$ and $b = e^{0.053} = 1.0544$. The given exponential function is equivalent to the exponential function $y = 25(1.0544)^t$. The annual percent growth rate is 5.44% and the continuous percent growth rate per year is 5.3% per year.

9. To convert to the form $Q = ae^{kt}$, we first say that the right sides of the two equations equal each other (since each equals Q), and then we solve for a and k. Thus, we have $ae^{kt} = 14(0.862)^{1.4t}$. At $t = 0$, we can solve for a:

$$\begin{aligned} ae^{k\cdot 0} &= 14(0.862)^0 \\ a \cdot 1 &= 14 \cdot 1 \\ a &= 14. \end{aligned}$$

Thus, we have $14e^{kt} = 14(0.862)^{1.4t}$, and we solve for k:

$$\begin{aligned} 14e^{kt} &= 14(0.862)^{1.4t} \\ e^{kt} &= \left(0.862^{1.4}\right)^t \\ \left(e^k\right)^t &= (0.812)^t \\ e^k &= 0.812 \\ \ln e^k &= \ln 0.812 \\ k &= -0.208. \end{aligned}$$

Therefore, the equation is $Q = 14e^{-0.208t}$.

13. We have $a = 230$, $b = 1.182$, $r = b - 1 = 18.2\%$, and $k = \ln b = 0.1672 = 16.72\%$.

17. Writing this as $Q = 12.1(10^{-0.11})^t$, we have $a = 12.1$, $b = 10^{-0.11} = 0.7762$, $r = b - 1 = -22.38\%$, and $k = \ln b = -25.32\%$.

PROBLEMS

21. Since the growth factor is $1.027 = 1 + 0.027$, the formula for the bank account balance, with an initial balance of a and time t in years, is

$$B = a(1.027)^t.$$

The balance doubles for the first time when $B = 2a$. Thus, we solve for t after putting B equal to $2a$ to give us the doubling time:

$$\begin{aligned} 2a &= a(1.027)^t \\ 2 &= (1.027)^t \\ \log 2 &= \log(1.027)^t \\ \log 2 &= t\log(1.027) \\ t &= \frac{\log 2}{\log(1.027)} = 26.017. \end{aligned}$$

So the doubling time is about 26 years.

25. The growth factor for Einsteinium-253 should be $1 - 0.03406 = 0.96594$, since it is decaying by 3.406% per day. Therefore, the decay equation starting with a quantity of a should be:

$$Q = a(0.96594)^t,$$

where Q is quantity remaining and t is time in days. The half-life will be the value of t for which Q is $a/2$, or half of the initial quantity a. Thus, we solve the equation for $Q = a/2$:

$$\begin{aligned} \frac{a}{2} &= a(0.96594)^t \\ \frac{1}{2} &= (0.96594)^t \\ \log(1/2) &= \log(0.96594)^t \\ \log(1/2) &= t\log(0.96594) \\ t &= \frac{\log(1/2)}{\log(0.96594)} = 20.002. \end{aligned}$$

So the half-life is about 20 days.

29. **(a)** To find the annual growth rate, we need to find a formula which describes the population, $P(t)$, in terms of the initial population, a, and the annual growth factor, b. In this case, we know that $a = 11{,}000$ and $P(3) = 13{,}000$. But $P(3) = ab^3 = 11{,}000b^3$, so

$$\begin{aligned} 13000 &= 11000b^3 \\ b^3 &= \frac{13000}{11000} \\ b &= \left(\frac{13}{11}\right)^{\frac{1}{3}} \approx 1.05726. \end{aligned}$$

Since b is the growth factor, we know that, each year, the population is about 105.726% of what it had been the previous year, so it is growing at the rate of 5.726% each year.

(b) To find the continuous growth rate, we need a formula of the form $P(t) = ae^{kt}$ where $P(t)$ is the population after t years, a is the initial population, and k is the rate we are trying to determine. We know that $a = 11{,}000$ and, in this case, that $P(3) = 11{,}000e^{3k} = 13{,}000$. Therefore,

$$e^{3k} = \frac{13000}{11000}$$

$$\ln e^{3k} = \ln\left(\frac{13}{11}\right)$$
$$3k = \ln\left(\frac{13}{11}\right) \quad \text{(because } \ln e^{3k} = 3k\text{)}$$
$$k = \frac{1}{3}\ln\left(\frac{13}{11}\right) \approx 0.05568.$$

So our continuous annual growth rate is 5.568%.

(c) The annual growth rate, 5.726%, describes the actual percent increase in one year. The continuous annual growth rate, 5.568%, describes the percentage increase of the population at any given instant, and so should be a smaller number.

33. We have

$$\begin{aligned}
\text{First investment} &= 5000(1.072)^t \\
\text{Second investment} &= 8000(1.054)^t \\
\text{so we solve}\quad 5000(1.072)^t &= 8000(1.054)^t \\
\frac{1.072^t}{1.054^t} &= \frac{8000}{5000} && \text{divide} \\
\left(\frac{1.072}{1.054}\right)^t &= 1.6 && \text{exponent rule} \\
t\ln\left(\frac{1.072}{1.054}\right) &= \ln 1.6 && \text{take logs} \\
t &= \frac{\ln 1.6}{\ln\left(\frac{1.072}{1.054}\right)} \\
&= 27.756,
\end{aligned}$$

so it will take almost 28 years.

37. We let N represent the amount of nicotine in the body t hours after it was ingested, and we let N_0 represent the initial amount of nicotine. The half-life tells us that at $t = 2$ the quantity of nicotine is $0.5N_0$. We substitute this point to find k:

$$\begin{aligned}
N &= N_0e^{kt} \\
0.5N_0 &= N_0e^{k(2)} \\
0.5 &= e^{2k} \\
\ln(0.5) &= 2k \\
k &= \frac{\ln(0.5)}{2} = -0.347.
\end{aligned}$$

The continuous decay rate of nicotine is 34.7% per hour.

41. **(a)** If $P(t)$ is the investment's value after t years, we have $P(t) = P_0e^{0.04t}$. We want to find t such that $P(t)$ is three times its initial value, P_0. Therefore, we need to solve:

$$\begin{aligned}
P(t) &= 3P_0 \\
P_0e^{0.04t} &= 3P_0 \\
e^{0.04t} &= 3 \\
\ln e^{0.04t} &= \ln 3 \\
0.04t &= \ln 3 \\
t &= (\ln 3)/0.04 \approx 27.465 \text{ years.}
\end{aligned}$$

With continuous compounding, the investment should triple in about $27\frac{1}{2}$ years.

(b) If the interest is compounded only once a year, the formula we will use is $P(t) = P_0b^t$ where b is the percent value of what the investment had been one year earlier. If it is earning 4% interest compounded once a year, it is 104% of what it had been the previous year, so our formula is $P(t) = P_0(1.04)^t$. Using this new formula, we will now solve

$$\begin{aligned}
P(t) &= 3P_0 \\
P_0(1.04)^t &= 3P_0 \\
(1.04)^t &= 3
\end{aligned}$$

$$\begin{aligned}\log(1.04)^t &= \log 3\\ t\log 1.04 &= \log 3\\ t &= \frac{\log 3}{\log 1.04} \approx 28.011 \text{ years.}\end{aligned}$$

So, compounding once a year, it will take a little more than 28 years for the investment to triple.

45. We have $P = ab^t$ and $2P = ab^{t+d}$. Using the algebra rules of exponents, we have

$$2P = ab^{t+d} = ab^t \cdot b^d = Pb^d.$$

Since P is nonzero, we can divide through by P, and we have

$$2 = b^d.$$

Notice that the time it takes an exponential growth function to double does not depend on the initial quantity a and does not depend on the time t. It depends only on the growth factor b.

49. **(a)** We see that the function $P = f(t)$, where t is years since 2012, is approximately exponential by looking at ratios of successive terms:

$$\begin{aligned}\frac{f(1)}{f(0)} &= \frac{122.09}{120.66} = 1.012\\ \frac{f(2)}{f(1)} &= \frac{123.52}{122.09} = 1.012\\ \frac{f(3)}{f(2)} &= \frac{124.94}{123.52} = 1.012\\ \frac{f(4)}{f(3)} &= \frac{126.39}{124.94} = 1.012.\end{aligned}$$

Since $P = 120.66$ when $t = 0$, the formula is $P = 120.66(1.012)^t$. Alternately, we could use exponential regression to find the formula. The world population age 80 or older is growing at an annual rate of 1.2% per year.

(b) We find k so that $e^k = 1.012$, giving $k = \ln(1.012) = 0.0119$. The continuous percent growth rate is 1.19% per year. The corresponding formula is $P = 120.66e^{0.0119t}$.

(c) We can use either formula to find the doubling time (and the results will differ slightly due to round off error.) Using the continuous version, we have

$$\begin{aligned}2(120.66) &= 120.66e^{0.0119t}\\ 2 &= e^{0.0119t}\\ \ln 2 &= 0.0119t\\ t &= \frac{\ln 2}{0.0119} = 58.248.\end{aligned}$$

The number of people in the world age 80 or older is doubling approximately every 58 years.

53. Since the half-life of carbon-14 is 5,728 years, and just a little more than 50% of it remained, we know that the man died nearly 5,700 years ago. To obtain a more precise date, we need to find a formula to describe the amount of carbon-14 left in the man's body after t years. Since the decay is continuous and exponential, it can be described by $Q(t) = Q_0e^{kt}$. We first find k. After 5,728 years, only one-half is left, so

$$Q(5{,}728) = \frac{1}{2}Q_0.$$

Therefore,

$$\begin{aligned}Q(5{,}728) = Q_0e^{5728k} &= \frac{1}{2}Q_0\\ e^{5728k} &= \frac{1}{2}\\ \ln e^{5728k} &= \ln\frac{1}{2}\\ 5728k &= \ln\frac{1}{2} = \ln 0.5\\ k &= \frac{\ln 0.5}{5728}\end{aligned}$$

So, $Q(t) = Q_0 e^{\frac{\ln 0.5}{5728}t}$.

If 46% of the carbon-14 has decayed, then 54% remains, so that $Q(t) = 0.54Q_0$.

$$\begin{aligned} Q_0 e^{\left(\frac{\ln 0.5}{5728}\right)t} &= 0.54Q_0 \\ e^{\left(\frac{\ln 0.5}{5728}\right)t} &= 0.54 \\ \ln e^{\left(\frac{\ln 0.5}{5728}\right)t} &= \ln 0.54 \\ \frac{\ln 0.5}{5728}t &= \ln 0.54 \\ t &= \frac{(\ln 0.54)\cdot(5728)}{\ln 0.5} = 5092.013 \end{aligned}$$

So the man died about 5092 years ago.

57. At an annual growth rate of 1%, the Rule of 70 tells us this investment doubles in $70/1 = 70$ years. At a 2% rate, the doubling time should be about $70/2 = 35$ years. The doubling times for the other rates are, according to the Rule of 70,

$$\frac{70}{5} = 14 \text{ years}, \qquad \frac{70}{7} = 10 \text{ years}, \qquad \text{and} \qquad \frac{70}{10} = 7 \text{ years}.$$

To check these predictions, we use logs to calculate the actual doubling times. If V is the dollar value of the investment in year t, then at a 1% rate, $V = 1000(1.01)^t$. To find the doubling time, we set $V = 2000$ and solve for t:

$$\begin{aligned} 1000(1.01)^t &= 2000 \\ 1.01^t &= 2 \\ \log(1.01^t) &= \log 2 \\ t\log 1.01 &= \log 2 \\ t &= \frac{\log 2}{\log 1.01} \approx 69.661. \end{aligned}$$

This agrees well with the prediction of 70 years. Doubling times for the other rates have been calculated and recorded in Table 5.1 together with the doubling times predicted by the Rule of 70.

Table 5.1 *Doubling times predicted by the Rule of 70 and actual values*

Rate (%)	1	2	5	7	10
Predicted doubling time (years)	70	35	14	10	7
Actual doubling time (years)	69.661	35.003	14.207	10.245	7.273

The Rule of 70 works reasonably well when the growth rate is small. The Rule of 70 does not give good estimates for growth rates much higher than 10%. For example, at an annual rate of 35%, the Rule of 70 predicts that the doubling time is $70/35 = 2$ years. But in 2 years at 35% growth rate, the \$1000 investment from the last example would be not worth \$2000, but only

$$1000(1.35)^2 = \$1822.50.$$

61. We have

$$\begin{aligned} e^{k(t-t_0)} &= e^{kt-kt_0} \\ &= e^{kt}\cdot e^{-kt_0} \\ &= \underbrace{e^{-kt_0}}_{a}\ \underbrace{\left(e^k\right)^t}_{b^t} \end{aligned}$$

$$\begin{aligned} \text{so} \quad \left(e^k\right)^t &= b^t \\ e^k &= b \\ k &= \ln b \end{aligned}$$

$$\text{and} \quad e^{-kt_0} = a$$
$$-kt_0 = \ln a$$
$$t_0 = -\frac{\ln a}{k}$$
$$= -\frac{\ln a}{\ln b} \quad \text{because } k = \ln b.$$

Solutions for Section 5.3

Skill Refresher

S1. $\log 0.0001 = \log 10^{-4} = -4 \log 10 = -4.$

S5. The equation $-\ln x = 12$ can be expressed as $\ln x = -12$, which is equivalent to $x = e^{-12}$.

S9. Rewrite the sum as $\ln x^3 + \ln x^2 = \ln(x^3 \cdot x^2) = \ln x^5$.

S13. **(a)** If x takes on positive values that approach 0, the values of y get smaller and approach $-\infty$.
(b) If x takes on negative values that approach 0, the values of y get smaller and approach $-\infty$.

EXERCISES

1. For $y = \ln x$, we have

$$\text{Domain is } x > 0$$
$$\text{Range is all } y.$$

The graph of $y = \ln(x - 3)$ is the graph of $y = \ln x$ shifted right by 3 units. Thus for $y = \ln(x - 3)$, we have

$$\text{Domain is } x > 3$$
$$\text{Range is all } y.$$

5. The rate of change is negative and decreases as we move right. The graph is concave down.

9. The graphs of $y = 10^x$ and $y = 2^x$ both have horizontal asymptotes, $y = 0$. The graph of $y = \log x$ has a vertical asymptote, $x = 0$.

13. See Figure 5.4. The graph of $y = \log(x - 4)$ is the graph of $y = \log x$ shifted to the right 4 units.

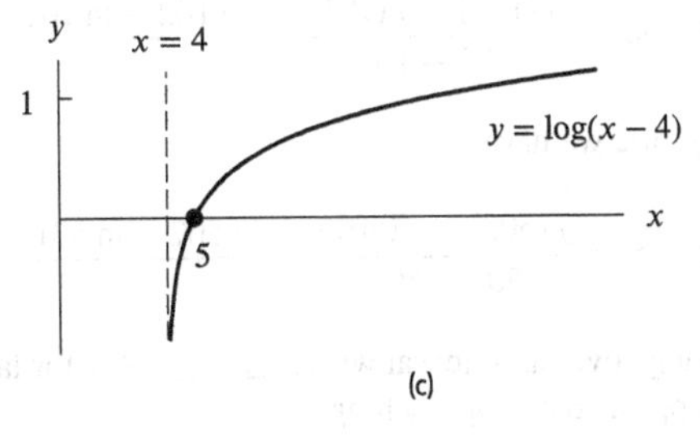

Figure 5.4

17. A graph of this function is shown in Figure 5.5. We see that the function has a vertical asymptote at $x = 2$. The domain is $(-\infty, 2)$.

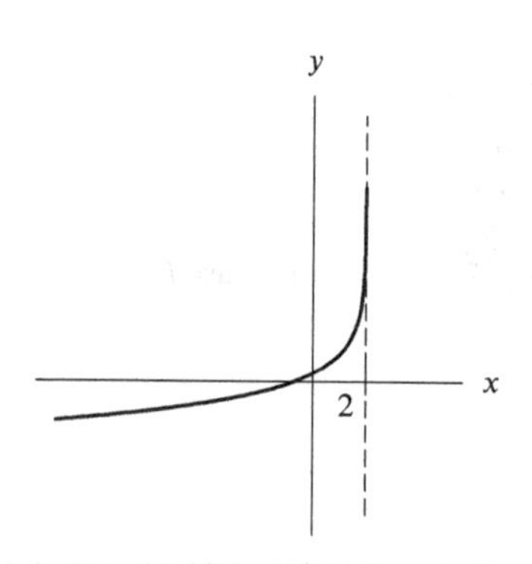

Figure 5.5

21. We know, by the definition of pH, that $8.3 = -\log[H^+]$. Therefore, $-8.3 = \log[H^+]$, and $10^{-8.3} = [H^+]$. Thus, the hydrogen ion concentration is $10^{-8.3} = 5.012 \times 10^{-9}$ moles per liter.

PROBLEMS

25. The log function is increasing but is concave down and so is increasing at a decreasing rate. It is not a compliment—growing exponentially would have been better. However, it is most likely realistic because after you are proficient at something, any increase in proficiency takes longer and longer to achieve.

29. This graph could represent exponential growth, with a y-intercept of 2. A possible formula is $y = 2b^x$ with $b > 1$.

33. **(a)** The graph in (III) is the only graph with a vertical asymptote.
(b) The graph in (I) has a horizontal asymptote that does not coincide with the x-axis, so it has a nonzero horizontal asymptote.
(c) The graph in (IV) has no horizontal or vertical asymptotes.
(d) The graph in (II) tends towards the x-axis as x gets larger and larger.

37. **(a)** (i) On the interval $1 \le x \le 2$ we have

$$\text{Rate of change} = \frac{f(2) - f(1)}{2 - 1} = \frac{\ln 2 - \ln 1}{1} = \frac{0.693 - 0}{1} = 0.693.$$

(ii) On the interval $11 \le x \le 12$ we have

$$\text{Rate of change} = \frac{f(12) - f(11)}{12 - 11} = \frac{\ln 12 - \ln 11}{1} = \frac{2.485 - 2.398}{1} = 0.087.$$

(iii) On the interval $101 \le x \le 102$ we have

$$\text{Rate of change} = \frac{f(102) - f(101)}{102 - 101} = \frac{\ln 102 - \ln 101}{1} = \frac{4.625 - 4.615}{1} = 0.01.$$

(iv) On the interval $501 \le x \le 502$ we have

$$\text{Rate of change} = \frac{f(502) - f(501)}{502 - 501} = \frac{\ln 502 - \ln 501}{1} = \frac{6.219 - 6.217}{1} = 0.002.$$

(b) Each time we take a rate of change over an interval with $n \le x \le n + 1$ for larger values of n, the value of the rate of change decreases. The values appear to be approaching 0.

41. Since the water in the stream has $[H^+] = 7.94(10^{-7})$, we have $pH = -\log(7.94(10^{-7})) = 6.10018$.

For water, in which rainbow trout start to die we have $[H^+] = 6$, so pH = 6. We want to know how much higher the second hydrogen ion concentration is compared to the first: $\frac{10^{-6}}{7.94(10^{-7})} = 1.25945$.

Therefore, the stream water would have to increase by 25.9% in acidity to cause the trout population to die. The corresponding change in pH would be from 6.10018 to 6, i.e. a decrease in pH value by 0.10018 points.

45. We use the formula $N = 10 \cdot \log\left(\frac{I}{I_0}\right)$, where N denotes the noise level in decibels and I and I_0 are sound's intensities in watts/cm^2. The intensity of a standard benchmark sound is $I_0 = 10^{-16}$ watts/cm^2, and the noise level when loss of hearing tissue occurs is $N = 180$ dB. Thus, we have $180 = 10 \cdot \log\left(\frac{I}{10^{-16}}\right)$. Solving for I, we have

$$\begin{aligned}
10\log\left(\frac{I}{10^{-16}}\right) &= 180 \\
\log\left(\frac{I}{10^{-16}}\right) &= 18 && \text{Dividing by 10} \\
\frac{I}{10^{-16}} &= 10^{18} && \text{Raising 10 to the power of both sides} \\
I &= 10^{18} \cdot 10^{-16} && \text{Multiplying both sides by } 10^{-16} \\
I &= 100 \text{ watts/cm}^2.
\end{aligned}$$

49. **(a)** We know that $D_1 = 10\log\left(\frac{I_1}{I_0}\right)$ and $D_2 = 10\log\left(\frac{I_2}{I_0}\right)$. Thus

$$\begin{aligned}
D_2 - D_1 &= 10\log\left(\frac{I_2}{I_0}\right) - 10\log\left(\frac{I_1}{I_0}\right) \\
&= 10\left(\log\left(\frac{I_2}{I_0}\right) - \log\left(\frac{I_1}{I_0}\right)\right) && \text{factoring} \\
&= 10\log\left(\frac{I_2/I_0}{I_1/I_0}\right) && \text{using a log property}
\end{aligned}$$

so

$$D_2 - D_1 = 10\log\left(\frac{I_2}{I_1}\right).$$

(b) Suppose the sound's initial intensity is I_1 and that its new intensity is I_2. Then here we have $I_2 = 2I_1$. If D_1 is the original decibel rating and D_2 is the new rating then

$$\begin{aligned}
\text{Increase in decibels} &= D_2 - D_1 \\
&= 10\log\left(\frac{I_2}{I_1}\right) && \text{using formula from part (a)} \\
&= 10\log\left(\frac{2I_1}{I_1}\right) \\
&= 10\log 2 \\
&\approx 3.01.
\end{aligned}$$

Thus, the sound increases by 3 decibels when it doubles in intensity.

53. Let $M_2 = 8.5$ and $M_1 = 5.1$, so

$$M_2 - M_1 = \log\left(\frac{W_2}{W_1}\right)$$

becomes

$$\begin{aligned}
8.5 - 5.1 &= \log\left(\frac{W_2}{W_1}\right) \\
3.4 &= \log\left(\frac{W_2}{W_1}\right)
\end{aligned}$$

so

$$\frac{W_2}{W_1} = 10^{3.4} \approx 2511.89.$$

Thus, the seismic waves of the 1883 Krakatoa earthquake were about 2512 times as large as those of the 2017 Alaska earthquake.

57. **(a)** The graph in (III) has a vertical asymptote at $x = 0$ and $f(x) \to -\infty$ as $x \to 0^+$.
(b) The graph in (IV) goes through the origin, so $f(x) \to 0$ as $x \to 0^-$.
(c) The graph in (I) goes upward without bound, that is $f(x) \to \infty$, as $x \to \infty$.
(d) The graphs in (I) and (II) tend toward the x-axis, that is $f(x) \to 0$, as $x \to -\infty$.

Solutions for Section 5.4

Skill Refresher

S1. 1.455×10^6

S5. 3.6×10^{-4}

S9. Since $0.1 < \frac{1}{3} < 1$, $10^{-1} < \frac{1}{3} < 1 = 10^0$.

S13. To solve for y, we "undo" the natural logarithm by raising e to each side of the equation:

$$\ln y = 5 + 2\ln x$$
$$e^{\ln y} = e^{5+2\ln x}$$
$$y = e^{5+2\ln x}$$

We now use properties of exponents and properties of logs to rewrite the right-hand side:

$$y = e^{5+2\ln x} = e^5 e^{2\ln x} = 148.413e^{\ln(x^2)} = 148.413x^2.$$

This is in the form $y = kx^p$ with $k = 148.413$ and $p = 2$.

EXERCISES

1. Using a linear scale, the wealth of everyone with less than a million dollars would be indistinguishable because all of them are less than one one-thousandth of the wealth of the average billionaire. A log scale is more useful.

5. **(a)**

Table 5.2

n	1	2	3	4	5	6	7	8	9
$\log n$	0	0.3010	0.4771	0.6021	0.6990	0.7782	0.8451	0.9031	0.9542

Table 5.3

n	10	20	30	40	50	60	70	80	90
$\log n$	1	1.3010	1.4771	1.6021	1.6990	1.7782	1.8451	1.9031	1.9542

(b) The first tick mark is at $10^0 = 1$. The dot for the number 2 is placed $\log 2 = 0.3010$ of the distance from 1 to 10. The number 3 is placed at $\log 3 = 0.4771$ units from 1, and so on. The number 30 is placed 1.4771 units from 1, the number 50 is placed 1.6989 units from 1, and so on.

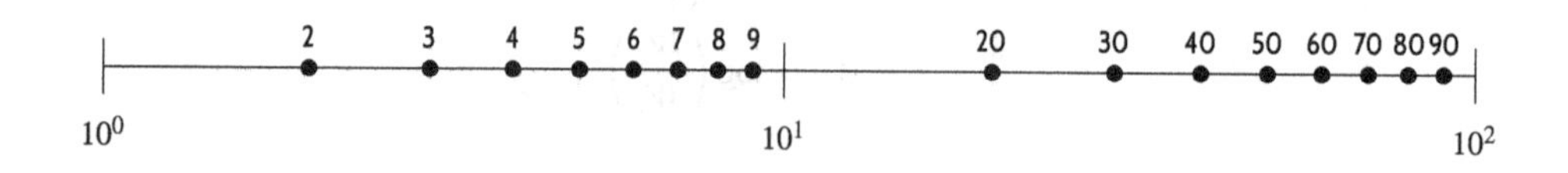

Figure 5.6

9. We have $\log 160 = 2.204$, so this lifespan would be marked at 2.2 inches.

13. **(a)** Run a linear regression on the data. The resulting function is $y = -3582.145 + 236.314x$, with $r \approx 0.7946$. We see from the sketch of the graph of the data that the estimated regression line provides a reasonable but not excellent fit. See Figure 5.7.

(b) If, instead, we compare x and $\ln y$ we get

$$\ln y = 1.568 + 0.200x.$$

We see from the sketch of the graph of the data that the estimated regression line provides an excellent fit with $r \approx 0.9998$. See Figure 5.8. Solving for y, we have

$$\begin{aligned} e^{\ln y} &= e^{1.568+0.200x} \\ y &= e^{1.568}e^{0.200x} \\ y &= 4.797e^{0.200x} \\ \text{or} \quad y &= 4.797(e^{0.200})^x \approx 4.797(1.221)^x. \end{aligned}$$

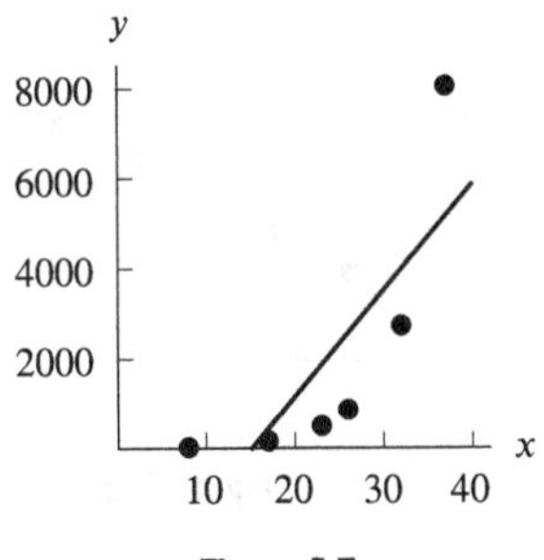

Figure 5.7

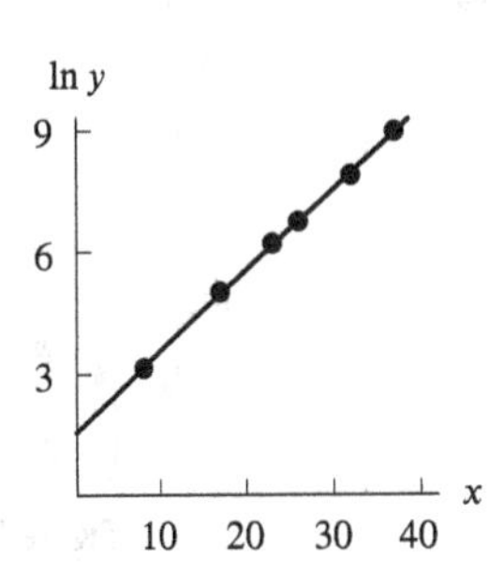

Figure 5.8

(c) The linear equation is a poor fit, and the exponential equation is a better fit.

PROBLEMS

17. **(a)** An appropriate scale is from 0 to 70 at intervals of 10. (Other answers are possible.) See Figure 5.9. The points get more and more spread out as the exponent increases.

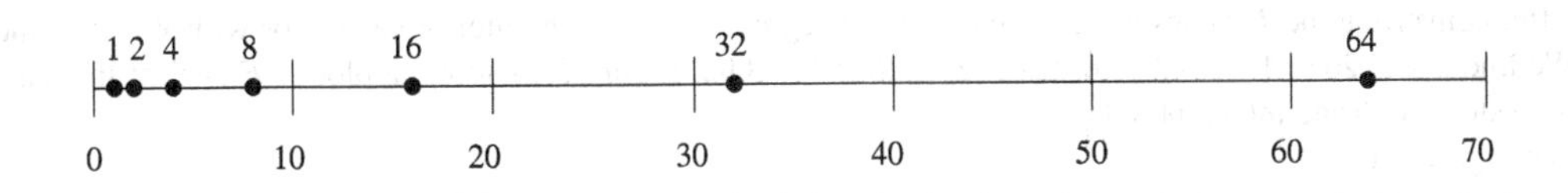

Figure 5.9

(b) If we want to locate 2 on a logarithmic scale, since $2 = 10^{0.3}$, we find $10^{0.3}$. Similarly, $8 = 10^{0.9}$ and $32 = 10^{1.5}$, so 8 is at $10^{0.9}$ and 32 is at $10^{1.5}$. Since the values of the logs go from 0 to 1.8, an appropriate scale is from 0 to 2 at intervals of 0.2. See Figure 5.10. The points are spaced at equal intervals.

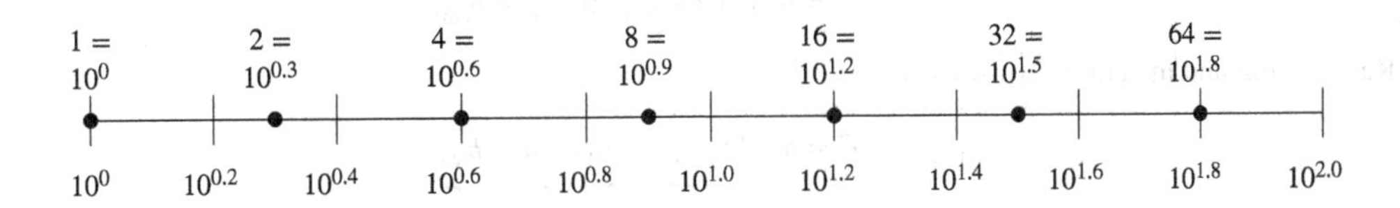

Figure 5.10

21. For the pack of gum, $\log(1.50) = 0.176$, so the pack of gum is plotted at 0.176. For the movie ticket, $\log(10) = 1$, so the ticket is plotted at 1, and so on. See Figure 5.11.

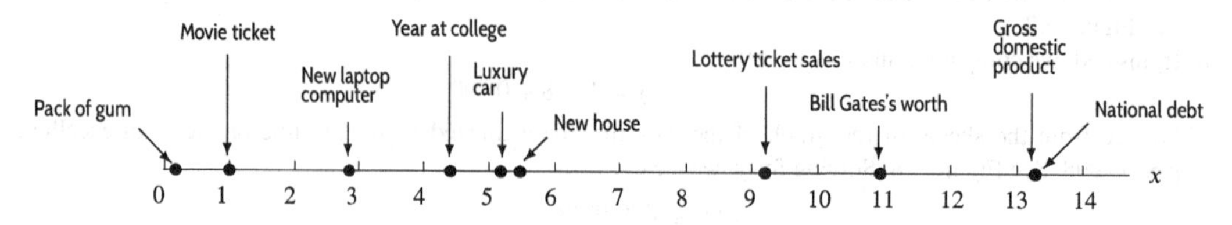

Figure 5.11: Log (dollar value)

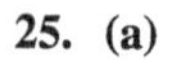

25. (a)

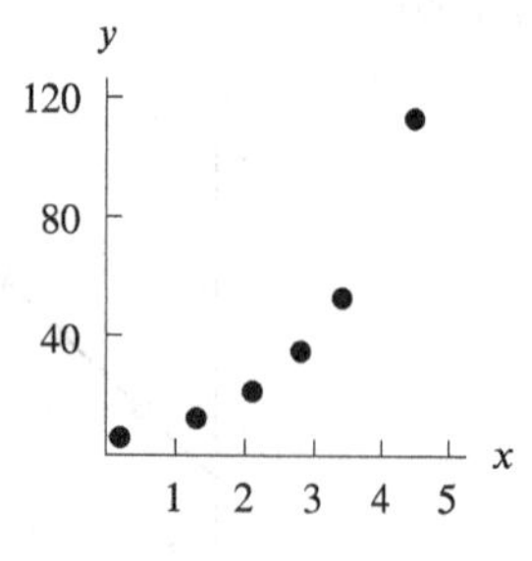

Figure 5.12

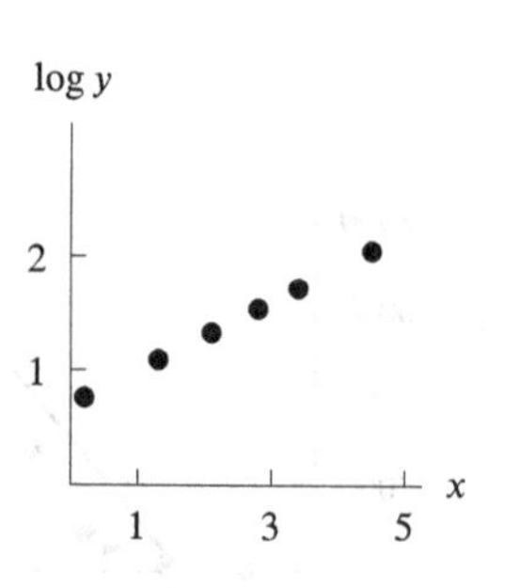

Figure 5.13

(b) The data appear to be exponential.
(c) See Figure 5.13. The data appear to be linear.

Table 5.4

x	0.2	1.3	2.1	2.8	3.4	4.5
$\log y$	.76	1.09	1.33	1.54	1.72	2.05

29. (a) After converting the I values to $\ln I$, we use linear regression on a computer or calculator with $x = \ln I$ and $y = F$. We find $a \approx 4.26$ and $b \approx 8.95$ so that $F = 4.26 \ln I + 8.95$. Figure 5.14 shows a plot of F against $\ln I$ and the line with slope 4.26 and intercept 8.95.
(b) See Figure 5.14.
(c) Figure 5.15 shows a plot of $F = 4.26 \ln I + 8.95$ and the data set in Table 5.22. The model seems to fit well.
(d) Imagine the units of I were changed by a factor of $\alpha > 0$ so that $I_{\text{old}} = \alpha I_{\text{new}}$.
Then

$$\begin{aligned} F &= a_{\text{old}} \ln I_{\text{old}} + b_{\text{old}} \\ &= a_{\text{old}} \ln(\alpha I_{\text{new}}) + b_{\text{old}} \\ &= a_{\text{old}}(\ln \alpha + \ln I_{\text{new}}) + b_{\text{old}} \\ &= a_{\text{old}} \ln \alpha + a_{\text{old}} \ln I_{\text{new}} + b_{\text{old}}. \end{aligned}$$

Rearranging and matching terms, we see:

$$F = \underbrace{a_{\text{old}} \ln I_{\text{new}}}_{a_{\text{new}} \ln I_{\text{new}}} + \underbrace{a_{\text{old}} \ln \alpha + b_{\text{old}}}_{b_{\text{new}}}$$

so

$$a_{\text{new}} = a_{\text{old}} \quad \text{and} \quad b_{\text{new}} = b_{\text{old}} + a_{\text{old}} \ln \alpha.$$

We can also see that if $\alpha > 1$ then $\ln \alpha > 0$ so the term $a_{\text{old}} \ln \alpha$ is positive and $b_{\text{new}} > b_{\text{old}}$. If $\alpha < 1$ then $\ln \alpha < 0$ so the term $a_{\text{old}} \ln \alpha$ is negative and $b_{\text{new}} < b_{\text{old}}$.

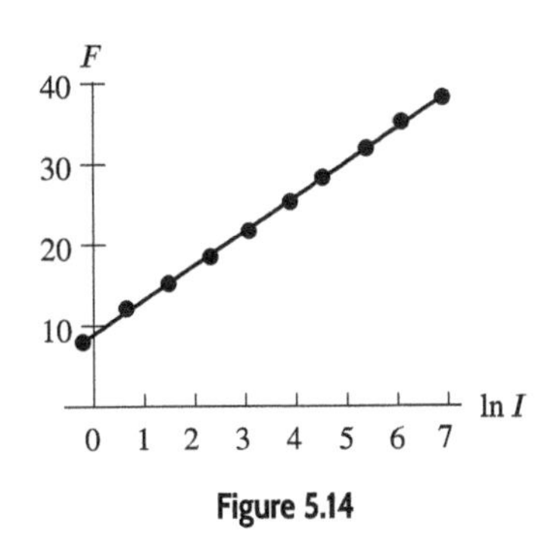

Figure 5.14

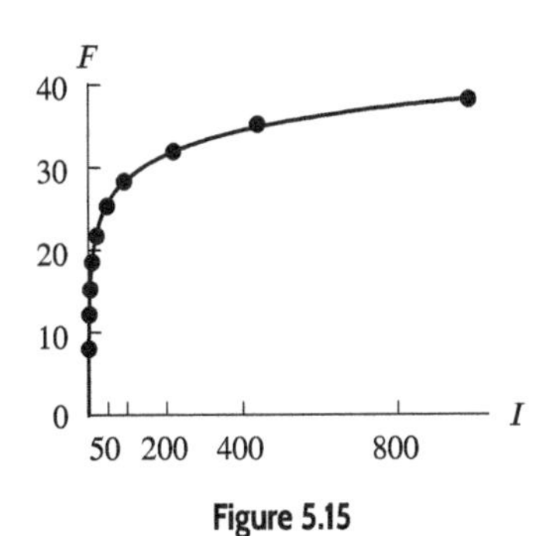

Figure 5.15

STRENGTHEN YOUR UNDERSTANDING

1. False. Since the $\log 1000 = \log 10^3 = 3$ we know $\log 2000 > 3$. Or use a calculator to find that $\log 2000$ is about 3.3.

5. True. Comparing the equation, we see $b = e^k$, so $k = \ln b$.

9. True. The log function outputs the power of 10 which in this case is n.

13. False. For example, $\log 10 = 1$, but $\ln 10 \approx 2.3$.

17. True. The two functions are inverses of one another.

21. False. Taking the natural log of both sides, we see $t = \ln 7.32$.

25. True. This is the definition of half-life.

29. True. Solve for t by dividing both sides by Q_0, taking the ln of both sides, and then dividing by k.

33. False. Since $26{,}395{,}630{,}000{,}000 \approx 2.6 \cdot 10^{13}$, we see that it would be between 13 and 14 on a log scale.

37. False. The fit will not be as good as $y = x^3$ but an exponential function can be found.

CHAPTER SIX

Solutions for Section 6.1

Skill Refresher

S1. Evaluating $f(x)$ at $x = 3$, we have $f(3) = e^3 = 20.086$.

S5. **(a)** $f(5) = 5^2 = 25$.
(b) $g(5) = -5^2 = -(5^2) = -25$.
(c) $h(5) = (-5)^2 = (-5) \cdot (-5) = 25$.
(d) $k(5) = -(-5)^2 = -[(-5) \cdot (-5)] = -(25) = -25$.

S9. **(a)** $f(-x) = 2(-x)^3 - 3 = -2x^3 - 3$
(b) $-f(x) = -\left(2x^3 - 3\right) = -2x^3 + 3$
Notice since $f(-x) \neq -f(x)$ and $f(-x) \neq -f(x)$, we see that $f(x) = 2x^3 - 3$ is neither an even nor an odd function.

EXERCISES

1. **(a)** The y-coordinate is unchanged, but the x-coordinate is the same distance to the left of the y-axis, so the point is $(-2, -3)$.
(b) The x-coordinate is unchanged, but the y-coordinate is the same distance above the x-axis, so the point is $(2, 3)$.

5. The negative sign reflects the graph of $Q(t)$ horizontally about the y-axis, so the domain of $y = Q(-t)$ is $t < 0$. A horizontal reflection of the graph of $Q(t)$ about the y-axis will not change the range. The range of $Q(-t)$ is therefore the same as the range of $Q(t)$, $-4 \leq Q(-t) \leq 7$.

9. To reflect about the y-axis, we substitute $-x$ for x in the formula, getting $y = e^{-x}$.

13. We have

$$\begin{aligned} y = m(-n) &= (-n)^2 - 4(-n) + 5 \\ &= n^2 + 4n + 5 \end{aligned}$$

To graph this function, reflect the graph of m across the y-axis. See Figure 6.1.

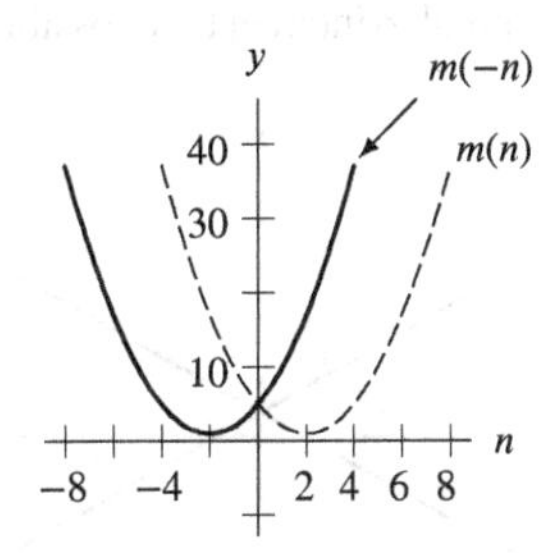

Figure 6.1: $y = m(-n)$

17. We have

$$y = k(-w) = 3^{-w}$$

To graph this function, reflect the graph of k across the y-axis. See Figure 6.2.

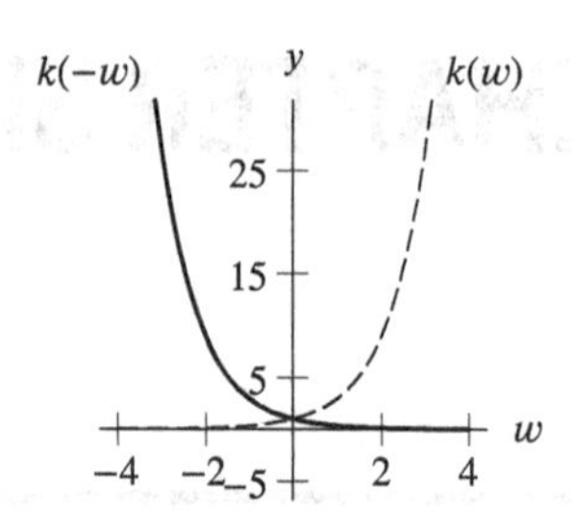

Figure 6.2: $y = k(-w)$

21. Since $f(-x) = 7(-x)^2 - 2(-x) + 1 = 7x^2 + 2x + 1$ is equal to neither $f(x)$ nor $-f(x)$, the function is neither even nor odd.

PROBLEMS

25. (a) See Figure 6.3.
    (b) See Figure 6.4.

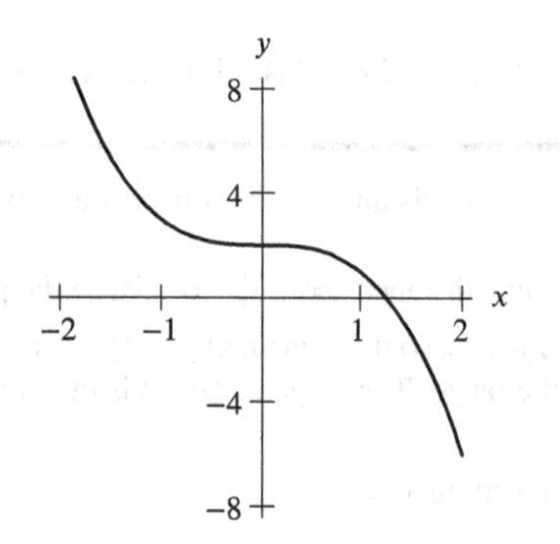

Figure 6.3: $y = -x^3 + 2$

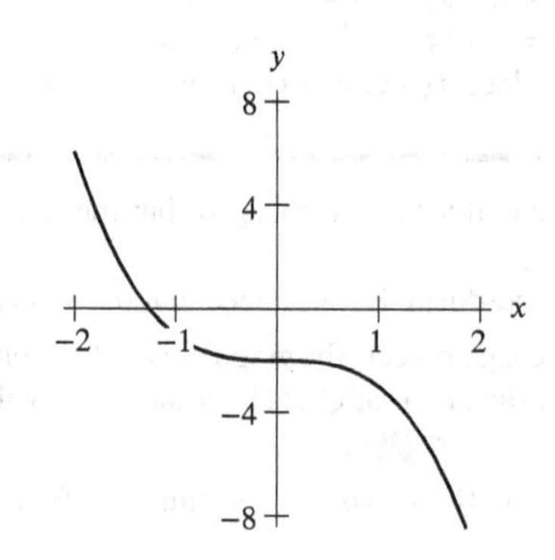

Figure 6.4: $y = -(x^3 + 2)$

    (c) The two functions are not the same.

29. The equation of the reflected line is

$$y = b + m(-x) = b - mx.$$

The reflected line has the same y-intercept as the original; that is b. Its slope is $-m$, the negative of the original slope, and its x-intercept is b/m, the negative of the original x-intercept. A possible graph is in Figure 6.5.

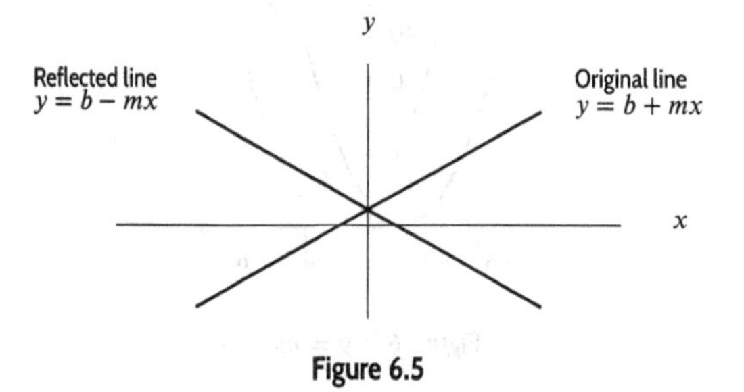

Figure 6.5

33. (a) If

$$H(t) = 20 + 54(0.91)^t$$

then $H(t) + k = (20 + 54(.91)^t) + k = (20 + k) + 54(0.91)^t$

(b) See Figure 6.6.

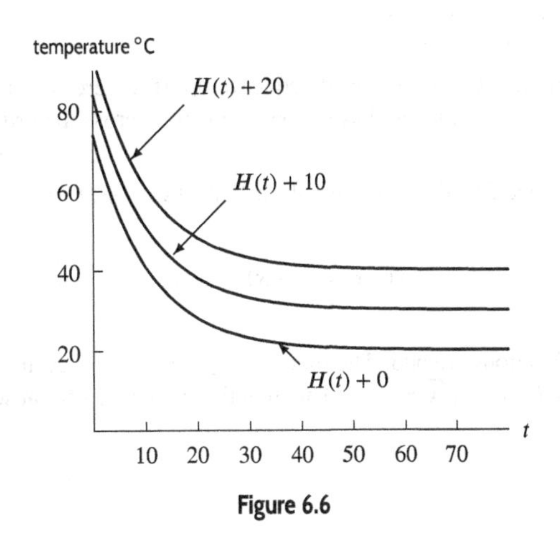

Figure 6.6

(c) As t gets very large, $H(t)+10$ approaches a final temperature of 30°C. Over time, the soup will gradually cool to the same temperature as its surroundings, so 30°C is the temperature of the room it is in. Similarly, $H(t)+k$ approaches a final temperature of $(20+k)$°C, which is the temperature of the room the bowl of soup is in.

37. (a) Figure 6.7 shows the graph of a function f that is symmetric across the y-axis.

(b) Figure 6.8 shows the graph of function f that is symmetric across the origin.

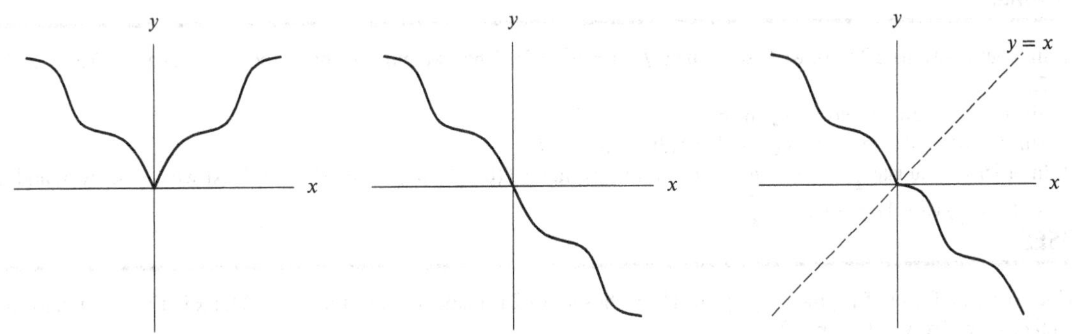

Figure 6.7: The graph of $f(x)$ that is symmetric across the y-axis

Figure 6.8: The graph of $f(x)$ that is symmetric across the origin

Figure 6.9: The graph of $f(x)$ that is symmetric across the line $y=x$

(c) Figure 6.9 shows the graph of function f that is symmetric across the line $y=x$.

41. See Figure 6.10.

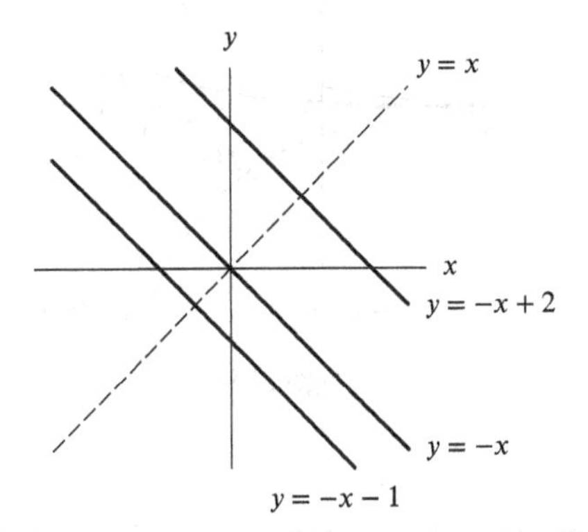

Figure 6.10: The graphs of $y=-x+2$, $y=-x$, and $y=-x-1$ are all symmetric across the line $y=x$.

Any straight line perpendicular to $y = x$ is symmetric across $y = x$. Its slope must be -1, so $y = -x + b$, for an arbitrary constant b, is symmetric across $y = x$.

Also, the line $y = x$ is symmetric about itself.

45. No, it is not possible for an odd function to be strictly concave up. If it were concave up in the first or second quadrants, then the fact that it is odd would mean it would have to be symmetric across the origin, and so would be concave down in the third or fourth quadrants.

49. To show that $f(x) = x^{1/3}$ is an odd function, we must show that $f(x) = -f(-x)$:

$$-f(-x) = -(-x)^{1/3} = x^{1/3} = f(x).$$

However, not all power functions are odd. The function $f(x) = x^2$ is an even function because $f(x) = f(-x)$ for all x. Another counterexample is $f(x) = \sqrt{x} = x^{1/2}$. This function is not odd because it is not defined for negative values of x.

Solutions for Section 6.2

Skill Refresher

S1. **(a)** In order to evaluate $2f(6)$, we first evaluate $f(6) = 6^2 = 36$. Then we multiply by 2, and thus we have $2f(6) = 2\cdot 36 = 72$.

(b) Since $f(6) = 36$, we have $-\frac{1}{2}f(6) = -\frac{1}{2}\cdot 36 = -18$.

(c) Since $f(6) = 36$, we have $5f(6) - 3 = 5(36) - 3 = 177$.

(d) In order to evaluate $\frac{1}{4}f(x-1)$ at $x = 6$, we first evaluate $f(6-1) = f(5) = 5^2 = 25$. Next we divide by 4, and thus we have $\frac{1}{4}f(6-1) = \frac{1}{4}\cdot 25 = \frac{25}{4}$.

EXERCISES

1. To increase by a factor of 10, multiply by 10. The right shift of 2 is made by substituting $x-2$ for x in the function formula. Together they give $y = 10f(x-2)$.

5. See Figure 6.11.

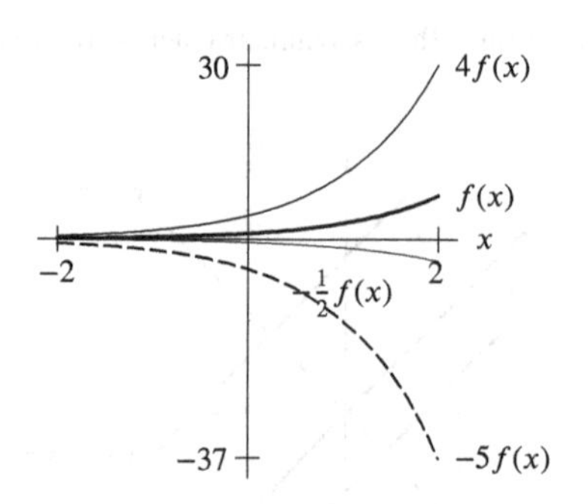

Figure 6.11

9. The function is $y = f(x+3)$. Since $f(x) = |x|$, we want $y = |x+3|$. The transformation shifts the graph of $f(x)$ by 3 units to the left. See Figure 6.12.

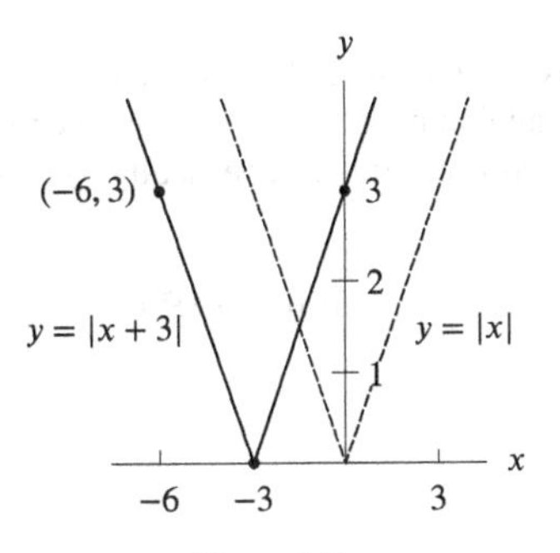

Figure 6.12

13. Since $h(x) = 2^x$, $3h(x) = 3 \cdot 2^x$. The graph of $h(x)$ is stretched vertically by a factor of 3. See Figure 6.13.

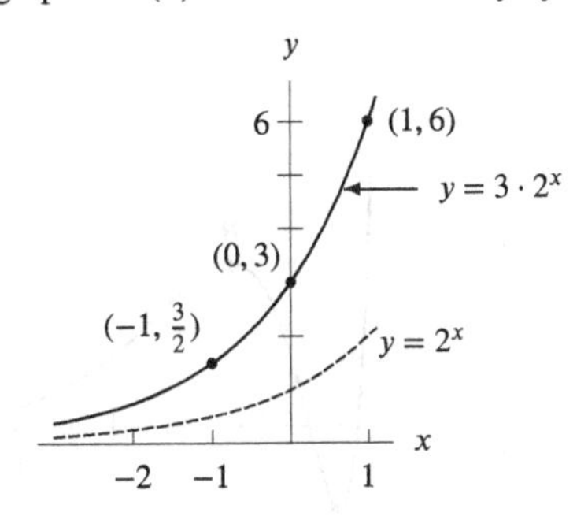

Figure 6.13

PROBLEMS

17. The graph of $f(t) = 1/(1 + t^2)$ resembles a bell-shaped curve. (It is not, however, a true "bell curve.") See Figure 6.14. The y-axis is the horizontal asymptote.

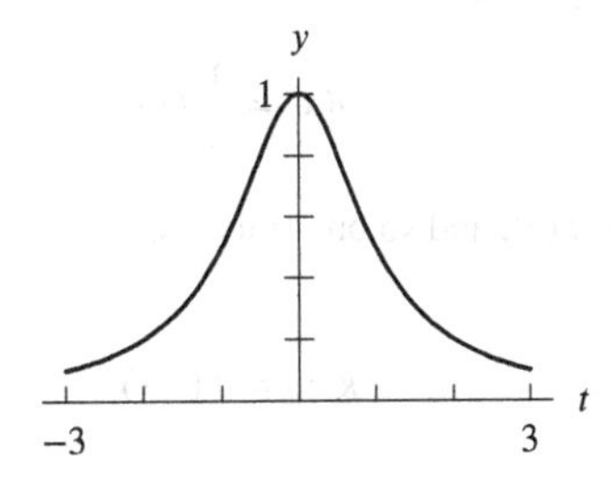

Figure 6.14

21. The graph of $f(t + 5) - 5$ is the graph of $f(t)$ shifted to the left by 5 units and then down by 5 units. See Figure 6.15. The horizontal asymptote is at $y = -5$.

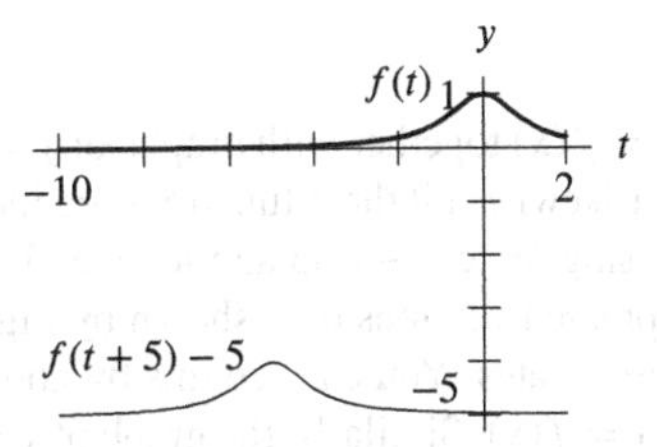

Figure 6.15

25. Since the domain of $R(n)$ is the same as the domain of $P(n)$, no horizontal transformations have been applied. Since the maximum value of $R(n)$ is -5 times the minimum value of $P(n)$, and the minimum value $R(n)$ approaches is -5 times the maximum value $P(n)$ approaches, $P(n)$ has been stretched vertically by a factor of 5 and reflected about the x-axis. Thus, we have

$$R(n) = -5P(n).$$

29. I is (b)
II is (d)
III is (c)
IV is (h)

33. See Figure 6.16. The graph is shifted to the right by 3 units.

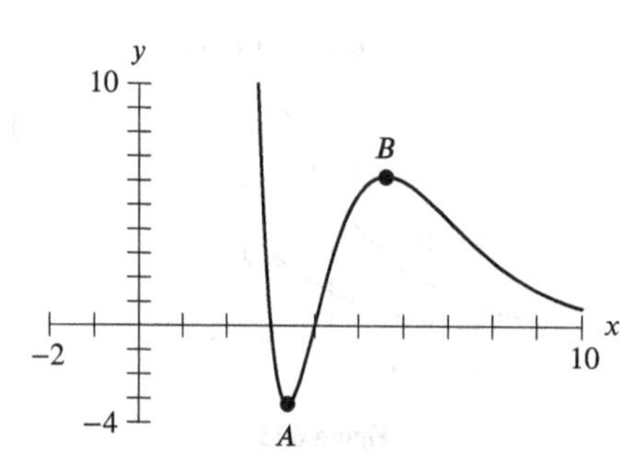

Figure 6.16

37. **(a)** Notice that the value of $h(x)$ at every value of x is one-half the value of $f(x)$ at the same x value. Thus, $f(x)$ has been compressed vertically by a factor of $1/2$, and

$$h(x) = \frac{1}{2}f(x).$$

(b) Observe that $k(-6) = f(6)$, $k(-4) = f(4)$, and so on. Thus, we have

$$k(x) = f(-x).$$

(c) The values of $m(x)$ are 4 less than the values of $f(x)$at the same x value. Thus, we have

$$m(x) = f(x) - 4.$$

41. Figure 6.17 gives a graph of a function $y = f(x)$ together with graphs of $y = \frac{1}{2}f(x)$ and $y = 2f(x)$. All three graphs cross the x-axis at $x = -2$, $x = -1$, and $x = 1$. Likewise, all three functions are increasing and decreasing on the same intervals. Specifically, all three functions are increasing for $x < -1.55$ and for $x > 0.21$ and decreasing for $-1.55 < x < 0.21$.

Even though the stretched and compressed versions of f shown by Figure 6.17 are increasing and decreasing on the same intervals, they are doing so at different rates. You can see this by noticing that, on every interval of x, the graph of $y = \frac{1}{2}f(x)$ is less steep than the graph of $y = f(x)$. Similarly, the graph of $y = 2f(x)$ is steeper than the graph of $y = f(x)$. This indicates that the magnitude of the average rate of change of $y = \frac{1}{2}f(x)$ is less than that of $y = f(x)$, and that the magnitude of the average rate of change of $y = 2f(x)$ is greater than that of $y = f(x)$.

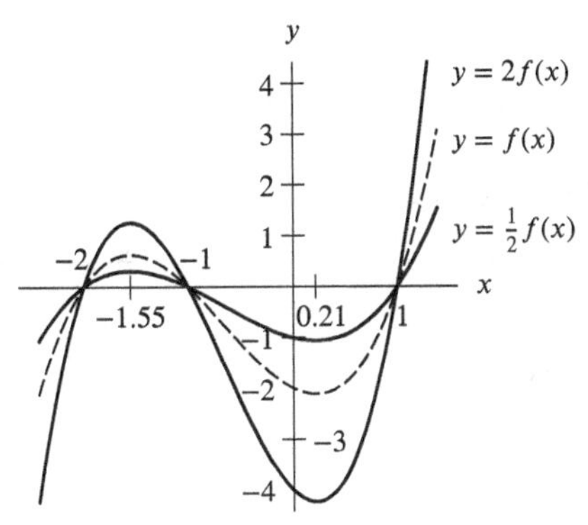

Figure 6.17: The graph of $y = 2f(x)$ and $y = \frac{1}{2}f(x)$ compared to the graph of $f(x)$

Solutions for Section 6.3

Skill Refresher

S1. **(a)** In order to evaluate $f(3x)$ at $x = 2$, we substitute $x = 2$ and find $f(3 \cdot 2) = f(6) = 6^2 = 36$.
(b) Since $f(2) = 2^2 = 4$, we have $3f(2) = 3 \cdot f(2) = 3 \cdot 2^2 = 3 \cdot 4 = 12$.
(c) At $x = 2$, we see that $5f(2x) = 5f(2 \cdot 2) = 5f(4) = 5 \cdot 4^2 = 5 \cdot 16 = 80$.
(d) At $x = 2$, we see that $2f(5x) = 2f(5 \cdot 2) = 2f(10) = 2 \cdot 10^2 = 2 \cdot 100 = 200$.

S5. $f\left(-\frac{1}{3}x\right) = \left(\frac{-x}{3}\right)^3 - 5 = -\frac{x^3}{27} - 5.$

S9. $Q(2t) + 11 = 4e^{6(2t)} + 11 = 4e^{12t} + 11.$

S13. Factoring -3 out from the left side of the equation, we have

$$-3(z - \frac{10}{3}) = -3(z - h).$$

Thus, we see $h = \frac{10}{3}$.

S17. We must write the expression $-1/3x - 9$ inside the function in the form $B(x - h)$. Factoring out the $-1/3$, we have $-1/3x - 9 = -1/3(x + 27)$. Thus, we have

$$y = 6f\left(-\frac{1}{3}x - 9\right) = 6f\left(-\frac{1}{3}(x + 27)\right).$$

We see from the form above that $A = 6$, $B = -1/3$, $h = -27$, and $k = 0$.

EXERCISES

1. **(a)** Since $y = -f(x) + 2$, we first need to reflect the graph of $y = f(x)$ over the x-axis and then shift it upward two units. See Figure 6.18.

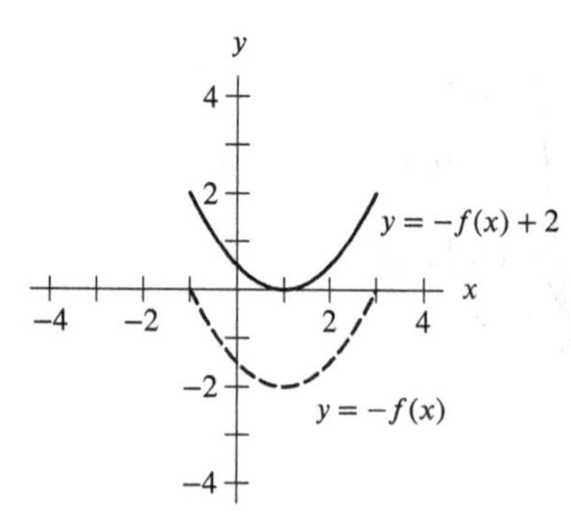

Figure 6.18

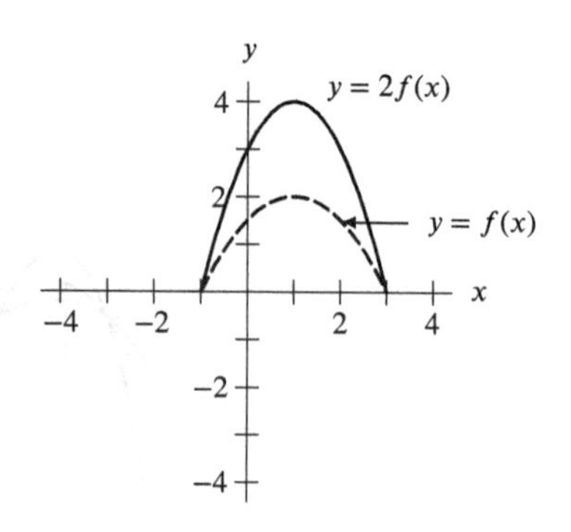

Figure 6.19

(b) We need to stretch the graph of $y = f(x)$ vertically by a factor of 2 in order to get the graph of $y = 2f(x)$. See Figure 6.19.

(c) In order to get the graph of $y = f(x-3)$, we will move the graph of $y = f(x)$ to the right by 3 units. See Figure 6.20.

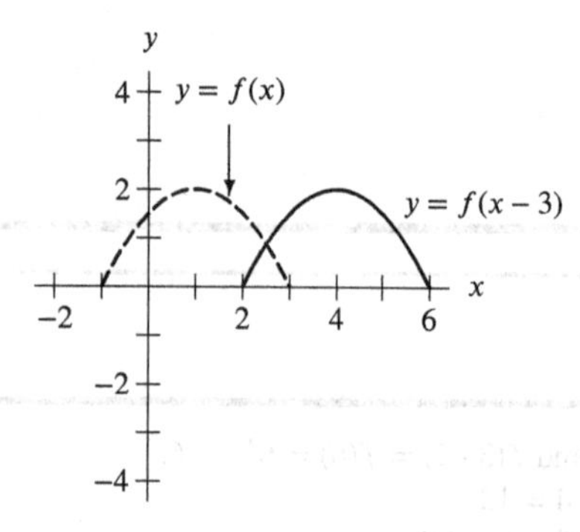

Figure 6.20

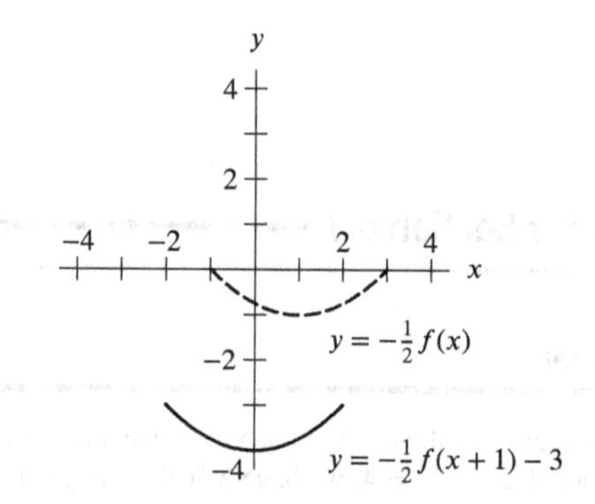

Figure 6.21

(d) To get the graph of $y = -\frac{1}{2}f(x+1) - 3$, first vertically compress the graph of $y = f(x)$ by a factor of $1/2$, then reflect about the x-axis, then horizontally shift left 1 unit, and finally vertically shift down 3 units. See Figure 6.21.

5. See Figure 6.22.

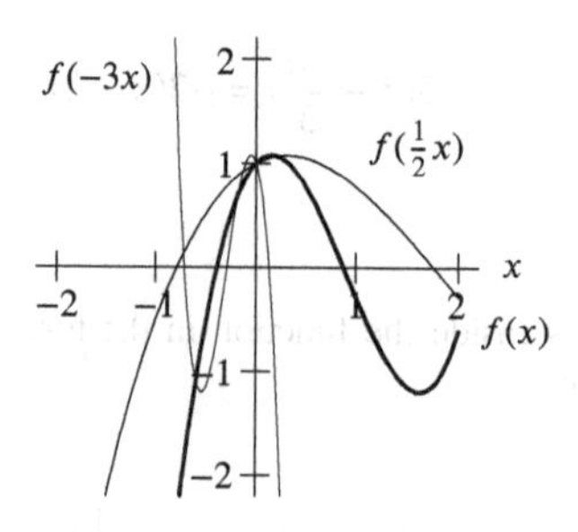

Figure 6.22

9. Since horizontal transformations do not affect features such as horizontal asymptotes or y-intercepts, each of the graphs in (a)-(c) will still have a horizontal asymptote at $y = 2$ and a y-intercept at $(0, -2)$.

(a) The graph of f has been compressed horizontally by a factor of $1/3$. The x-intercepts $(-1, 0)$ and $(3, 0)$ on the graph of f are moved to $(-\frac{1}{3}, 0)$ and $(1, 0)$ respectively on the graph of $y = f(3x)$ in Figure 6.23.

(b) The graph of f has been compressed horizontally by a factor of $1/2$ and reflected about the y-axis. The x-intercepts $(-1,0)$ and $(3,0)$ on the graph of f are moved to $(\frac{1}{2},0)$ and $(-\frac{3}{2},0)$ respectively on the graph of $y = f(-2x)$ in Figure 6.24.

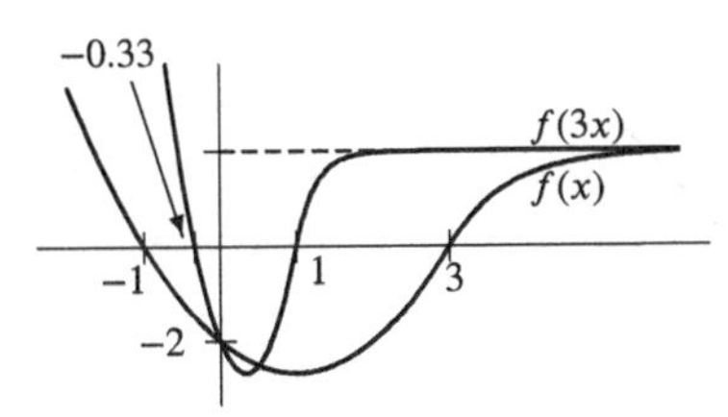

Figure 6.23

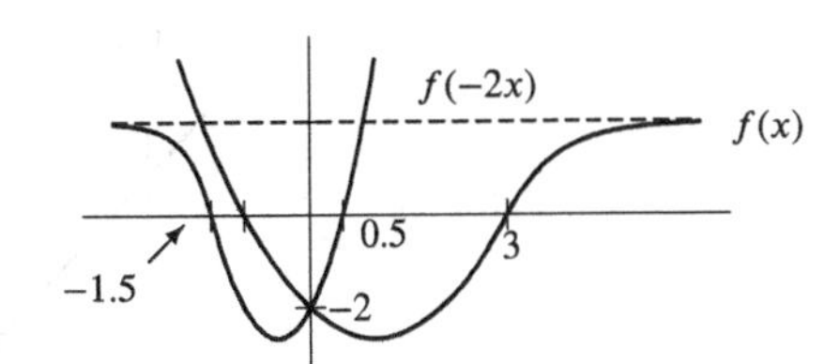

Figure 6.24

(c) The graph of f has been stretched horizontally by a factor of 2. The x-intercepts $(-1,0)$ and $(3,0)$ on the graph of f are moved to $(-2,0)$ and $(6,0)$ respectively on the graph of $y = f(\frac{1}{2}x)$ in Figure 6.25.

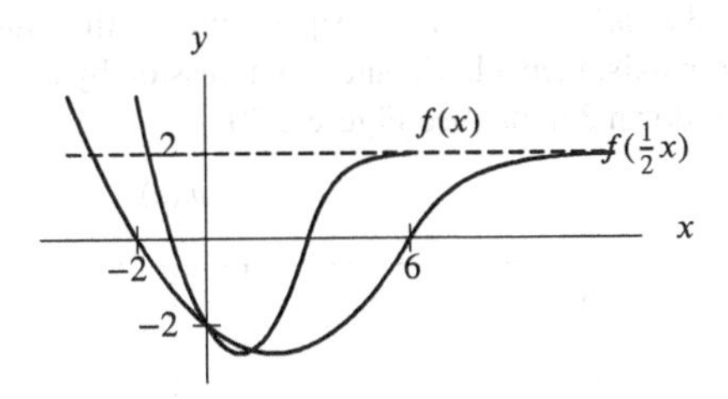

Figure 6.25

13. **(a)** This is a vertical shift of one unit upward. See Figure 6.26.

(b) Writing $h(x) = |x+1|$ as $h(x) = |x-(-1)|$, we see that this is a horizontal shift of one unit to the left. See Figure 6.27.

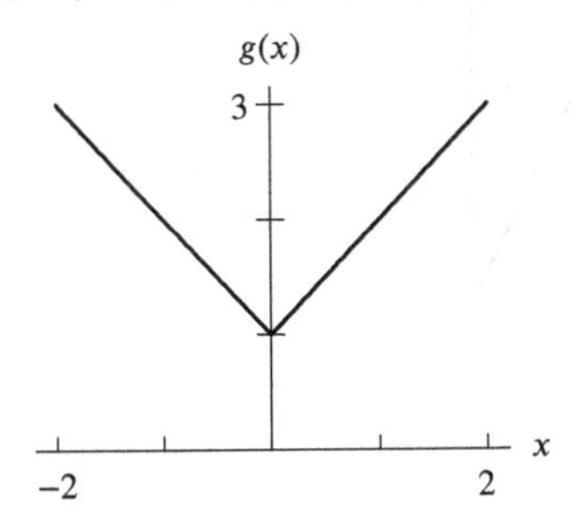

Figure 6.26

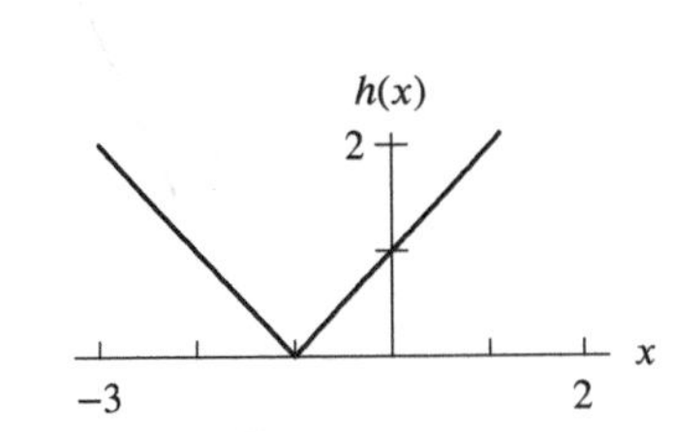

Figure 6.27

(c) Writing $j(x) = |2x+1| - 3$ as $j(x) = |2(x+\frac{1}{2})| - 3$, we see that this is a horizontal compression by a factor of $1/2$, then a horizontal shift of $1/2$ unit to the left, and finally a vertical shift down 3 units. See Figure 6.28.

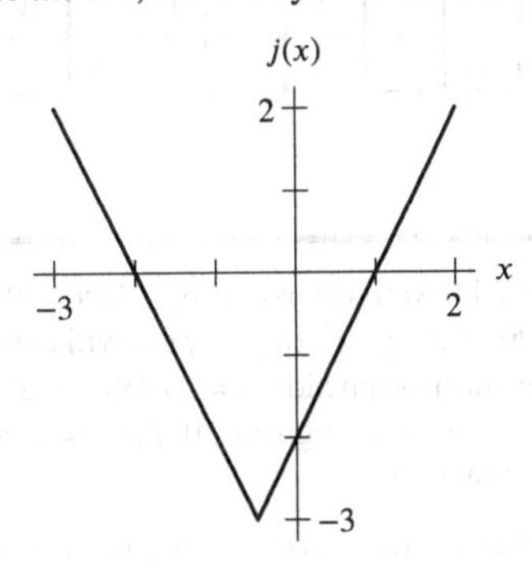

Figure 6.28

(d) Writing $k(x) = \frac{1}{2}|2x - 4| + 1$ as $k(x) = \frac{1}{2}|2(x - 2)| + 1$, we see that this is a vertical compression by a factor of $1/2$, then a horizontal compression by a factor of $1/2$, then a horizontal shift right 2 units, and finally a vertical shift up 1 unit. See Figure 6.29.

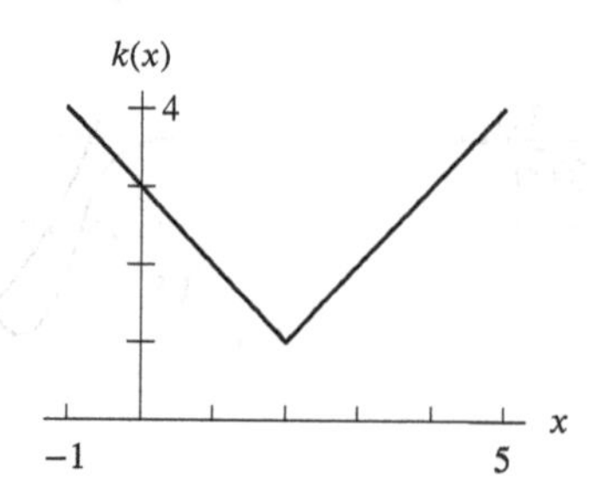

Figure 6.29

(e) Writing $m(x) = -\frac{1}{2}|4x + 12| - 3$ as $m(x) = -\frac{1}{2}|4(x + 3)| - 3$, we see that this is a vertical compression by a factor of $1/2$, then a reflection about the x-axis, then a horizontal compression by a factor of $1/4$, then a horizontal shift left 3 units, and finally a vertical shift down 3 units. See Figure 6.30.

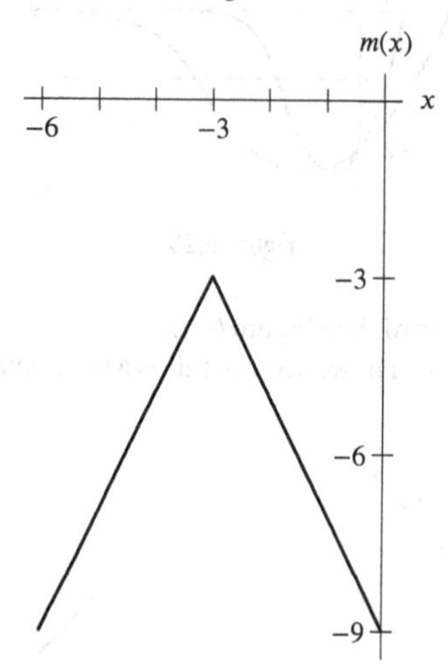

Figure 6.30

17.

Table 6.1

x	-3	-2	-1	0	1	2	3
$f(x)$	-4	-1	2	3	0	-3	-6
$f(\frac{1}{2}x)$	–	2	–	3	–	0	–
$f(2x)$	–	–	-1	3	-3	–	–

PROBLEMS

21. If $d > 1$ and $c > 0$, the graph is compressed toward the y-axis by a factor of $1/d$, and then translated upward by c units. After these transformations, the point (a, b) on the graph of $y = f(x)$ will correspond to the point $(a/d, b+c)$ on the graph of $y = f(dx) + c$. In other words, the transformed function $y = f(dx) + c$ gives an output of $b + c$ when the input is a/d.

One can also see this algebraically in the following way: If $f(a) = b$ and $h(x) = f(dx) + c$, then we can evaluate $h(x)$ when $dx = a$, hence when $x = a/d$. We have

$$h(a/d) = f((a/d)/d) + c = f(a) + c = b + c.$$

Thus $(a/d, b + c)$ is on the graph of $y = f(dx) + c$.

25. **(a)** (ii) The \$5 tip is added to the fare $f(x)$, so the total is $f(x) + 5$.
(b) (iv) There were 5 extra miles so the trip was $x + 5$. I paid $f(x + 5)$.
(c) (i) Each trip cost $f(x)$ and I paid for 5 of them, or $5f(x)$.
(d) (iii) The miles were 5 times the usual so $5x$ is the distance, and the cost is $f(5x)$.

29. **(a)** Since I is horizontally stretched compared to one graph and compressed compared to another, it should be $f(x)$.
(b) The most horizontally compressed of the graphs are III and II, so they should be $f(-2x)$ and $f(2x)$. Since III appears to be a compressed version of I reflected across the y-axis, it should be $f(-2x)$.
(c) The most horizontally stretched of the graphs should be $f(-\frac{1}{2}x)$, which is IV.
(d) The most horizontally compressed of the graphs are III and II, so they should be $f(-2x)$ and $f(2x)$. Since II appears to be a compressed version of I, it should be $f(2x)$.

33. We have:

First step:	$y = -f(x)$	reflect
Second step:	$y = -f(x) + 2$	shift up

Therefore, our final formula is $y = -f(x) + 2$.

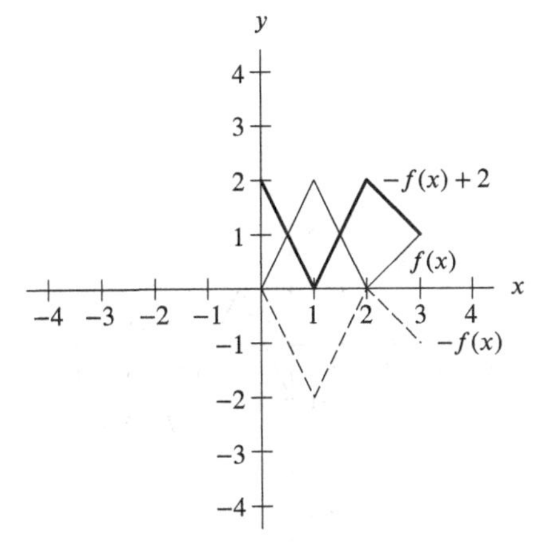

Figure 6.31

37. We have:

First step:	$y = f(x) - 2$	shift down
Second step:	$y = 2(f(x) - 2)$	stretch vertically

Multiplying out, this gives $y = 2f(x) - 4$.

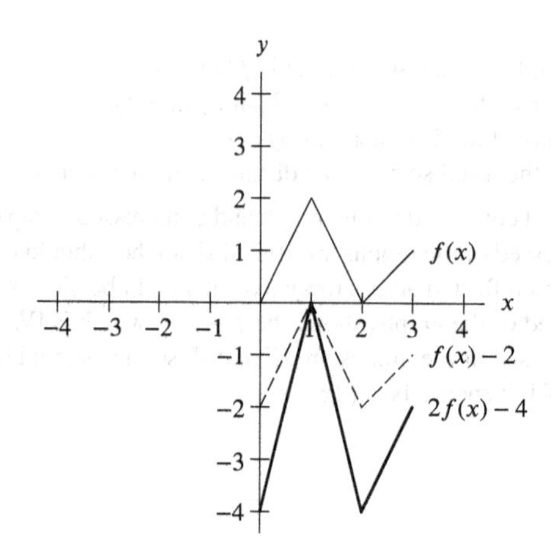

Figure 6.32

41. We have:

First step:	$y = f(-x)$	reflect
Second step:	$y = f(-(x-2))$	shift right

After simplifying, our answer is $y = f(-x+2)$.

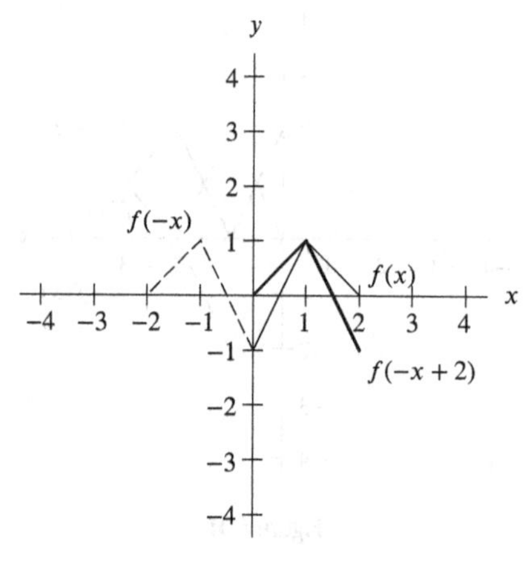

Figure 6.33

45. The function f has been reflected over the x-axis and the y-axis and stretched horizontally by a factor of 2. Thus, $y = -f(-\frac{1}{2}x)$.

49. **(a)** Since $f(x)$ has been shifted left by 2 units, the entire domain of $f(x)$ is shifted to the left by 2. Thus, the domain of $k(x) = f(x+2)$ is $-8 \le x \le 0$.

(b) Since the average rate of change of $f(x)$ over $-6 \le x \le 2$ is given as -3, we have

$$\begin{aligned}
\frac{f(2) - f(-6)}{2 - (-6)} &= \text{Average rate of change of } f(x) \\
\frac{f(2) - f(-6)}{8} &= -3 \\
f(2) - f(-6) &= 8(-3) = -24.
\end{aligned}$$

We now use the fact that $f(2) - f(-6) = -24$ to calculate the average rate of change of $k(x)$ over its domain $-8 \leq x \leq 0$.

$$\begin{aligned}
\text{Average rate of change of } k(x) &= \frac{\Delta y}{\Delta x} = \frac{k(0) - k(-8)}{0 - (-8)} \\
&= \frac{f(0+2) - f(-8+2)}{8} \quad \text{(since } k(x) = f(x+2)\text{)} \\
&= \frac{f(2) - f(-6)}{8} \\
&= \frac{-24}{8} \quad \text{(since } f(2) - f(-6) = -24\text{)} \\
&= -3.
\end{aligned}$$

53. We have

$$\begin{aligned}
y &= f(x+4) && \text{shift left 4 units} \\
y &= -f(x+4) && \text{then flip vertically} \\
y &= -f(x+4) + 2 && \text{then shift up 2 units} \\
y &= 3\,(-f(x+4) + 2) && \text{then stretch by 3} \\
\text{so} \quad g(x) &= -3f(x+4) + 6.
\end{aligned}$$

57. **(a)** The building is kept at 60° F until 5 am when the heat is turned up. The building heats up at a constant rate until 7 am when it is 68° F. It stays at that temperature until 3 pm when the heat is turned down. The building cools at a constant rate until 5 pm. At that time, the temperature is 60° F and it stays that level through the end of the day.

(b) Since $c(t) = 142 - d(t) = -d(t) + 142$, the graph of $c(t)$ will look like the graph of $d(t)$ that has been first vertically reflected across the t-axis and then vertically shifted up 142 units.

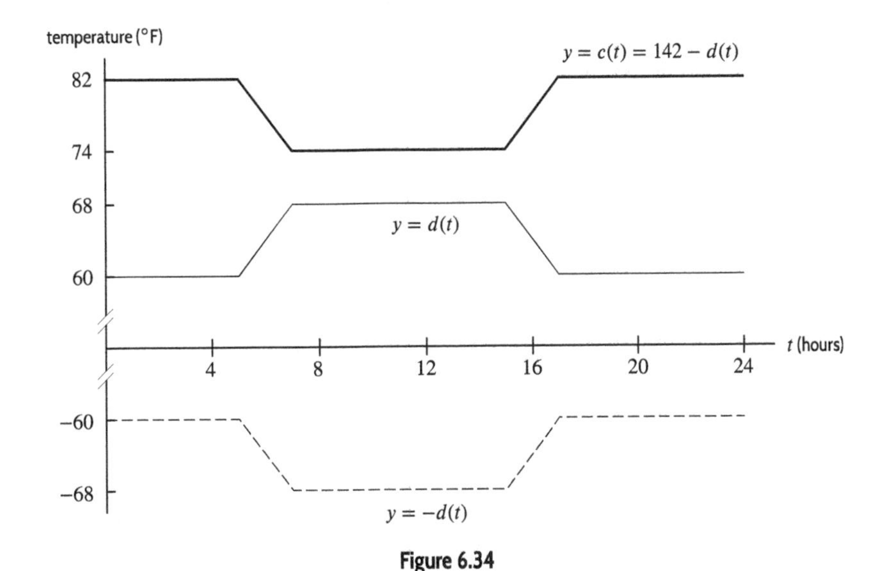

Figure 6.34

(c) This could describe the cooling schedule in the summer months when the temperature is kept at 82° F at night and cooled down to 74° during the day.

61. **(a)** Graphing f and g shows that there is a vertical shift up 1 unit. See Figure 6.35.

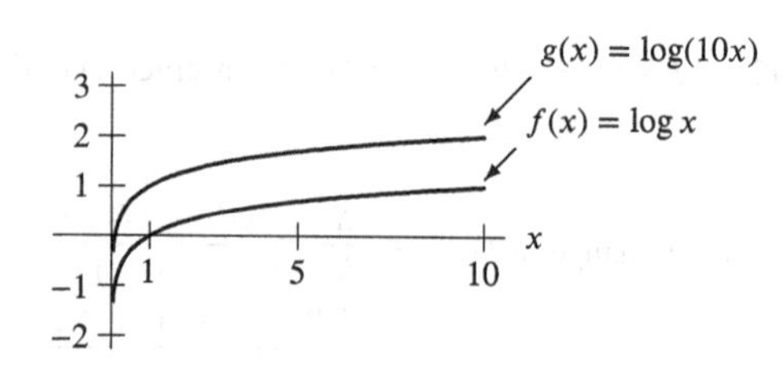

Figure 6.35: A vertical shift of +1

(b) Using the property that $\log(ab) = \log a + \log b$, we have

$$g(x) = \log(10x) = \log 10 + \log x = 1 + f(x).$$

Thus, $g(x)$ is $f(x)$ shifted vertically upward by 1.

(c) Using the same property of logarithms,

$$\log(ax) = \log a + \log x \qquad \text{so} \qquad k = \log a.$$

STRENGTHEN YOUR UNDERSTANDING

1. True. The graph is shifted down by $|k|$ units.

5. False. The graphs of odd functions are symmetric about the origin.

9. False. Substituting $(x - 2)$ in to the formula for g gives $g(x - 2) = (x - 2)^2 + 4 = x^2 - 4x + 4 + 4 = x^2 - 4x + 8$.

13. True.

17. False. Consider $f(x) = x^2$. Shifting up first and then compressing vertically gives the graph of $g(x) = \frac{1}{2}(x^2+1) = \frac{1}{2}x^2 + \frac{1}{2}$. Compressing first and then shifting gives the graph of $h(x) = \frac{1}{2}x^2 + 1$.

CHAPTER SEVEN

Solutions for Section 7.1

Skill Refresher

S1. **(a)** $f(5)$ is the height, in meters, of the apple 5 seconds after being dropped. Since $f(5) = 59.5$, after 5 seconds, the apple is 59.5 m from the ground.

(b) $f(5)$ is the height of the apple 5 seconds after being dropped and $f(1)$ is the height of the apple 1 second after being dropped. Therefore, $f(5) - f(1)$ is the change in height of the apple between $t = 1$ and $t = 5$ seconds. Since $f(5) - f(1) = -117.6$, the change in height of the apple between $t = 1$ and $t = 5$ seconds is -117.6 m. It makes sense that this is negative since the apple is dropping.

(c) $f(5)$ is the height of the apple 5 seconds after being dropped and $f(1)$ is the height of the apple 1 second after being dropped. Therefore, $(f(5) - f(1))/(5 - 1)$ is the average change in height of the apple between $t = 1$ and $t = 5$ seconds. Since $(f(5) - f(1))/(5 - 1) = -29.4$, the average rate of change of height of the apple between $t = 1$ and $t = 5$ seconds is -29.4 m/s.

S5. **(a)** The diameter is 126 ft and the platform is 2 ft high, so the minimum height is 2 ft, the same as the platform, and the maximum height is $126 + 2 = 128$ ft.

(b) The distance from the ground to the center of the Ferris wheel is equal to the halfway point between the minimum and maximum heights, or $(128 + 2)/2 = 65$ ft.

(c) The radius is half the diameter, so $126/2 = 63$ ft.

(d) Since it take 10 minutes to make quarter of a revolution, it will take 40 minutes to complete a full revolution.

EXERCISES

1. Graphs (I), (II), and (IV) appear to describe periodic functions.

(I) This function appears periodic. The rapid variation overlays a slower variation that appears to repeat every 8 units. (It almost appears to repeat every 4 units, but there is subtle difference between consecutive 4-second intervals. Do you see it?)

(II) This function also appears periodic, again with a period of about 4 units. For instance, the x-intercepts appear to be evenly spaced, at approximately $-11, -7, -3, 1, 5, 9$, and the peaks are also evenly spaced, at $-9, -5, -1, 3, 7, 11$.

(III) This function does not appear periodic. For instance, the x-intercepts grow increasingly close together (when read from left to right).

(IV) At first glance this function might appear to vary unpredictably. But on closer inspection we see that the graph repeats the same pattern on the interval $-12 \le x \le 0$ and $0 \le x \le 12$.

(V) This function does not appear periodic. The peaks of the graph appear to rise slowly (when read from left to right), and the troughs appear to fall slowly.

(VI) This function does not appear to be periodic. The peaks and troughs of its graph seem to vary unpredictably, although they are more or less evenly spaced.

5. In the 9 o'clock position, the person is midway between the top and bottom of the wheel. Since the diameter is 150 m, the radius is 75 m, so the person is 75 m below the top, or $165 - 75 = 90$ m above the ground, as in the 3 o'clock position.

9. The graph appears to have a period of b. Every change in the x value of b brings us back to the same y value.

PROBLEMS

13. After 24 minutes, the person is three-fourths of the way through one rotation. Since the wheel is turning clockwise, this means she is in the 3 o'clock position, midway between the top and bottom of the wheel. Since the diameter is 150 m, the radius is 75 m, so the person is 75 m below the top, or $165 - 75 = 90$ m above the ground.

17. The wheel will complete two full revolutions after 20 minutes, and the height ranges from $h = 50$ to $h = 650$. So the function is graphed on the interval $0 \le t \le 20$. See Figure 7.1.

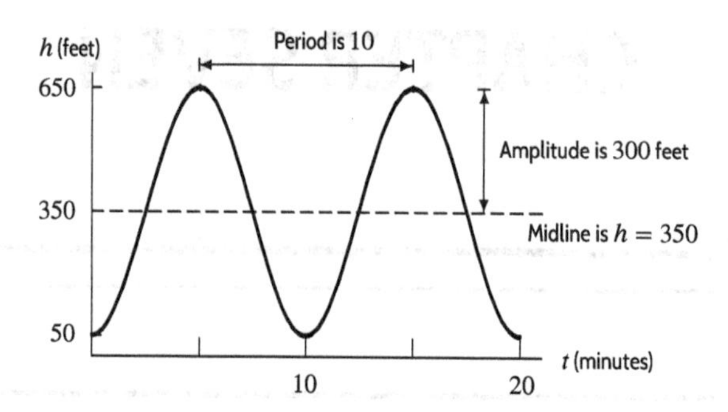

Figure 7.1: Graph of $h = f(t)$, $0 \le t \le 20$

21. The ride is completed in 30 minutes, so the average vertical speed in the last 5 minutes is given by

$$\frac{h(30) - h(25)}{5}.$$

Even though the table does not show this data, we know, due to symmetry, that the passenger is as high 5 minutes before the ride ends as she is 5 minutes into the ride. In other words:

$$h(25) = h(5) = 160 \text{ ft.}$$

Since we are back at the lowest point after 30 minutes, we have $h(30) = h(0) = 30$. Thus:

$$\frac{h(30) - h(25)}{5} = \frac{30 - 160}{5} = -26 \text{ ft/min.}$$

We note the sign is negative since we are moving down.

25. See Figure 7.2.

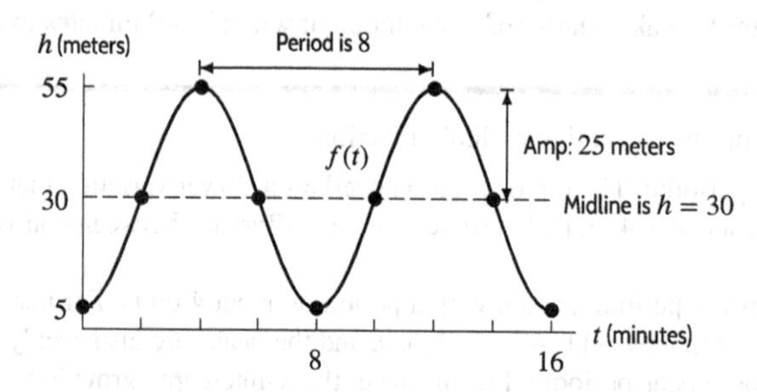

Figure 7.2: Graph of $h = f(t)$, $0 \le t \le 16$

29. Your initial position is twelve o'clock, since at $t = 0$, the value of h is at its maximum of 35. The period is 4 because the wheel completes one cycle in 4 minutes. The diameter is 30 meters and the boarding platform is 5 meters above ground. Because you go through 2.5 cycles, the length of time spent on the wheel is 10 minutes.

33. **(a)** Weight B, because the midline is $d = 10$, compared to $d = 20$ for weight A. This means that when the spring is not oscillating, weight B is 10 cm from the ceiling, while weight A is 20 cm from the ceiling.

(b) Weight A, because its amplitude is 10 cm, compared to the amplitude of 5 cm for weight B.

(c) Weight A, because its period is 0.5, compared to the period of 2 for weight B. This means that it takes weight A only half a second to complete one oscillation, whereas weight B completes one oscillation in 2 seconds.

37. **(a)** Looking at the input values for f, we see the quotient on the left gives the average rate of change in height as we move from 5 o'clock to 4 o'clock. The quotient on the right gives the average rate of change as we move from 4 o'clock to 3 o'clock. On average, we move upward faster when our motion is mostly vertical than when our motion is mostly horizontal. This suggests the quotient on the right is bigger than the one on the left. We use values in the table to confirm:

$$\frac{f(35) - f(32.5)}{35 - 32.5} = \frac{150 - 76.795}{2.5} = 29.282 \text{ ft/min.}$$

$$\frac{f(37.5) - f(35)}{37.5 - 35} = \frac{250 - 150}{2.5} = 40 \text{ ft/min.}$$

(b) The quotient on the left gives the average rate of change in height as we move from 6 o'clock to 5 o'clock. The quotient on the right gives the average rate of change as we move from 12 o'clock to 11 o'clock. Using values from the table, we find

$$\frac{f(32.5)-f(30)}{32.5-30} = \frac{76.795-50}{2.5} = 10.718 \text{ ft/min}$$
$$\frac{f(47.5)-f(45)}{47.5-45} = \frac{423.205-450}{2.5} = -10.718 \text{ ft/min.}$$

These values are equal in magnitude and opposite in sign. The equal magnitudes tell us that, on average, we travel the same vertical distance per minute as we lift off from the ground and when we start to descend from the highest point. The negative sign tells us that, on average, we are moving downward as we go from the top of the London Eye to the 11 o'clock position.

41. Both quantities describe our average rate of change in height, in meters per minute, over a 15 minute time interval while riding in the London Eye (first difference quotient) and the High Roller (second difference quotient).

The first quotient gives our average change in vertical height as we travel from the half-way up point (on our way up) to our half-way down point (on our way down) in the London Eye ($t = 7.5$ to $t = 22.5$).

The second quotient describes the same change, but in the High Roller.

In each wheel, we are equally far from the boarding platform at each half-way point: 60 meters away in the London Eye, and 80 meters away in the High Roller. Thus, our average change in height is 0 in both cases, that is:

$$\frac{f(22.5)-f(7.5)}{15} = 0 = \frac{h(22.5)-h(7.5)}{15}.$$

Solutions for Section 7.2

Skill Refresher

S1. The angle lies in quadrant I.

S5. The angle is between 90° and 180° so it terminates in quadrant II.

S9. The angle is between 0° and 90°, so it terminates in quadrant I.

S13. Since $40° - 180° = -140°$, the terminal side will rotate clockwise more than 90° but less than 180° ending in quadrant III.

S17. Since the angle is $3(40°) - 100° = 20°$, the terminal side will rotate counterclockwise less than 90° ending in quadrant I.

EXERCISES

1. See Figure 7.3 for the positions of the angles. The coordinates of the points are found using the sine and cosine functions on a calculator.

(a) $(\cos 100°, \sin 100°) = (-0.174, 0.985)$
(b) $(\cos 200°, \sin 200°) = (-0.940, -0.342)$
(c) $(\cos(-200°), \sin(-200°)) = (-0.940, 0.342)$
(d) $(\cos(-45°), \sin(-45°)) = (0.707, -0.707)$
(e) $(\cos 1000°, \sin 1000°) = (0.174, -0.985)$
(f) $(\cos -720°, \sin -720°) = (1, 0)$

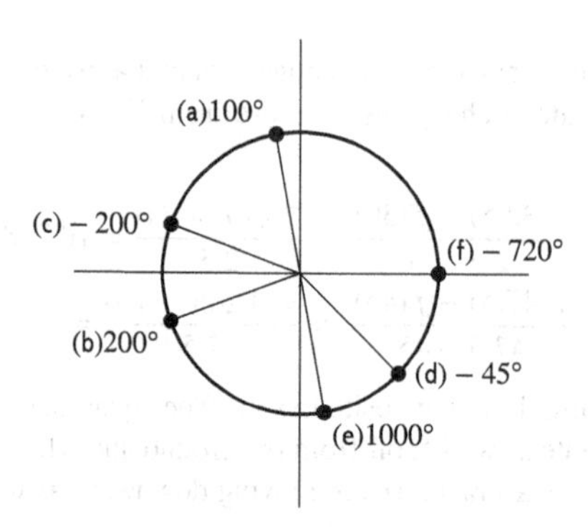

Figure 7.3

5. Since 100° is in the second quadrant, its reference angle is measured from the negative x-axis, corresponding to 180°. We have

$$\text{Reference angle} = 180^\circ - 100^\circ = 80^\circ.$$

9. The car on the Ferris wheel starts at the 3 o'clock position. Let's suppose that you see the wheel rotating counterclockwise. (If not, move to the other side of the wheel.)

The angle $\phi = 420^\circ$ indicates a counterclockwise rotation of the Ferris wheel from the 3 o'clock position all the way around once (360°), and then two-thirds of the way back up to the top (an additional 60°). This leaves you in the 1 o'clock position, or at the angle 60°.

A negative angle represents a rotation in the opposite direction, that is, clockwise. The angle $\theta = -150^\circ$ indicates a rotation from the 3 o'clock position in the clockwise direction, past the 6 o'clock position and two-thirds of the way up to the 9 o'clock position. This leaves you in the 8 o'clock position, or at the angle 210°. (See Figure 7.4.)

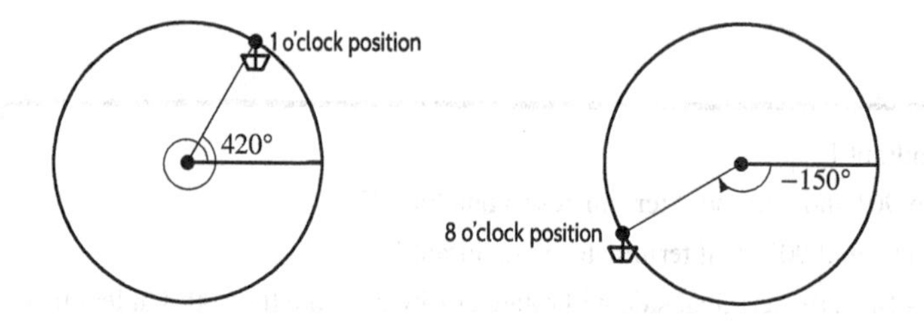

Figure 7.4: The positions and displacements on the Ferris wheel described by 420° and −150°

13. Since the x-coordinate is $r\cos\theta$ and the y-coordinate is $r\sin\theta$ and $r = 3.8$ and $\theta = -270^\circ$, the point is $(3.8\cos(-270^\circ), 3.8\sin(-270^\circ)) = (0, 3.8)$.

17. Since the x-coordinate is $r\cos\theta$ and the y-coordinate is $r\sin\theta$ and $r = 3.8$, the point is $(3.8\cos(-10^\circ), 3.8\sin(-10^\circ)) = (3.742, -0.660)$.

PROBLEMS

21. The angles 70°, $180^\circ - 70^\circ = 110^\circ$, $180^\circ + 70^\circ = 250^\circ$, and $360 - 70^\circ = 290^\circ$ are in different quadrants and all have the same reference angle 70°. Other answers are possible.

25. **(a)** As we see from Figure 7.5, the angle 135° specifies a point P' on the unit circle directly across the y-axis from the point P. Thus, P' has the same y-coordinate as P, but its x-coordinate is opposite in sign to the x-coordinate of P. Therefore, $\sin 135^\circ = 0.707$, and $\cos 135^\circ = -0.707$.

(b) As we see from Figure 7.6, the angle 285° specifies a point Q' on the unit circle directly across the x-axis from the point Q. Thus, Q' has the same x-coordinate as Q, but its y-coordinate is opposite in sign to the y-coordinate of Q. Therefore, $\sin 285^\circ = -0.966$, and $\cos 285^\circ = 0.259$.

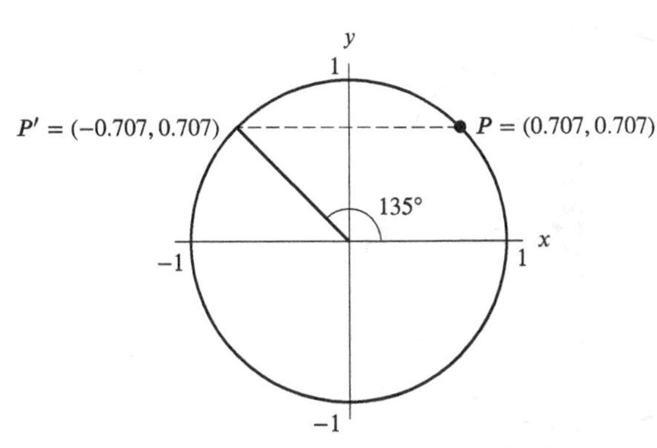

Figure 7.5: The sine and cosine of 135° can be found by referring to the sine and cosine of 45°

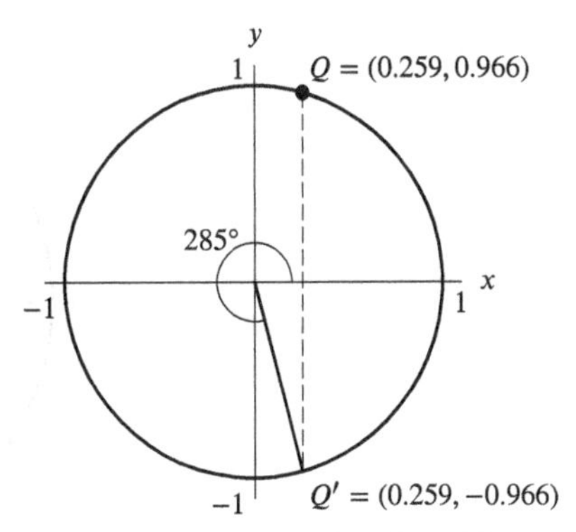

Figure 7.6: The sine and cosine of 285° can be found by referring to the sine and cosine of 75°

29. **(a)** $\sin(\theta + 360°) = \sin\theta = a$, since the sine function is periodic with a period of 360°.
(b) $\sin(\theta + 180°) = -a$. (A point on the unit circle given by the angle $\theta + 180°$ diametrically opposite the point given by the angle θ. So the y-coordinates of these two points are opposite in sign, but equal in magnitude.)
(c) $\cos(90° - \theta) = \sin\theta = a$. This is most easily seen from the right triangles in Figure 7.7.

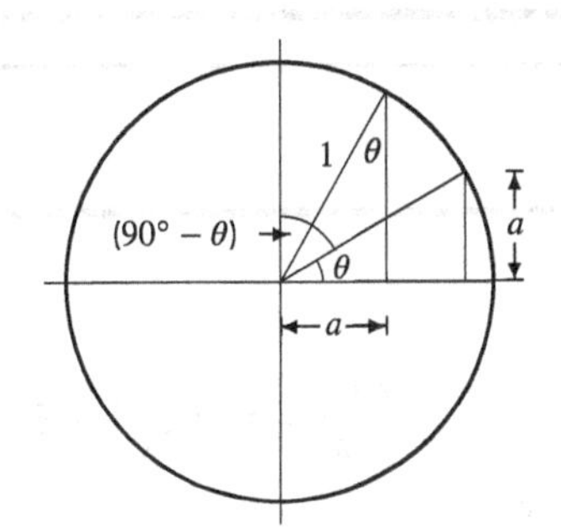

Figure 7.7

(d) $\sin(180° - \theta) = a$. (A point on the unit circle given by the angle $180° - \theta$ has a y-coordinate equal to the y-coordinate of the point on the unit circle given by θ.)
(e) $\sin(360° - \theta) = -a$. (A point on the unit circle given the the angle $360° - \theta$ has a y-coordinate of the same magnitude as the y-coordinate of the point on the unit circle given by θ, but is of opposite sign.)
(f) $\cos(270° - \theta) = -\sin\theta = -a$.

33. **(a)** The five panels split the circle into five equal parts, so the angle between each panel is $360°/5 = 72°$.
(b) Point B is directly across the circle from D, so 180°.
(c) The angle from A to D is the same as the angle from B to C, and the BC angle is the angle between panels, which is 72°. So moving the panel between A and D gives an angle of $(72°)/2 = 36°$. The panel then goes from point D to point B, spanning another 180°. Thus in total the panel traveled $36° + 180° = 216°$.

37. See Figure 7.8. Since the diameter is 120 mm, the radius is 60 mm. The coordinates of the outer edge point, A, on the x-axis is $(60, 0)$. Similarly, the inner edge at point B has coordinates $(7.5, 0)$.

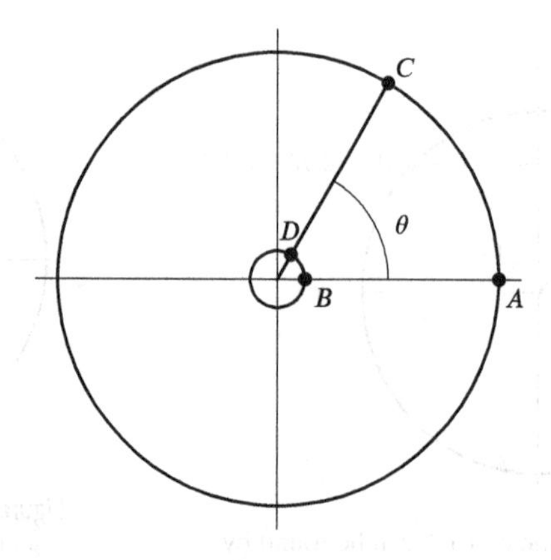

Figure 7.8

Points C and D are at an angle θ from the x-axis and have coordinates of the form $(r\cos\theta, r\sin\theta)$. For the outer edge, $r = 60$ so $C = (60\cos\theta, 60\sin\theta)$. The inner edge has $r = 7.5$, so $D = (7.5\cos\theta, 7.5\sin\theta)$.

Solutions for Section 7.3

Skill Refresher

S1. Dividing by 2π, we have $r = \dfrac{C}{2\pi}$.

S5. Multiplying both sides by x gives

$$x\left(\frac{360}{\pi}\right) = x\left(\frac{2}{3x}\right) = \frac{2}{3}.$$

Then, multiplying both sides by $\dfrac{\pi}{360}$ gives

$$x\left(\frac{360}{\pi}\right)\left(\frac{\pi}{360}\right) = \frac{2}{3}\left(\frac{\pi}{360}\right), \text{ so}$$

$$x = \frac{\pi}{540}.$$

S9. The angle lies in quadrant I, and item (a) is the only interval that lies within quadrant I.

EXERCISES

1. To convert 60° to radians, multiply by $\pi/180°$:

$$60°\left(\frac{\pi}{180°}\right) = \left(\frac{60°}{180°}\right)\pi = \frac{\pi}{3}.$$

We say that the radian measure of a 60° angle is $\pi/3$.

5. In order to change from degrees to radians, we multiply the number of degrees by $\pi/180$, so we have $150 \cdot \pi/180$, giving $\frac{5}{6}\pi$ radians.

9. In order to change from radians to degrees, we multiply the number of radians by $180/\pi$, so we have $\frac{7}{2}\pi \cdot 180/\pi$, giving 630 degrees.

13. In order to change from radians to degrees, we multiply the number of radians by $180/\pi$, so we have $45 \cdot 180/\pi$, giving $8100/\pi \approx 2578.310$ degrees.

17. If we go around twice, we make two full circles, which is $2\pi \cdot 2 = 4\pi$ radians. Since we're going around in the negative direction, we have -4π radians.

21. The arc length, s, corresponding to an angle of θ radians in a circle of radius r is $s = r\theta$. In order to change from degrees to radians, we multiply the number of degrees by $\pi/180$, so we have $45 \cdot \pi/180$, giving $\frac{\pi}{4}$ radians. Thus, our arc length is $6.2\pi/4 \approx 4.869$.

PROBLEMS

25. The reference angle between the ray and the negative x-axis is 30°. Since both sine and cosine are negative in the third quadrant, we have

$$x = r\cos\theta = 10\cos 210° = 10(-\cos 30°) = 10(-\sqrt{3}/2) = -5\sqrt{3}$$

and

$$y = r\sin\theta = 10\sin 210° = 10(-\sin 30°) = 10(-1/2) = -5,$$

so the coordinates of W are $(-5\sqrt{3}, -5)$.

29. Since the x-coordinate is $r\cos\theta$ and the y-coordinate is $r\sin\theta$ and $r = 5$, the point is $(5\cos 135°, 5\sin 135°) = (-5\sqrt{2}/2, 5\sqrt{2}/2)$.

33. The reference angle for 300° is $360° - 300° = 60°$, so $\sin 300° = -\sin 60° = -\sqrt{3}/2$.

37. The reference angle for 210° is $210° - 180° = 30°$, so $\sin 210° = -\sin 30° = -1/2$.

41. The reference angle for 405° is $405° - 360° = 45°$, so $\sin 405° = \sin 45° = \frac{1}{\sqrt{2}}$.

45. The reference angle for $11\pi/6$ is $2\pi - 11\pi/6 = \pi/6$, so $\cos(11\pi/6) = \cos(\pi/6) = \sqrt{3}/2$.

49.

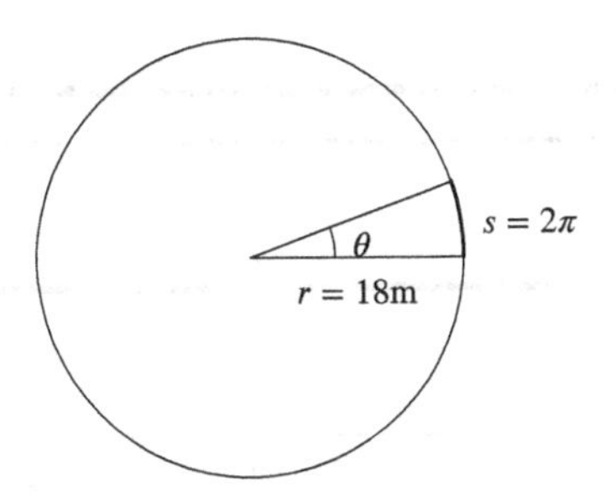

Figure 7.9

In Figure 7.9, we have $s = 2\pi$ and $r = 18$. Therefore,

$$\theta = \frac{s}{r} = \frac{2\pi}{18} = \frac{\pi}{9}.$$

Now,

$$\frac{\pi}{9}\text{ radians} = \frac{\pi}{9}\left(\frac{180°}{\pi}\right) = 20°.$$

Therefore, an arc of length 2π m on a circle of radius 18 m determines an angle of $\pi/9$ radians or 20°.

53. We have $\theta = 1.3$ rad or, in degrees,

$$1.3\left(\frac{180°}{\pi}\right) = 74.4845°.$$

We also have $r = 12$, so

$$s = 12(1.3) = 15.6,$$

and $P = (r\cos\theta, r\sin\theta) = (3.2100, 11.5627)$.

57. **(a)** Negative
(b) Negative
(c) Positive
(d) Positive

61. The angle spanned by the arc shown is $\theta = s/r = 4/5$ radians, so $m = r\cos\theta = 5\cos(4/5)$ and $n = r\sin\theta = 5\sin(4/5)$. By the Pythagorean theorem,

$$\begin{aligned} p^2 &= n^2 + (5-m)^2 \\ &= n^2 + m^2 - 10m + 25 \\ &= 25\sin^2(4/5) + 25\cos^2(4/5) - 10m + 25 \\ &= 50 - 10m \end{aligned}$$

so

$$\begin{aligned} p &= \sqrt{50 - 10m} \\ &= \sqrt{50 - 50\cos(4/5)} \\ &= 5\sqrt{2(1-\cos(4/5))}. \end{aligned}$$

65. We can approximate this angle by using $s = r\theta$. The arc length is approximated by the moon diameter; and the radius is the distance to the moon. Therefore $\theta = s/r = 2160/238{,}860 \approx 0.009$ radians. Change this to degrees to get $\theta = 0.009(180/\pi) \approx 0.516°$. Note that we could also consider the radius to cut across the moon's center, in which case the radius would be $r = 238{,}860 + 2160/2 = 239{,}940$. The difference in the two answers is negligible.

69. Using $s = r\theta$, we know the arc length $s = 600$ and $r = 3960 + 500$. Therefore $\theta = 600/4460 \approx 0.1345$ radians.

Solutions for Section 7.4

Skill Refresher

S1.

$0 \quad \frac{\pi}{4} \quad \frac{\pi}{2} \quad \frac{3\pi}{4} \quad \pi \quad \frac{5\pi}{4} \quad \frac{3\pi}{2} \quad \frac{7\pi}{4} \quad 2\pi \quad x$

S5. Since the graph is obtained by shifting the graph of f up 2 units, the formula is $y = f(x) + 2$, so our answer is (i).

S9. The graph of $y = -f(x)$ is obtained by reflecting the graph of $y = f(x)$ about the x-axis. See Figure 7.10.

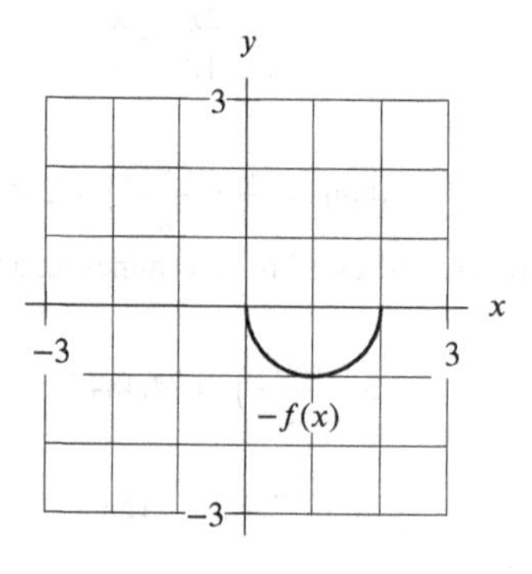

Figure 7.10

S13. This is a vertical stretch of $f(x)$ by a factor of two.

EXERCISES

1. The midline is $y = 3$ and the amplitude is 2.

5. The midline of $f(t)$ is $y = 4$ and the amplitude is $|-2| = 2$. Since the amplitude measures the vertical distance of the minimum and maximum value of $f(t)$ from its midline, the range of $f(t)$ will be $2 \le y \le 6$.

9. Judging from the figure:

- The midline is the dashed horizontal line $y = -1$.
- The vertical distance from the first peak to the midline is 0.5, so the amplitude is 0.5.
- The function starts at its minimum value, so it must be a shift and stretch of $\cos t$.

13. For a function of the form $A \sin t + k$ or $A \cos t + k$, $y = k$ is the midline and $|A|$ is the amplitude. This means possible formulas for this function are $g(t) = 2\sin t + 3$, $g(t) = -2\sin t + 3$, $g(t) = 2\cos t + 3$, or $g(t) = -2\cos t + 3$. The only one of these to have a maximum at $t = 0$ is $2\cos t + 3$ and so we have $g(t) = 2\cos t + 3$.

PROBLEMS

17. $f(x) = (\sin x) + 1$
$g(x) = (\sin x) - 1$

21. Since $\sin\theta$ is the y-coordinate of a point on the unit circle, its height above the x-axis can never be greater than 1. Otherwise the point would be outside the circle. See Figure 7.11.

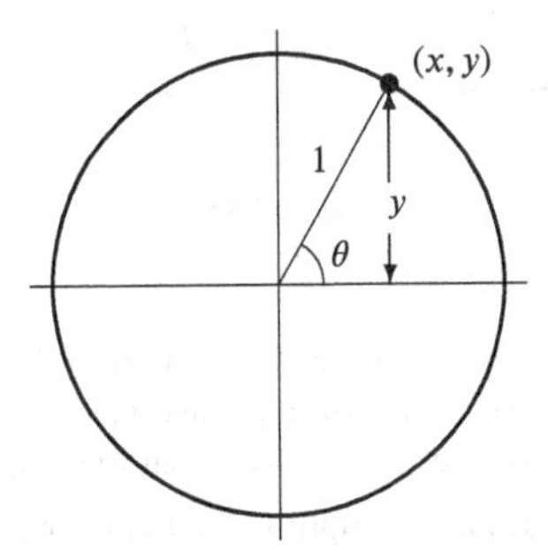

Figure 7.11

25. The midline of $y = \frac{1}{2}\sin x$ is $y = 0$, and the midline of $y = \sin x + \frac{1}{2}$ is $y = \frac{1}{2}$. In the figure, the graph of $g(x)$ has a midline above the x-axis, so $f(x) = \frac{1}{2}\sin x$.

Since $x = a$ is the first positive x-value for which $f(x) = 0$, we have $a = \pi$.

The first minimum value of $f(x)$ with a positive x-value occurs at $x = b$, so $b = 3\pi/2$.

The minimum value of $f(x)$ is $x = c$, so $c = -\frac{1}{2}$.

The maximum value of $g(x)$ is a maximum value of the function $g(x) = \sin x + \frac{1}{2}$, so $d = \frac{3}{2}$.

29. This function resembles a sine graph that has been reflected across the y-axis and has an amplitude of 2 and a midline of $y = 1$. Thus $y = -2f(x) + 1$.

33. **(a)** (i) See Figure 7.12. The function $f(\theta)$ has 4 zeros.

(ii) See Figure 7.13. The function $g(\theta)$ has no zeros.

(iii) See Figure 7.14. The function $h(\theta)$ has 2 zeros.

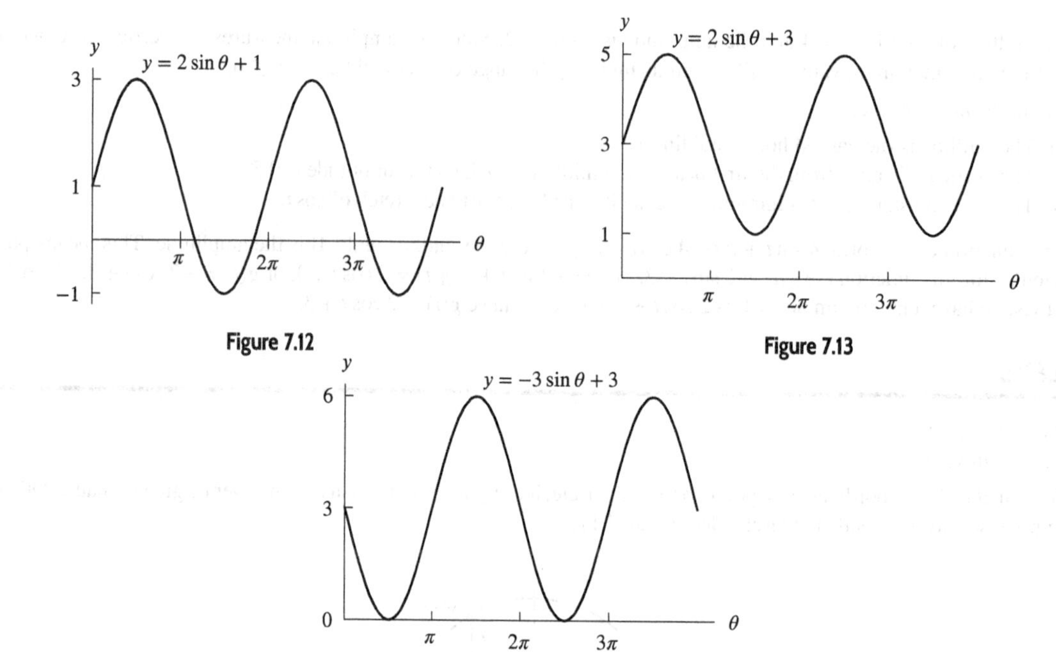

Figure 7.12

Figure 7.13

Figure 7.14

(b) In order to have no zeros, the y-value on the midline $y = k$ must be larger than the amplitude, that is, $|A| < k$.

(c) In order to have exactly two zeros, $f(\theta) = A\sin\theta + k$ must have its zeros at the minimum or maximum values of $f(\theta)$. This means the y-value on the midline $y = k$ must equal the amplitude, that is, $|A| = k$.

37. (a) The midline of the function is 50 meters and the amplitude is 45 meters. So the highest point on the ride occurs when the person on the Ferris wheel is in the 12 o'clock position, where $\theta = 90^\circ$. Since $\sin 90^\circ = 1$, $f(90) = 50{+}45\sin 90^\circ = 95$ m. Similarly, the lowest point occurs when $\theta = 270^\circ$, which is in the 6 o'clock position: $f(270) = 50{+}45\sin 270^\circ = 5$ m, since $\sin 270^\circ = -1$.

(b) Since the amplitude of the height function is 45 meters, this is also the radius of the Ferris wheel. The midline of the function is 50 meters, so the center of the wheel is 50 meters off the ground. This means that as the Ferris wheel rotates, the lowest point on the wheel will be 5 meters above the ground.

Solutions for Section 7.5

Skill Refresher

S1. Remove the common factor of three, and we get

$$3x - 6 = 3(x - 2).$$

S5. Remove the common factor of $\frac{\pi}{6}$, and we get

$\frac{\pi}{6}(x - 3)$.

S9. Since the graph is obtained by shifting the graph of f left by 2, the formula is $y = f(x + 2)$, so our answer is (iv).

S13. The graph of $y = f(2x)$ is obtained by horizontally compressing the graph of $y = f(x)$ by a factor of 2. See Figure 7.15.

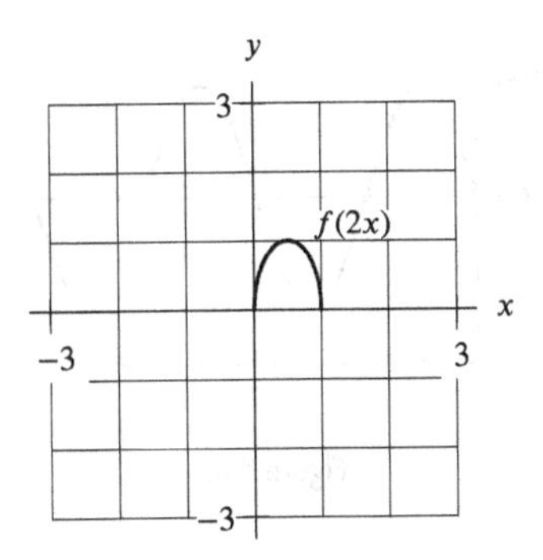

Figure 7.15

S17. This transformation will take $f(x)$ and shift it right three units.

EXERCISES

1. The midline is $y = 0$. The amplitude is 6. The period is 2π.

5. Judging from the figure:

- The curve looks the same from $t = 0$ to $t = 8$ as from $t = 8$ to $t = 16$, so it repeats with a period of 8.
- The midline is the dashed horizontal line $y = 30$.
- The vertical distance from the first peak to the midline is 20, so the amplitude is 20.

9. Judging from the figure:

- The curve looks the same from $t = 0$ to $t = 0.5$ as from $t = 0.5$ to $t = 1$, so it repeats with a period of 0.5.
- The midline is the dashed horizontal line $y = 0.5$.
- The vertical distance from the first peak to the midline is 0.5, so the amplitude is 0.5.

13. This function resembles a sine curve in that it passes through the origin and then proceeds to grow from there. We know that the smallest value it attains is −4, and the largest it attains is 4; thus its amplitude is 4, with a midline of 0. It has a period of 1. Thus in the equation

$$g(t) = A\sin(Bt)$$

we know that $A = 4$ and

$$1 = \text{period} = \frac{2\pi}{B}.$$

So $B = 2\pi$, and then

$$h(t) = 4\sin(2\pi t).$$

17. The midline is $y = 4000$. The amplitude is $8000 - 4000 = 4000$. The period is 60, so B is $2\pi/60$. The graph at $x = 0$ rises from its midline, so we use the sine. Thus,

$$y = 4000 + 4000\sin\left(\frac{2\pi}{60}x\right).$$

PROBLEMS

21. See Figure 7.16.

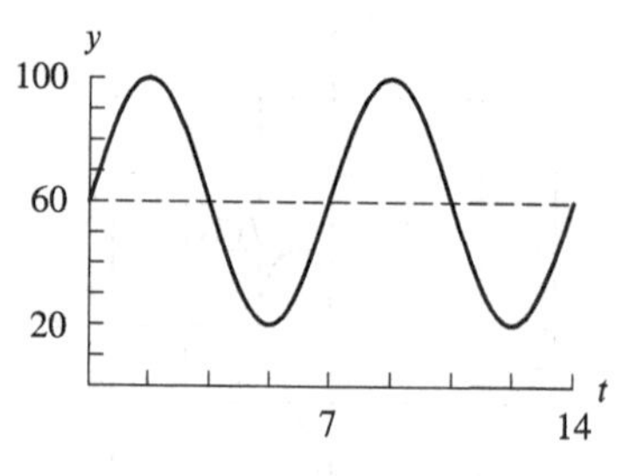

Figure 7.16

25. Because the period of $\sin x$ is 2π, and the period of $\sin 2x$ is π, so from the figure in the problem we see that

$$f(x) = \sin x.$$

The points on the graph are $a = \pi/2$, $b = \pi$, $c = 3\pi/2$, $d = 2\pi$, and $e = 1$.

29. $f(t) = 14 + 10\sin\left(\pi t + \frac{\pi}{2}\right)$

33. The data given describe a trigonometric function shifted vertically because all the $g(x)$ values are greater than 1. Since the maximum is approximately 3 and the minimum approximately 1, the midline value is 2. We choose the sine function over the cosine function because the data tell us that at $x = 0$ the function takes on its midline value, and then increases. Thus our function will be of the form

$$g(x) = A\sin(Bx) + k.$$

We know that A represents the amplitude, k represents the vertical shift, and the period is $2\pi/B$.

We have already noted the midline value is $k \approx 2$. This means $A = \max - k = 1$. We also note that the function completes a full cycle after 1 unit. Thus

$$1 = \frac{2\pi}{B}$$

so

$$B = 2\pi.$$

Thus

$$g(x) = \sin(2\pi x) + 2.$$

37. This function has an amplitude of 2 and a period of 1 and a midline of $y = -3$, and resembles a cosine graph. Thus $y = 2g(x) - 3$.

41. **(a)** See Figure 7.17.

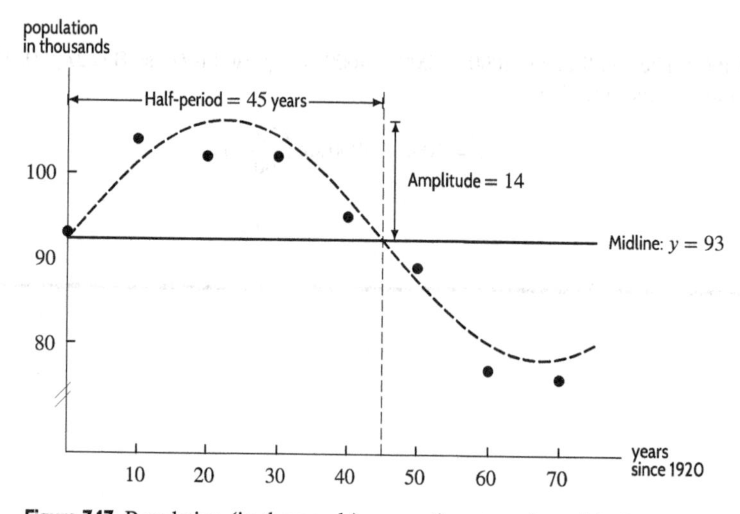

Figure 7.17: Population (in thousands) versus time, together with sine curve.

(b) The variation is possibly sinusoidal, but not necessarily. The population rises first, then falls. From the graph, it appears it could soon rise again, but this need not be the case.

(c) See Figure 7.17. There are many possible answers.

(d) For the graph drawn, the amplitude of the population function is $107-93=14$. The average value of P from the data is 93. The graph of $P=f(t)$ behaves as the graph of $\sin t$, for $3/4$ of a period. Therefore, we look for a reasonable approximation to the data of the form $P=f(t)=14\sin(Bt)+93$. To determine B, we assume that 45 years is half the period of f. Thus, the period equals 90 years and so $B=2\pi/90=\pi/45$. Hence an approximation to the data is

$$P=f(t)=14\sin\left(\frac{\pi}{45}t\right)+93.$$

(e) $P=f(-10)\approx 84.001$, which means that our formula predicts a population of about 84,000. This is not too far off the mark, but not all that close, either.

45. We see that the phase shift is 4; the shift is to the left. To find the horizontal shift, we factor out a 3 within the cosine function, giving

$$y=2\cos\left(3\left(t+\frac{4}{3}\right)\right)-5.$$

Thus, the horizontal shift is $-4/3$.

49. The amplitude and midline are 20 and the period is 5. The graph is a sine curve shifted half a period to the right (or left), so the phase shift is $\frac{1}{2}(2\pi)=\pi$. Thus

$$h=20+20\sin\left(\frac{2\pi}{5}t-\pi\right).$$

Solutions for Section 7.6

Skill Refresher

S1. Vertical lines have the form $x=k$, where k is the x-coordinate of the points on the line. Thus, the equation of the line is $x=3$.

S5. The slope of the line is defined as

$$m=\frac{\text{change in } y}{\text{change in } x}=\frac{3}{4}$$

S9. The slope of the line is defined as

$$m=\frac{\text{change in } y}{\text{change in } x}=\frac{-4}{2}=-2$$

EXERCISES

1. $\sin 0^\circ=0$, $\cos 0^\circ=1$, $\tan 0^\circ=\sin 0^\circ/\cos 0^\circ=0/1=0$.

5. We see that $\tan 540^\circ=\tan 180^\circ=\dfrac{\sin 180^\circ}{\cos 180^\circ}=\dfrac{0}{-1}=0.$

9. Using the exact values of sine and cosine for special angles, we have

$$\tan\frac{\pi}{3}=\frac{\sin(\pi/3)}{\cos(\pi/3)}=\frac{\sqrt{3}/2}{1/2}=\sqrt{3}$$

13. The graph of $\tan\theta$ has asymptotes at $\theta=\pm\pi/2,\pm 3\pi/2,\pm 5\pi/2,\ldots$. Since the graph of $h(\theta)$ is the graph of $\tan\theta$ vertically shifted up by 2 units, it has asymptotes at the same places. Of these, the only ones in the interval $0\le\theta\le 2\pi$ are $\theta=\pi/2$ and $\theta=3\pi/2$.

17. See Figure 7.18. The graph has been shifted up by 1.

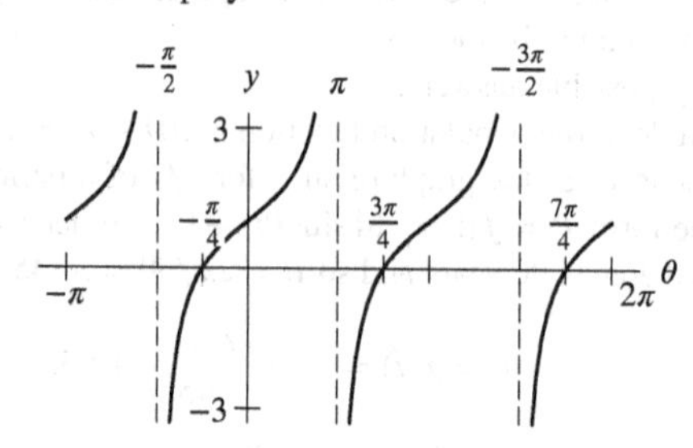

Figure 7.18

PROBLEMS

21. We have:

$$\text{Slope} = \frac{\Delta y}{\Delta x} = \frac{y-0}{x-0} = \frac{\sin(6\pi/7)}{\cos(6\pi/7)} = \tan(6\pi/7) = -0.482.$$

So, the slope of this line is -0.482. This makes sense since $6\pi/7$ is more than $90°$ but less than $180°$ and the line's slope is then negative.

25. Since $(0,0)$ and $(p,5)$ are points on the line,

$$\text{Slope of line } = \frac{\Delta y}{\Delta x} = \frac{5-0}{p-0} = \frac{5}{p}.$$

Since the line forms an angle of $15°$ we also have

$$\text{Slope of line} = \tan(15°)$$

The two expressions together give

$$\frac{5}{p} = \tan(15°) \qquad \text{which means} \qquad p = \frac{5}{\tan(15°)} = 18.660.$$

29. This looks like a tangent graph. At $\pi/4$, we have $\tan\theta = 1$. On this graph $y = 1/2$ if $\theta = \pi/4$, and since it appears to have the same period as $\tan\theta$ without a horizontal or vertical shift, a possible formula is $y = \frac{1}{2}\tan\theta$.

33. **(a)** For a possible graph see Figure 7.19.

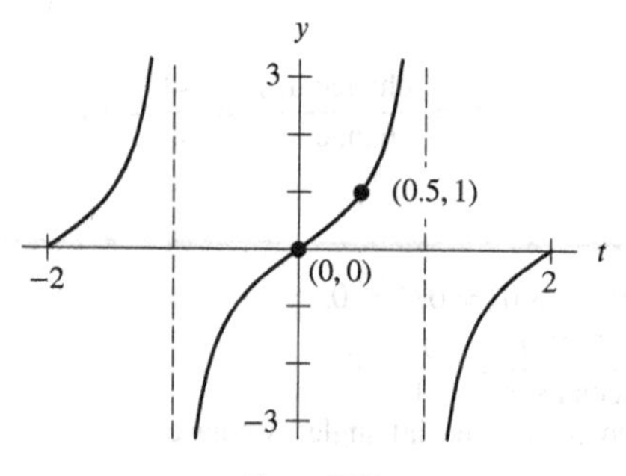

Figure 7.19

(b) We can see from the graph that this function has period 2. Since the graph passes through $(0,0)$, which lies directly in the middle of the period, there is no vertical shift. Therefore a formula is of the form $f(t) = A\tan\left(\frac{\pi}{2}t\right)$ for some A. Since $f(t)$ passes through $(0.5, 1)$, we have

$$f(0.5) = A\tan\left(\frac{\pi}{4}\right) = A = 1.$$

Therefore, a possible formula for $f(t)$ is $f(t) = \tan\left(\frac{\pi}{2}t\right)$.

37. Your friend is incorrect. The function $f(t) = \tan t$ is an example of a function that is increasing everywhere it is defined, yet it is periodic.

Solutions for Section 7.7

Skill Refresher

S1. By the Pythagorean Theorem, a right triangle with side lengths a, b, c and c the hypotenuse, satisfies $a^2 + b^2 = c^2$. Thus,

$$\begin{aligned} 3^2 + 4^2 &= c^2, \\ 9 + 16 &= c^2, \\ 25 &= c^2, \\ \sqrt{25} &= c, \\ 5 &= c. \end{aligned}$$

S5. The reciprocal of $\frac{a}{b}$ is $\frac{b}{a}$, so the reciprocal is $\frac{1}{5}$.

S9. We have $\frac{x}{2} \cdot \frac{3}{x} = \frac{3x}{2x} = \frac{3}{2}$.

S13. We have $\frac{y-3}{5} = \frac{y}{5} - \frac{3}{5}$.

EXERCISES

1. $-\sqrt{3}$

5. Since $\cot(5\pi/3) = 1/\tan(5\pi/3) = 1/(\sin(5\pi/3)/\cos(5\pi/3)) = \cos(5\pi/3)/\sin(5\pi/3)$, we know that $\cot(5\pi/3) = (1/2)/(-\sqrt{3}/2) = -1/\sqrt{3}$.

9. Since $\csc(3\pi/4) = 1/\sin(3\pi/4)$, we know that $\csc(3\pi/4) = 1/(1/\sqrt{2}) = \sqrt{2}$.

13. Writing $\sec t = 1/\cos t$, we have

$$\sec t \cos t = 1.$$

17. Expanding the square and combining terms gives

$$(\sin x - \cos x)^2 + 2\sin x \cos x = \sin^2 x - 2\sin x \cos x + \cos^2 + 2\sin x \cos x = \sin^2 x + \cos^2 x = 1.$$

PROBLEMS

21. Since $\sec\theta = 1/\cos\theta$, we have $\sec\theta = 1/(1/2) = 2$.
To find $\tan\theta$, we first find $\sin\theta$ from the identity $\cos^2\theta + \sin^2\theta = 1$, so

$$\begin{aligned} \left(\frac{1}{2}\right)^2 + \sin^2\theta &= 1 \\ \sin^2\theta &= 1 - \frac{1}{4} = \frac{3}{4} \\ \sin\theta &= \pm\frac{\sqrt{3}}{2}. \end{aligned}$$

Since $0 \le \theta \le \pi/2$, we know that $\sin\theta \ge 0$, so $\sin\theta = \sqrt{3}/2$ and

$$\tan\theta = \frac{\sin\theta}{\cos\theta} = \frac{\sqrt{3}/2}{1/2} = \sqrt{3}.$$

25. Since $y = \sin\theta$, we can construct the following triangle:

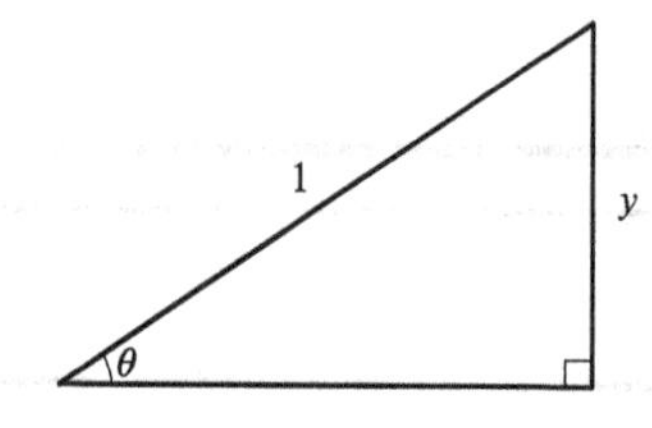

Figure 7.20

The adjacent side, using the Pythagorean theorem, has length $\sqrt{1-y^2}$. So, $\cos\theta = \frac{\text{adj}}{\text{hyp}} = \frac{\sqrt{1-y^2}}{1} = \sqrt{1-y^2}$.

29.

$$\cos^2\theta = 1 - \sin^2\theta = 1 - \left(\frac{x}{3}\right)^2 = 1 - \frac{x^2}{9} = \frac{9-x^2}{9},$$

so

$$\cos\theta = \sqrt{\frac{9-x^2}{9}} = \frac{\sqrt{9-x^2}}{3}.$$

$$\tan\theta = \frac{\sin\theta}{\cos\theta} = \frac{x}{3} \cdot \frac{3}{\sqrt{9-x^2}} = \frac{x}{\sqrt{9-x^2}}$$

33. **(a)** (i) This is an identity. The graph of $y = \sin t$ is the graph of $y = \cos t$ shifted to the right by $\pi/2$. This means that value of $\cos\left(t - \frac{\pi}{2}\right)$ equals the value of $\sin t$ for all values of t.

(ii) This is not an identity. In Figure 7.21, we see that the graph of $y = \sin 2t$ has a period of π and an amplitude of 1, whereas the graph of $y = 2\sin t$ has a period of 2π and an amplitude of 2. Instead of being true for all values of t, this equation is true only where the graphs of these two functions intersect.

(b) From Figure 7.21, we see there are three solutions to the equation $\sin 2t = 2\sin t$ on the interval $0 \le t \le 2\pi$, at $t = 0, \pi$, and 2π.

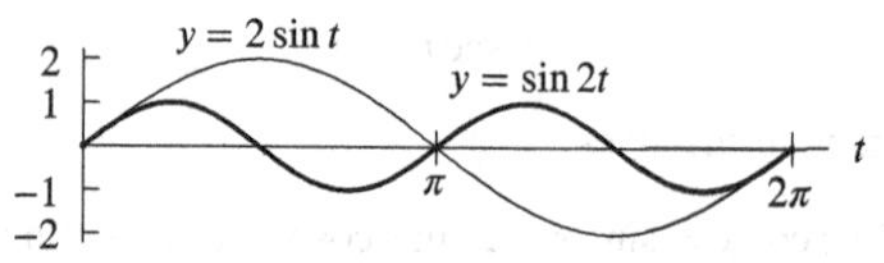

Figure 7.21: The graphs of $y = \sin 2t$ and $y = 2\sin t$ are different, so $\sin 2t = 2\sin t$ is not an identity. It is an equation with three solutions on the interval shown

37. Divide both sides of $\cos^2\theta + \sin^2\theta = 1$ by $\sin^2\theta$. For $\sin\theta \neq 0$,

$$\frac{\cos^2\theta}{\sin^2\theta} + \frac{\sin^2\theta}{\sin^2\theta} = \frac{1}{\sin^2\theta}$$

$$\left(\frac{\cos\theta}{\sin\theta}\right)^2 + 1 = \left(\frac{1}{\sin\theta}\right)^2$$

$$\cot^2\theta + 1 = \csc^2\theta.$$

Solutions for Section 7.8

Skill Refresher

S1. The variable p is the value you put into the sine function and the variable q is the result. So, p is the input and q is the output.

S5. By definition, if (x, y) is a point on the unit circle specified by angle θ, then $\cos\theta = x$. Since $\cos\theta = 0.1$, we have $x > 0$. The only other quadrant with $x > 0$ is the fourth quadrant, so the other angle α for which $\cos\alpha = 0.1$ must be in the fourth quadrant.

S9. By definition, if (x, y) is a point on the unit circle specified by angle θ, then $\tan\theta = y/x$. Since $\tan\theta = 1.7$, we have either both $x, y > 0$ or both $x, y < 0$. In the third quadrant, we have both $x, y < 0$, so the other angle α must be in the quadrant where both $x, y > 0$, so the first quadrant.

S13. We have

$$\begin{aligned} 5\cos\theta + 8 &= 4 \\ 5\cos\theta &= -4 \\ \cos\theta &= -\frac{4}{5}. \end{aligned}$$

EXERCISES

1. (a) We are looking for the value of the sine of $\left(\frac{1}{2}\right)^\circ$. Using a calculator or computer, we have $\sin\left(\frac{1}{2}\right)^\circ = 0.009$.

(b) We are looking for the angle between $-\pi/2$ and $\pi/2$ whose sine is $\frac{1}{2}$. Therefore, we have $\sin^{-1}\left(\frac{1}{2}\right) = \pi/6 = 30^\circ$.

(c) We are looking for the reciprocal of the sine of $\left(\frac{1}{2}\right)^\circ$. We have

$$\begin{aligned} (\sin x)^{-1} &= \left(\sin\frac{1}{2}\right)^{-1} \\ &= \frac{1}{\sin\frac{1}{2}} \\ &= 114.593. \end{aligned}$$

5. We have

$$\begin{aligned} 6\cos\theta - 2 &= 3 \\ 6\cos\theta &= 5 \\ \cos\theta &= \left(\frac{5}{6}\right) \\ \theta &= \cos^{-1}\left(\frac{5}{6}\right) \\ \theta &= 0.58569 = 33.557^\circ. \end{aligned}$$

9. We use the inverse sine function on a calculator to get $\theta = 0.608$.

13. Since the angles have a negative cosine they must be in the second or third quadrant. An angle in the second quadrant with a reference angle of $\pi/3$ must measure

$$\pi - \frac{\pi}{3} = \frac{2\pi}{3}.$$

An angle in the third quadrant with this same reference angle must measure

$$\pi + \frac{\pi}{3} = \frac{4\pi}{3}.$$

So $\theta = 2\pi/3, 4\pi/3$ are two of many possible answers.

PROBLEMS

17. Since $\cos t = -1$, we have $t = \pi$.

21. Since $\tan t = \sqrt{3}$, we have $t = \pi/3$ and $t = 4\pi/3$.

25. To find the angles when the height of the ant is 18 cm, we substitute 18 for y in the formula for the height of the ant:

$$\begin{aligned} 12 + 12\sin\theta &= 18 \\ 12\sin\theta &= 6 \\ \sin\theta &= \frac{1}{2}. \end{aligned}$$

Since $\sin\theta = 1/2$, one solution to the equation is

$$\theta = \sin^{-1}\left(\frac{1}{2}\right) = 30^\circ.$$

By symmetry, a second solution corresponds to the angle at which the ant has a height of 18 cm and is on the way back down to its starting height. At that time, the reference angle is 30°, so $\theta = 180^\circ - 30^\circ = 150^\circ$ (see Figure 7.22). Our answers are therefore $\theta = 30^\circ$ and $\theta = 150^\circ$.

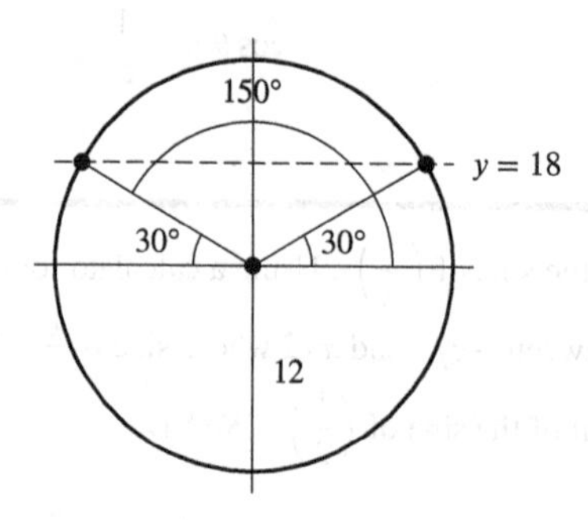

Figure 7.22

29. The letter k represents the angle and the letter a represents the value of the function.

33. The angle is n; the value is p.

37. **(a)** A sketch of the graph of h with respect to t is shown in Figure 7.23. The points labeled P, Q, R, and S represent the points on the graph corresponding to the times when the rider is at a height of 100 meters above the ground; we estimate these times to be 6, 10, 22, and 26 minutes.

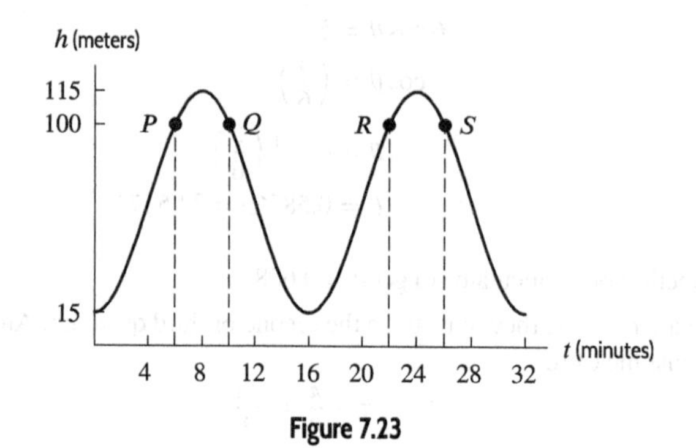

Figure 7.23

(b) To answer this question, we first find an equation for the height of the rider as a function of time. This function is periodic and achieves its minimum value at $t = 0$; therefore, we can write

$$h = A\cos(Bt) + k$$

for some constants A, B, and k. From Figure 7.23, we see that the period of the function is 16 minutes, so

$$\begin{aligned}\frac{2\pi}{B} &= 16\\ 2\pi &= 16B\\ B &= \frac{\pi}{8}.\end{aligned}$$

Similarly, the midline is $(115+15)/2 = 65$ meters, and the amplitude is $115-65 = 50$ meters, so a possible formula for the function is

$$h = f(t) = -50\cos\left(\frac{\pi}{8}t\right) + 65.$$

The times when the rider is 100 meters above the ground are the solutions to the equation $f(t) = 100$ that lie in the interval $0 \le t \le 32$. We obtain one solution to this equation as follows:

$$\begin{aligned}-50\cos\left(\frac{\pi}{8}t\right) + 65 &= 100\\ \cos\left(\frac{\pi}{8}t\right) &= -\frac{7}{10}\\ t &= \frac{8}{\pi}\cos^{-1}\left(-\frac{7}{10}\right)\\ &= 5.9745 \text{ minutes}\end{aligned}$$

Referring to Figure 7.23, we use symmetry to determine that $t = 16 - 5.9745 = 10.0255$ minutes is a second solution to $f(t) = 100$. The remaining two solutions occur one period to the right of the previous two, so they are given by $t = 16 + 5.9745 = 21.9745$ minutes and $t = 16 + 10.0255 = 26.0255$ minutes.

Finally, we wish to express our answers in terms of minutes and seconds. Since $0.9745 \cdot 60 = 58.47$ and $0.0255 \cdot 60 = 1.53$, we see that 0.9745 of a minute and 0.0255 of a minute correspond to about 58 and 2 seconds, respectively. Therefore, the four times at which the rider is at a height of 100 meters are about 5 minutes and 58 seconds, 10 minutes and 2 seconds, 21 minutes and 58 seconds, and 26 minutes and 2 seconds.

41. **(a)** $\sin^{-1} x$ is the angle between $-\pi/2$ and $\pi/2$ whose sine is x. When $x = 0.5$, $\sin^{-1}(0.5) = \pi/6$.
(b) $\sin(x^{-1})$ is the sine of $1/x$. When $x = 0.5$, $\sin(0.5^{-1}) = \sin(2) \approx 0.909$.
(c) $(\sin x)^{-1} = \frac{1}{\sin x}$. When $x = 0.5$, $(\sin 0.5)^{-1} = \frac{1}{\sin(0.5)} \approx 2.086$.

45. Statement II is always true, because $\arcsin x$ is an angle whose sine is x, and thus the sine of $\arcsin x$ will necessarily equal x. Statement I could be true or false. For example,

$$\arcsin\left(\sin\frac{\pi}{4}\right) = \arcsin\left(\frac{\sqrt{2}}{2}\right) = \frac{\pi}{4}.$$

On the other hand,

$$\arcsin(\sin\pi) = \arcsin(0) = 0,$$

which is not equal to π.

STRENGTHEN YOUR UNDERSTANDING

1. True, because $\sin x$ is an odd function.

5. The function $\sin(3x)$ is a horizontal transformation of $\sin x$, while $3\sin x$ is a vertical stretch. Therefore, the statement is false.

9. We know that $\csc\left(\frac{x}{2}\right)$ is a horizontal transformation and $\frac{\csc}{2}$ is a vertical transformation.
Therefore, this statement is false.

13. True. The point $(1, 0)$ on the unit circle is the starting point to measure angles so $\theta = 0°$.

17. False. One radian is about 57 degrees.

21. False. The point is $(2\cos 240°, 2\sin 240°) = (-1, -\sqrt{3})$.

25. False. Since both angles are negative they are measured in the clockwise direction. Going clockwise beyond 180° takes the point into the second quadrant.

29. False. The angle 315° is in the fourth quadrant, so the cosine value is positive. The correct value is $\sqrt{2}/2$.

33. False. A parabola is the graph a quadratic function and can never repeat its values more than twice.

37. True. Since sine is an odd function.

41. True. The midline is $y = (\text{maximum} + \text{minimum})/2 = (6+2)/2 = 4$.

45. False. The maximum y-value is when $\cos x = 1$. It is 35.

49. False, since $\sin(\pi x)$ has period $2\pi/\pi = 2$.

53. False. The period is $\frac{1}{3}$ that of $y = \cos x$.

57. False. Numerically, we could check the equation at $x = 0$ to find $y = -0.5\cos(2\cdot 0+\frac{\pi}{3})+1 = 0.75$, but the graph at $x = 0$ shows $y = 1$.

61. True. Let $\theta = 5x$ in the identity $\sin^2\theta + \cos^2\theta = 1$.

65. False. The reciprocal of the sine function is the cosecant function.

69. False. $y = \sin^{-1} x = \arcsin x \neq 1/\sin x$.

73. False; for instance, $\cos^{-1}(\cos(2\pi)) = 0$.

77. False. For example, $\cos(\pi/4) = \cos(-\pi/4)$.

81. True. This is true since $\cos\frac{\pi}{3} = 0.5$ and $0 < \pi/3 < \pi$.

85. False. Any angle $\theta = \frac{\pi}{4} + n\pi$, or $\theta = -\frac{\pi}{4} + n\pi$ with n an integer, will have the same cosine value.

CHAPTER EIGHT

Solutions for Section 8.1

Skill Refresher

S1. We see that x is the hypotenuse of this right triangle, so by the Pythagorean Theorem:

$$\begin{aligned} 6^2 + 8^2 &= x^2 \\ x &= \sqrt{6^2 + 8^2} = \sqrt{100} = 10. \end{aligned}$$

We see that $x = 10$. In this case, we could have also found x using the fact that this is a 3-4-5 right triangle, with each length multiplied by 2 to give a 6-8-10 right triangle.

S5. We see that x is one of the sides of this right triangle, so by the Pythagorean Theorem:

$$\begin{aligned} x^2 + 3^2 &= 7^2 \\ x &= \sqrt{7^2 - 3^2} = \sqrt{40} = 6.325. \end{aligned}$$

We see that $x = \sqrt{40} = 6.325$.

S9. We use the fact that the angles of a triangle always add to 180°. Since a right angle is 90°, we have:

$$\begin{aligned} \theta + 25 + 90 &= 180 \\ \theta &= 180 - 90 - 25 = 65. \end{aligned}$$

We see that $\theta = 65°$.

EXERCISES

1. By the Pythagorean theorem, the hypotenuse has length $\sqrt{1^2 + 2^2} = \sqrt{5}$.

(a) $\tan\theta = \dfrac{\text{opposite}}{\text{adjacent}} = \dfrac{2}{1} = 2.$

(b) $\sin\theta = \dfrac{\text{opposite}}{\text{hypotenuse}} = \dfrac{2}{\sqrt{5}}.$

(c) $\cos\theta = \dfrac{\text{adjacent}}{\text{hypotenuse}} = \dfrac{1}{\sqrt{5}}.$

5. We use the Pythagorean theorem to find the length of the hypotenuse:

$$\begin{aligned} \text{Hypotenuse}^2 &= (0.1)^2 + (0.2)^2 = 0.01 + 0.04 = 0.05 \\ \text{Hypotenuse} &= \sqrt{0.05}. \end{aligned}$$

(a) We have

$$\sin\theta = \frac{\text{Side opposite}}{\text{Hypotenuse}} = \frac{0.1}{\sqrt{0.05}} = 0.447.$$

(b) We have

$$\cos\theta = \frac{\text{Side adjacent}}{\text{Hypotenuse}} = \frac{0.2}{\sqrt{0.05}} = 0.894.$$

9. Since $\cos 30° = x/10$, we have $x = 10(\sqrt{3}/2) = 5\sqrt{3}$.

13. Since $\sin 12° = 4/r$, we have $r = 4/\sin 12°$. Similarly, since $\tan 12° = 4/x$, we have $x = 4/\tan 12°$.

17. Since $\sin 22° = \lambda/r$, we have $r = \lambda/\sin 22°$. Similarly, since $\tan 22° = \lambda/y$, we have $y = \lambda/\tan 22°$.

21. In a 45°-45°-90° triangle, the two legs are equal and the hypotenuse is $\sqrt{2}$ times the length of a leg. So each leg is $7/\sqrt{2}$.

25. We have $\tan\theta = 54.169$, so $\theta = \tan^{-1} 54.169 = 88.942°$.

29. We have

$$\begin{aligned} b &= \sqrt{c^2 - a^2} = \sqrt{384} \approx 19.596 \\ \sin A &= \frac{a}{c} \\ A &= \sin^{-1}\frac{a}{c} = 45.585° \\ B &= 90° - A = 44.415°. \end{aligned}$$

PROBLEMS

33. In a right triangle, we have

$$\sin\theta = \frac{\text{Opposite}}{\text{Hypotenuse}} \quad \text{and} \quad \cos\theta = \frac{\text{Adjacent}}{\text{Hypotenuse}} \quad \text{and} \quad \tan\theta = \frac{\text{Opposite}}{\text{Adjacent}}.$$

Thus, rewriting and canceling, we have

$$\frac{\sin\theta}{\cos\theta} = \frac{\text{Opposite/Hypotenuse}}{\text{Adjacent/Hypotenuse}} = \frac{\text{Opposite}}{\text{Hypotenuse}} \cdot \frac{\text{Hypotenuse}}{\text{Adjacent}} = \frac{\text{Opposite}}{\text{Adjacent}} = \tan\theta.$$

37. A right triangle whose hypotenuse is twice the length of a leg is a 30°-60°-90° triangle. In such triangles the length of the longer leg is $\sqrt{3}$ times the length of the shorter leg, so the third side is $7\sqrt{3}$.

41. See Figure 8.20. Since the tangent is the length of the opposite side divided by the length of the adjacent side,

$$\begin{aligned} \tan 58° &= \frac{d}{50} \\ d &= 50\tan 58° \approx 80.017 \end{aligned}$$

The width of the river is about 80 meters.

45. Using right triangle trigonometry, we have

$$\begin{aligned} \tan\theta &= \frac{240}{130} \\ \theta &= (\tan)^{-1}\left(\frac{240}{130}\right) \\ &= 61.557°. \end{aligned}$$

49. **(a)** Since the grade of the ramp is $7\% = 7/100$, this means that a 7-foot height difference occurs over a horizontal distance of 100 feet. So we have $\tan\theta = \frac{7}{100}$. Using the $\tan^{-1}$ button on the calculator, we get

$$\theta = \tan^{-1}\left(\frac{7}{100}\right) = 4.00417°.$$

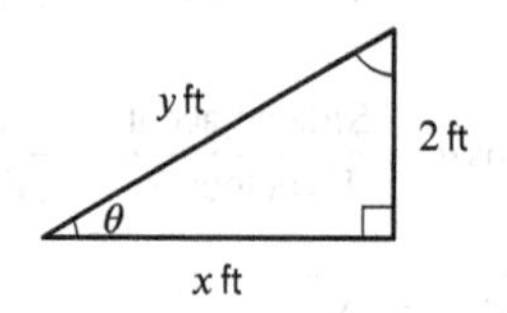

Figure 8.1

(b) From the right triangle representing the ramp we see that one leg represents the height difference between the path and the deck, which is 2 ft. The other leg represents the path, of which we would like to find the length x. Then $2/x = \tan 4.00417°$. Solving this equation for x, we have

$$x = \frac{2}{\tan 4.00417°} = 28.57.$$

So the path has to be 28.57 feet long. (We can also use similar triangles: $2/7 = x/100$.)

(c) The ramp is represented by the hypotenuse y of the right triangle. Using the Pythagorean theorem we have $y = \sqrt{28.57^2 + 2^2} = 28.64$ feet. (We can also use $\sin 4.00417° = 2/y$.)

53. To solve for the distance x, we use $\tan 53° = \frac{954}{x}$ and solve for x:

$$x = 954/\tan 53° = 718.891 \text{ ft}.$$

To solve for the height of the Seafirst Tower, we can use $\tan 37° = y/x$ and solve for y:

$$y = 718.891 \tan 37° = 541.723 \text{ ft}.$$

(The actual height of the Seafirst Tower is 543 ft.)

57. In the figure, we see that $\tan\alpha = 2/10 = 0.2$ and $\tan\beta = (2+1)/10 = 0.3$. Thus,

$$\begin{aligned} \tan\alpha &= 0.2, \quad \text{so} \quad \alpha = \tan^{-1}(0.2) \approx 11.310° \\ \tan\beta &= 0.3, \quad \text{so} \quad \beta = \tan^{-1}(0.3) \approx 16.699° \end{aligned}$$

Solutions for Section 8.2

Skill Refresher

S1. We have:

$$x = \frac{\sin(40°)}{\sin(10°)} = 3.702.$$

S5. We cross-multiply and solve for x:

$$\begin{aligned} \frac{\sin(70°)}{x} &= \frac{\sin(40°)}{8} \\ 8\sin(70°) &= x\sin(40°) \\ x &= \frac{8\sin(70°)}{\sin(40°)} = 11.695. \end{aligned}$$

S9. We have

$$\sin\theta = \frac{1}{2}\cdot\frac{\sin(40°)}{2} = \frac{\sin(40°)}{4} = 0.25\sin(40°) = 0.161.$$

EXERCISES

1. We have

$$\sin 30° = \frac{1}{2}.$$

5. We have

$$\cos 120° = -\cos(180-120)° = -\cos 60° = -\frac{1}{2}.$$

9. In Figure 8.2, we have

$$\beta = 180^\circ - 90^\circ - 38^\circ$$
$$\beta = 52^\circ.$$

$$\sin 38^\circ = \frac{4}{c}$$
$$c = \frac{4}{\sin 38^\circ} \approx 6.497.$$

$$c^2 = 4^2 + b^2$$
$$b = \sqrt{c^2 - 16} \approx 5.120.$$

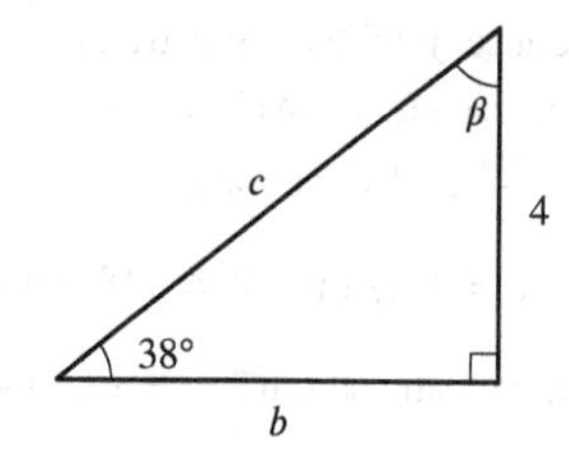

Figure 8.2

13. Using the Law of Sines in Figure 8.3, we have

$$\frac{\sin\theta}{5} = \frac{\sin 20^\circ}{6}$$
$$\theta = \sin^{-1}\left(\frac{5\sin 20^\circ}{6}\right) = 16.560^\circ$$

This is correct since $\theta < 90^\circ$ in the triangle. We expect $\theta < 20^\circ$ because θ is opposite a side which is shorter than 6. Therefore

$$\psi = 180^\circ - 16.560^\circ - 20^\circ = 143.440^\circ.$$
$$\frac{\sin 143.440^\circ}{a} = \frac{\sin 20^\circ}{6}$$
$$a = \frac{6\sin 143.440^\circ}{\sin 20^\circ}$$
$$a = 10.450.$$

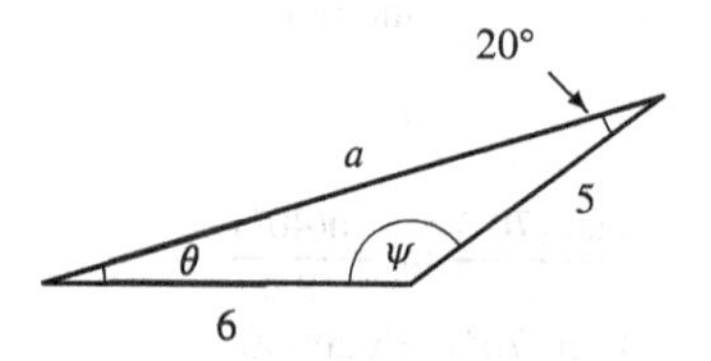

Figure 8.3

17. Using the Law of Cosines, we have

$$b^2 = 20^2 + 28^2 - 2\cdot 20\cdot 28\cos 81^\circ$$
$$b^2 = 1008.793$$
$$b = 31.762.$$

Using the Law of Sines, we have

$$\frac{\sin A}{20} = \frac{\sin 81^\circ}{31.762}$$
$$\sin A = 0.622$$
$$A = \sin^{-1}(0.622) \text{ or } 180 - \sin^{-1}(0.622)$$
$$A = 38.458^\circ \text{ or } 141.542^\circ.$$

We can discard the larger answer, as it does not fit in a triangle with $B = 81^\circ$. Thus we have $C = 180^\circ - A - B = 60.542^\circ$.

21. By the Law of Cosines,

$$\begin{aligned} c^2 &= 5^2 + 11^2 - 2 \cdot 5 \cdot 11 \cos 32^\circ \\ c &= 7.2605. \end{aligned}$$

Angle A is acute (since $a < b$ and thus $A < B$), so we solve for it using the Law of Sines. We have

$$\begin{aligned} \frac{\sin A}{a} &= \frac{\sin C}{c} \\ \frac{\sin A}{5} &= \frac{\sin 32^\circ}{7.2605} \\ \sin A &= \frac{5 \sin 32^\circ}{7.2605} \\ A &= \sin^{-1}\left(\frac{5\sin 32^\circ}{7.2605}\right) \\ &= 21.403^\circ. \end{aligned}$$

This gives $B = 180 - A - C = 126.597^\circ$.

25. We begin by finding the angle B, which is $180^\circ - 35^\circ - 92^\circ = 53^\circ$.

We can now use the law of sines to find the other two sides.

$$\begin{aligned} \frac{a}{\sin 92^\circ} &= \frac{9}{\sin 35^\circ} \\ a &= \sin 92^\circ \cdot \frac{9}{\sin 35^\circ} \\ a &= 15.681. \end{aligned}$$

Similarly,

$$\begin{aligned} \frac{b}{\sin 53^\circ} &= \frac{9}{\sin 35^\circ} \\ b &= \sin 53^\circ \cdot \frac{9}{\sin 35^\circ} \\ b &= 12.531. \end{aligned}$$

29. First, we recognize that it is possible that there are two triangles, since we may have the ambiguous case. However, we know that angle C must be less than 75°, since the side across from it is shorter than 7. Thus, we begin by finding the angle C using the law of sines:

$$\begin{aligned} \frac{\sin C}{2} &= \frac{\sin 75^\circ}{7} \\ \sin C &= 2 \cdot \frac{\sin 75^\circ}{7} \\ C &= \sin^{-1} 0.276 \\ C &= 16.020^\circ. \end{aligned}$$

We can now solve for A, which is $180^\circ - 75^\circ - 16.020^\circ = 88.980^\circ$.

Using the law of sines, we can solve for side a:

$$\begin{aligned} \frac{a}{\sin 88.980} &= \frac{7}{\sin 75^\circ} \\ a &= \sin 88.980 \cdot \frac{7}{\sin 75^\circ} \\ a &= 7.246. \end{aligned}$$

33. In Figure 8.4, use the Law of Sines:

$$\frac{\sin(30^\circ)}{259} = \frac{\sin\beta}{510}$$

to obtain $\sin\beta \approx 0.9846$ and use $\sin^{-1}$ to find $\beta_1 \approx 79.917^\circ$ or $\beta_2 \approx 100.083^\circ$. We then know $\alpha_1 = 180^\circ - 30^\circ - 79.917^\circ \approx 70.083^\circ$, or $\alpha_2 = 180^\circ - 30^\circ - 100.083^\circ \approx 49.917^\circ$. We can use the value of α and the Law of Sines to find the length of side a:

$$\frac{a_1}{\sin(70.083^\circ)} = \frac{259}{\sin 30^\circ}, \quad \text{or} \quad \frac{a_2}{\sin(49.917^\circ)} = \frac{259}{\sin 30^\circ}.$$

$$a_1 \approx 487.016 \text{ ft} \qquad a_2 \approx 396.330 \text{ ft}$$

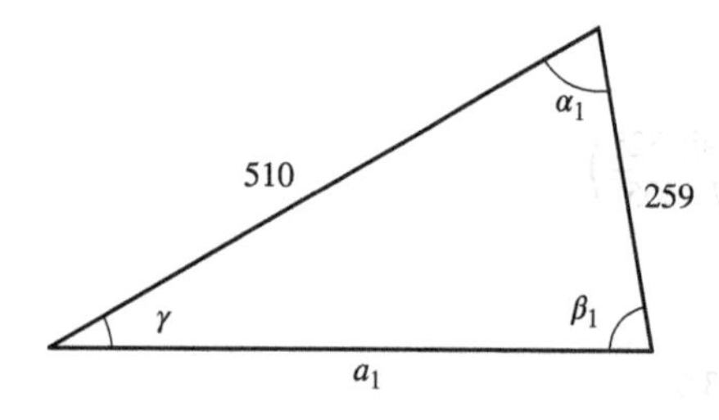

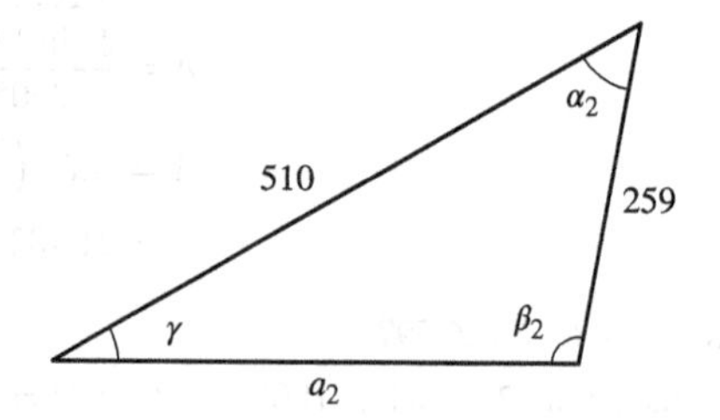

Figure 8.4

PROBLEMS

37. In Figure 8.5, the fire stations are at A and B and the forest fire is at C. The angle at C is $180^\circ - 54^\circ - 58^\circ = 68^\circ$. Solving for a and b using the Law of Sines, we get

$$\frac{56.7}{\sin 68^\circ} = \frac{a}{\sin 54^\circ} \qquad \frac{56.7}{\sin 68^\circ} = \frac{b}{\sin 58^\circ}$$

$$a = \frac{56.7 \sin 54^\circ}{\sin 68^\circ} \qquad b = \frac{56.7 \sin 58^\circ}{\sin 68^\circ}$$

$$a = 49.4738 \qquad b = 51.8606.$$

The fire station at point B is closer by $51.8606 - 49.4738 = 2.387$ miles.

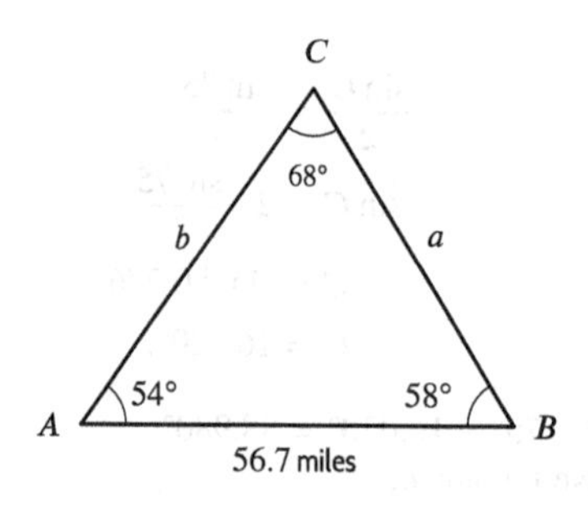

Figure 8.5

41. The horizontal and vertical displacements of the image are given by

$$\Delta x = 12\cos 25^\circ = 10.876$$
$$\Delta y = 12\sin 25^\circ = 5.071.$$

Thus, the new coordinates are $(x, y) = (8 + \Delta x, 5 + \Delta y) = (18.876, 10.071)$.

45. **(a)** See Figure 8.6.

Let x be the distance from the pitcher's rubber to first base. Then, by the Law of Cosines,

$$x^2 = 60.5^2 + 90^2 - 2(60.5)(90)\cos 45^\circ$$
$$x = 63.717.$$

To find the distance from the pitcher's rubber to second base, let y be the distance from home plate to second base. Then

$$y^2 = 90^2 + 90^2$$
$$y = 127.279.$$

Then we find the distance from the pitcher's rubber to second:

$$\text{Distance} = 127.279 - 60.5 = 66.779.$$

From the pitcher's rubber to first base is closer by $66.779 - 63.717 = 3.062$ feet.

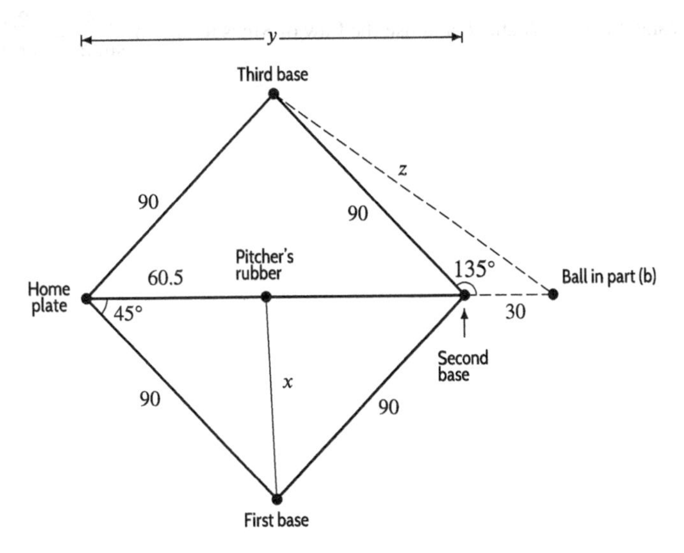

Figure 8.6

(b) Using a result from part (a), the distance from 30 feet past second base to home plate is given by

$$\text{Distance} = 30 + 127.279 = 157.279 \text{ feet.}$$

Let z be the distance from 30 feet past second base to third base:

$$z^2 = 30^2 + 90^2 - 2(30)(90)\cos 135^\circ$$
$$z = 113.218 \text{ feet.}$$

49. Using the Law of Sines on triangle DEF in Figure 8.7, we have

$$\frac{105.2}{\sin 29^\circ} = \frac{ED}{\sin 68^\circ}$$
$$ED = \frac{105.2\sin 68^\circ}{\sin 29^\circ} = 201.192 \text{ feet.}$$

$$\text{Total amount of wire needed} = 201.192 + 145.3 + 23.5 + 20 = 389.992 \text{ feet.}$$

Since wire is sold in 100-foot rolls, 4 rolls of wire are needed.

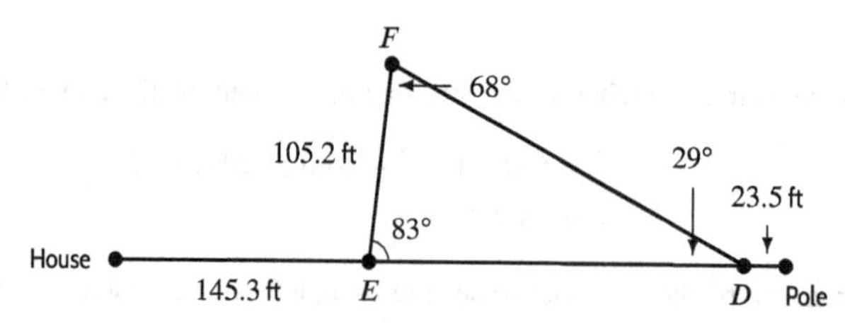

Figure 8.7

STRENGTHEN YOUR UNDERSTANDING

1. False. Both acute angles are 45 degrees, and $\sin 45° = \sqrt{2}/2$.

5. True. By the Law of Cosines, we have $p^2 = n^2 + r^2 - 2nr\cos P$, so $\cos P = (n^2 + r^2 - p^2)/(2nr)$.

9. True. Identify the opposite angles as B and L and use the Law of Sines to obtain $\frac{LA}{\sin B} = \frac{BA}{\sin L}$. Thus $\frac{LA}{BA} = \frac{\sin B}{\sin L}$.

CHAPTER NINE

Solutions for Section 9.1

Skill Refresher

S1. We have

$$2\cos\theta\sin\theta + 8\sin\theta = 2\sin\theta(\cos\theta + 4).$$

S5. **(a)** We have

$$\begin{aligned}11\cdot\frac{x}{4} + 3 &= 5\\ \frac{11x}{4} &= 2\\ 11x &= 8\\ x &= \frac{8}{11}.\end{aligned}$$

(b) We have

$$\begin{aligned}11\cdot\frac{\cos\theta}{4} + 3 &= 5\\ \frac{11\cos\theta}{4} &= 2\\ 11\cos\theta &= 8\\ \cos\theta &= \frac{8}{11}.\end{aligned}$$

S9. We have

$$\begin{aligned}8\cdot\frac{\cos\theta - 1}{5} + 1 &= \frac{1}{5}\\ 8\cdot\frac{\cos\theta - 1}{5} &= \frac{1}{5} - 1 = \frac{1}{5} - \frac{5}{5} = -\frac{4}{5}\\ \frac{\cos\theta - 1}{5} &= \frac{1}{8}\cdot\left(-\frac{4}{5}\right) = -\frac{4}{40} = -\frac{1}{10}\\ \cos\theta - 1 &= 5\cdot\left(-\frac{1}{10}\right) = -\frac{5}{10} = -\frac{1}{2}\\ \cos\theta &= 1 - \frac{1}{2} = \frac{2}{2} - \frac{1}{2} = \frac{1}{2}.\end{aligned}$$

EXERCISES

1. We draw a graph of $y = \cos t$ for $-\pi \le t \le 3\pi$ and trace along it on a calculator to find points at which $y = 0.4$. We read off the t-values at the points t_0, t_1, t_2, t_3 in Figure 9.1. If t is in radians, we find $t_0 = -1.159$, $t_1 = 1.159$, $t_2 = 5.124$, $t_3 = 7.442$. We can check these values by evaluating:

$$\cos(-1.159) = 0.40, \quad \cos(1.159) = 0.40, \quad \cos(5.124) = 0.40, \quad \cos(7.442) = 0.40.$$

Notice that because the cosine function is periodic, the equation $\cos t = 0.4$ has infinitely many solutions. The symmetry of the graph suggests that the solutions are related.

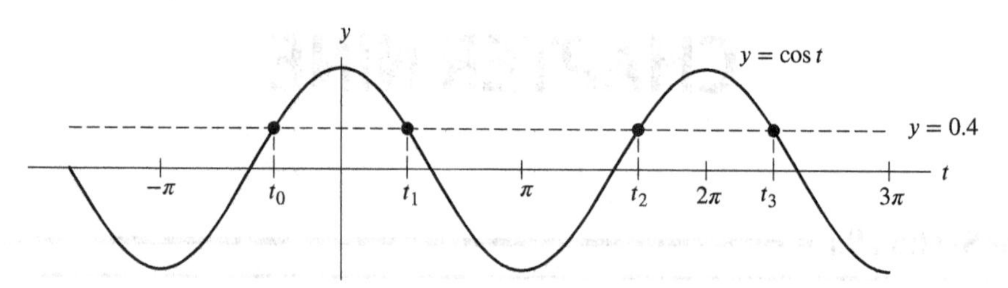

Figure 9.1: The points t_0, t_1, t_2, t_3 are solutions to the equation $\cos t = 0.4$

5. Graph $y = \cos t$ on $0 \leq t \leq 2\pi$ and locate the two points with y-coordinate -0.24. The t-coordinates of these points are approximately $t = 1.813$ and $t = 4.473$. See Figure 9.2.

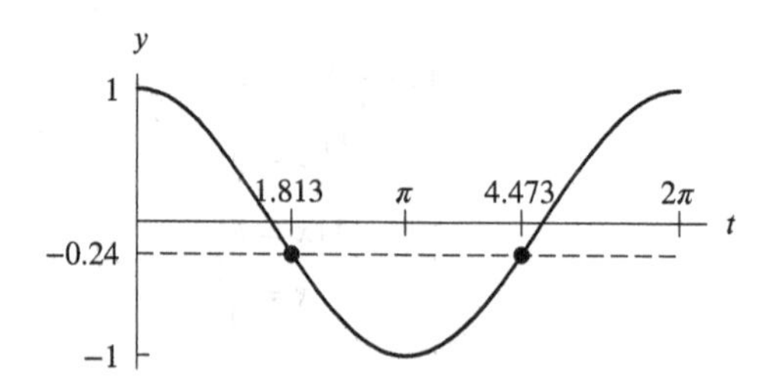

Figure 9.2

9. We have

$$\begin{aligned}
3\sin\theta + 3 &= 5\sin\theta + 2 \\
1 &= 2\sin\theta \\
\frac{1}{2} &= \sin\theta \\
\sin^{-1}\left(\frac{1}{2}\right) &= \theta \\
0.5236 = 30^\circ &= \theta.
\end{aligned}$$

13. We use the inverse tangent function on a calculator to get $5\theta + 7 = -0.236$. Solving for θ, we get $\theta = -1.447$.

PROBLEMS

17. We have $x = \cos^{-1}(0.6) = 0.927$. A graph of $\cos x$ shows that the second solution is $x = 2\pi - 0.927 = 5.356$.

21. Since $\sin(x - 1) = 0.25$, we know

$$\begin{aligned}
x - 1 &= \sin^{-1}(0.25) = 0.253. \\
x &= 1.253.
\end{aligned}$$

Another solution for $x - 1$ is given by

$$\begin{aligned}
x - 1 &= \pi - 0.253 = 2.889 \\
x &= 3.889.
\end{aligned}$$

25. By sketching a graph, we see that there are four solutions (see Figure 9.3). The first solution is given by $x = \cos^{-1}(0.6) = 0.927$, which is equivalent to the length labeled "b" in Figure 9.3. Next, note that, by the symmetry of the graph of the cosine function, we can obtain a second solution by subtracting the length b from 2π. Therefore, a second solution to the

equation is given by $x = 2\pi - 0.927 = 5.356$. Similarly, our final two solutions are given by $x = 2\pi + 0.927 = 7.210$ and $x = 4\pi - 0.927 = 11.639$.

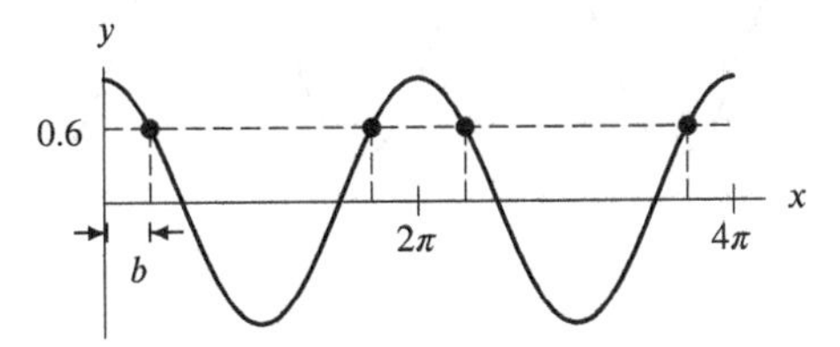

Figure 9.3

29. One solution is $\theta = \sin^{-1}(-\sqrt{2}/2) = -\pi/4$, and a second solution is $5\pi/4$, since $\sin(5\pi/4) = -\sqrt{2}/2$. All other solutions are found by adding integer multiples of 2π to these two solutions. See Figure 9.4.

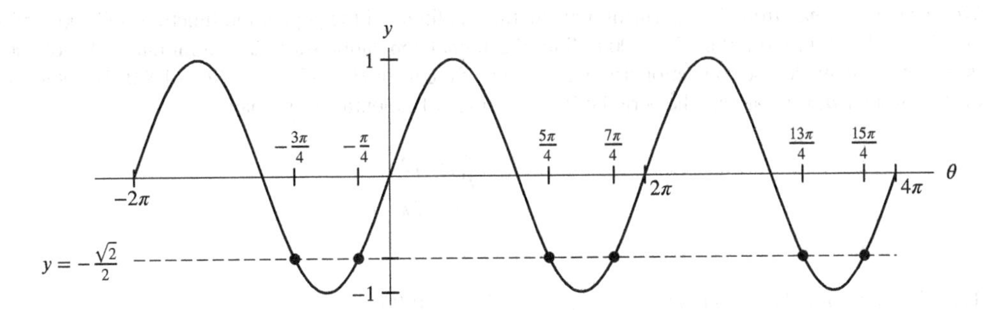

Figure 9.4

33. From Figure 9.5 we can see that the solutions lie on the intervals $\frac{\pi}{8} < t < \frac{\pi}{4}$, $\frac{3\pi}{4} < t < \frac{7\pi}{8}$, $\frac{9\pi}{8} < t < \frac{5\pi}{4}$ and $\frac{7\pi}{4} < t < \frac{15\pi}{8}$. Using the trace mode on a calculator, we can find approximate solutions $t = 0.52$, $t = 2.62$, $t = 3.67$ and $t = 5.76$.

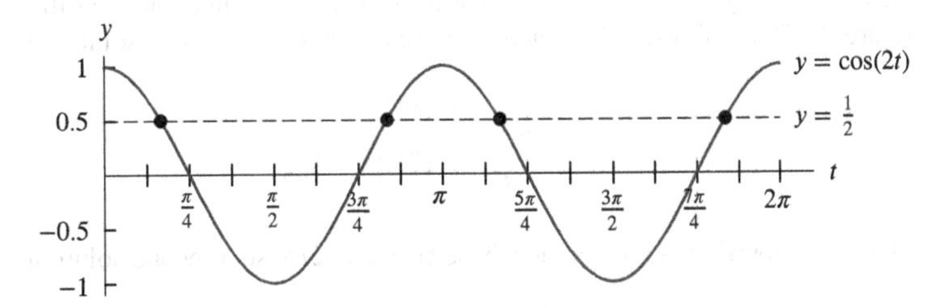

Figure 9.5

For a more precise answer we solve $\cos(2t) = \frac{1}{2}$ algebraically, giving $2t = \arccos(1/2)$. One solution is $2t = \pi/3$. But $2t = 5\pi/3, 7\pi/3$, and $11\pi/3$ are also angles that have a cosine of $1/2$. Thus $t = \pi/6, 5\pi/6, 7\pi/6$, and $11\pi/6$ are the solutions between 0 and 2π.

37. Graph $y = 12 - 4\cos(3t)$ on $0 \le t \le 2\pi/3$ and locate the two points with y-coordinate 14. (See Figure 9.6.) These points have t-coordinates of approximately $t = 0.698$ and $t = 1.396$. There are six solutions in three cycles of the graph between 0 and 2π.

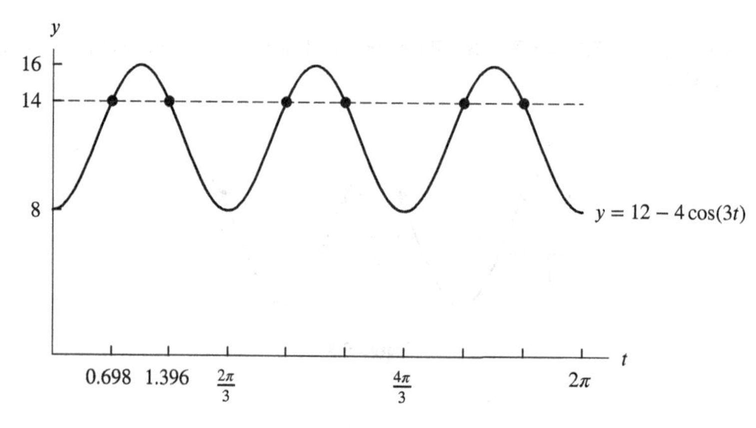

Figure 9.6

41. **(a)** We first observe that, from the information given, the amplitude of the population function is $(1500 - 500)/2 = 500$ and the midline is $(1500 + 500)/2 = 1000$. Since the animal population starts at its minimum value of 500 animals, it is reasonable to model the population using a function of the form $P = -500\cos(Bt) + 1000$. To determine the value of the constant B, we note that the period of the function is 12 months, so we have:

$$\begin{aligned}\frac{2\pi}{B} &= 12\\ 12B &= 2\pi\\ B &= \frac{\pi}{6}.\end{aligned}$$

Therefore, our formula is given by $P = -500\cos((\pi/6)t) + 1000$.

(b) Substituting $P = 800$ into our formula from part (a), we have:

$$\begin{aligned}-500\cos\left(\frac{\pi}{6}t\right) + 1000 &= 800\\ \cos\left(\frac{\pi}{6}t\right) &= \frac{-200}{-500} = \frac{2}{5}.\end{aligned}$$

Our reference angle is $\cos^{-1}(2/5) = 1.159$, and we know that the cosine function is positive in the first and the fourth quadrants. Therefore, 1.159 is a first-quadrant angle whose cosine is $2/5$, so one solution to our equation is

$$\begin{aligned}\frac{\pi}{6}t &= 1.159\\ t &= \frac{1.159(6)}{\pi} = 2.2.\end{aligned}$$

Similarly, $2\pi - 1.159$ is a fourth-quadrant angle whose cosine is $2/5$, so a second solution to our equation is

$$\begin{aligned}\frac{\pi}{6}t &= 2\pi - 1.159 = 5.124\\ t &= \frac{5.124(6)}{\pi} = 9.8.\end{aligned}$$

Therefore, the animal population reaches 800 at $t = 2.2$ months and at $t = 9.8$ months.

45. Figure 9.7 shows a graph of the height h of the block as a function of the time t for the first two seconds of the block's motion. Note that there appear to be four times at which the height of the block is 40 cm. To find these times, we will first find a formula for h as a function of t. From the graph, we can see that that the midline is the line $h = 30$, and the amplitude is 20. Since the height starts at the midline and decreases to its minimum value, we know that $h = -20\sin(Bt) + 30$ for some constant B. We can also see that the period of the function is 1 second, so we have

$$\begin{aligned}\frac{2\pi}{B} &= 1\\ B &= 2\pi.\end{aligned}$$

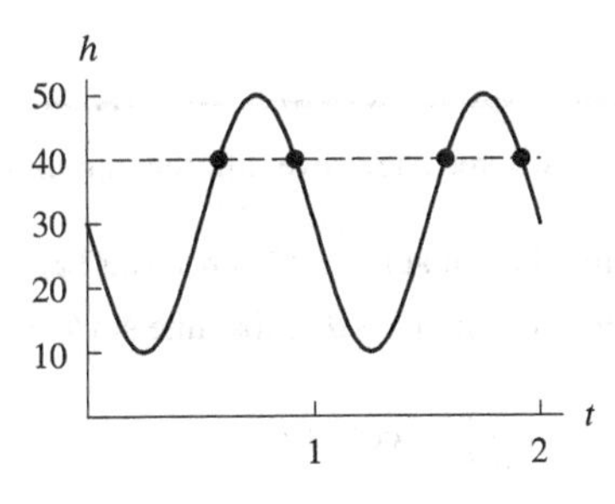

Figure 9.7

Therefore, $h = -20\sin(2\pi t) + 30$. To find the times when the height of the block is 40 cm, we substitute 40 for h in our formula:

$$-20\sin(2\pi t) + 30 = 40$$
$$\sin(2\pi t) = \frac{10}{-20} = -\frac{1}{2}.$$

Our reference angle is $\sin^{-1}(1/2) = \pi/6$, and we know that the sine function is negative in the third and fourth quadrants. Therefore, $\pi + (\pi/6) = 7\pi/6$ is a third-quadrant angle whose sine is $-1/2$, so one solution to our equation is

$$2\pi t = \frac{7\pi}{6}$$
$$t = \frac{7}{12} \text{ sec.}$$

Similarly, $2\pi - (\pi/6) = 11\pi/6$ is a fourth-quadrant angle whose sine is $-1/2$, so a second solution is

$$2\pi t = \frac{11\pi}{6}$$
$$t = \frac{11}{12} \text{ sec.}$$

Therefore, the height of the block is 40 cm at $7/12$ sec and at $11/12$ sec. Since the period of the height function is 1 sec, the height of the block will also be 40 cm at $1 + (7/12) = 19/12$ sec and at $1 + (11/12) = 23/12$ sec.

Solutions for Section 9.2

Skill Refresher

S1. Finding a common denominator, we have

$$\frac{y}{5x} + \frac{y}{x} = \frac{y}{5x} + \frac{5y}{5x} = \frac{6y}{5x}.$$

S5. We have

$$x^2 = 1 + y^2$$
$$x^2 - y^2 = 1,$$

so the equation $x^2 = 1 + y^2$ is equivalent to equation (ii).

S9. We have

$$x^2 - y^2 = -(1 + x^2)$$
$$x^2 - y^2 = -1 - x^2$$
$$2x^2 - y^2 = -1$$
$$-2x^2 + y^2 = 1$$
$$y^2 - 2x^2 = 1,$$

so $x^2 - y^2 = -(1 + x^2)$ is equivalent to equation (iv).

EXERCISES

1. No, these expressions are not equal everywhere. They have different amplitudes (1 and 3) and different periods ($2\pi/3$ and 2π).

The value of the two functions are different at $t = \pi/2$, since $\sin(3\pi/2) = -1$ and $3\sin(\pi/2) = 3$.

5. We use the relationship $\sin^2\theta + \cos^2\theta = 1$ to find $\cos\theta$. Substitute $\sin\theta = 1/4$:

$$\begin{aligned}\left(\frac{1}{4}\right)^2 + \cos^2\theta &= 1\\ \frac{1}{16} + \cos^2\theta &= 1\\ \cos^2\theta &= 1 - \frac{1}{16} = \frac{15}{16}\\ \cos\theta &= \pm\sqrt{\frac{15}{16}} = \pm\frac{\sqrt{15}}{4}.\end{aligned}$$

Because θ is in the first quadrant, $\cos\theta$ is positive, so $\cos\theta = \sqrt{15}/4$. To find $\tan\theta$, use the relationship

$$\tan\theta = \frac{\sin\theta}{\cos\theta} = \frac{1/4}{\sqrt{15}/4} = \frac{1}{\sqrt{15}}.$$

9. We have:

$$\begin{aligned}\tan t\cos t - \frac{\sin t}{\tan t} &= \frac{\sin t}{\cos t}\cdot\cos t - \frac{\sin t}{\left(\frac{\sin t}{\cos t}\right)} \quad \text{because } \tan t = \tfrac{\sin t}{\cos t}\\ &= \sin t - \sin t\cdot\frac{\cos t}{\sin t}\\ &= \sin t - \cos t.\end{aligned}$$

13. Writing $\sin 2\alpha = 2\sin\alpha\cos\alpha$, we have

$$\frac{\sin 2\alpha}{\cos\alpha} = \frac{2\sin\alpha\cos\alpha}{\cos\alpha} = 2\sin\alpha.$$

17. Combining terms and using $\cos^2\phi + \sin^2\phi = 1$, we have

$$\frac{\cos\phi - 1}{\sin\phi} + \frac{\sin\phi}{\cos\phi + 1} = \frac{(\cos\phi - 1)(\cos\phi + 1) + \sin^2\phi}{\sin\phi(\cos\phi + 1)} = \frac{\cos^2\phi - 1 + \sin^2\phi}{\sin\phi(\cos\phi + 1)} = \frac{0}{\sin\phi(\cos\phi + 1)} = 0$$

21. We have: $\dfrac{\sin\sqrt{\theta}}{\cos\sqrt{\theta}} = \tan\sqrt{\theta}.$

25. We have

$$\begin{aligned}2\sin\left(\frac{2}{k+3}\right)\cdot\frac{5}{3\cos\left(\frac{2}{k+3}\right)} &= \frac{10}{3}\cdot\frac{\sin\left(\frac{2}{k+3}\right)}{\cos\left(\frac{2}{k+3}\right)}\\ &= \frac{10}{3}\tan\left(\frac{2}{k+3}\right).\end{aligned}$$

PROBLEMS

29. Since $\cos 2t$ has period π and $\sin t$ has period 2π, if the result we want holds for $0 \le t \le 2\pi$, it holds for all t. So let's concentrate on the interval $0 \le t \le 2\pi$.

Solving $\cos 2t = 0$ gives $t = \pi/4, 3\pi/4, 5\pi/4, 7\pi/4$.

From the graph in Figure 9.8, we see $\cos 2t > 0$ for $0 \le t < \pi/4$, $3\pi/4 < t < 5\pi/4$, $7\pi/4 < t \le 2\pi$.

Solving $1 - 2\sin^2 t = 0$ gives

$$\sin^2 t = \frac{1}{2}$$
$$\sin t = \pm\frac{1}{\sqrt{2}},$$

so $t = \pi/4, 3\pi/4, 5\pi/4, 7\pi/4$.

From the graph of $y = \sin t$ and the lines $y = 1/\sqrt{2}$ and $y = -1/\sqrt{2}$ in Figure 9.9, we see that $-1/\sqrt{2} < \sin t < 1/\sqrt{2}$ on the same intervals that $\cos 2t > 0$.

Now if $-1/\sqrt{2} < \sin t < 1/\sqrt{2}$, then

$$\sin^2 t < \frac{1}{2}$$
$$1 - 2\sin^2 t > 0.$$

Thus, $\cos 2t$ and $1 - 2\sin^2 t$ have the same sign for all t.

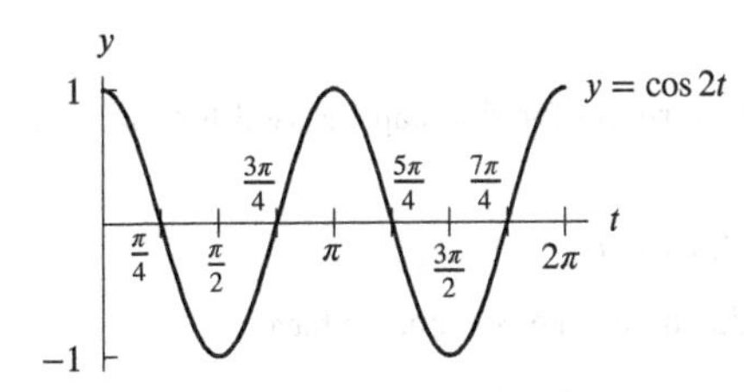

Figure 9.8

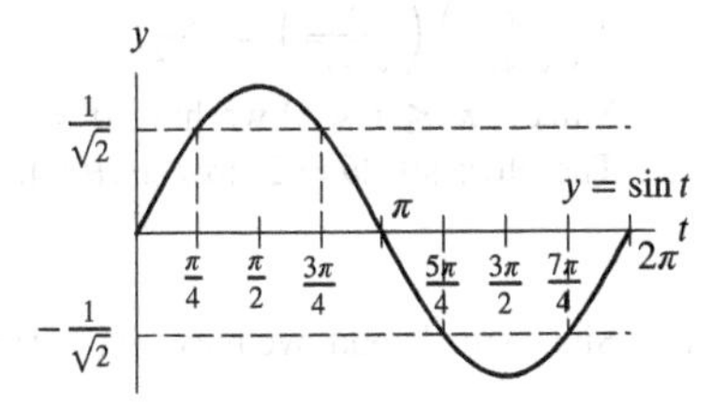

Figure 9.9

33. Multiply the denominator by $1 + \cos t$ to get $\sin^2 t$:

$$\begin{aligned}\frac{\sin t}{1-\cos t} &= \frac{\sin t}{(1-\cos t)}\frac{(1+\cos t)}{(1+\cos t)}\\ &= \frac{\sin t(1+\cos t)}{1-\cos^2 t}\\ &= \frac{\sin t(1+\cos t)}{\sin^2 t}\\ &= \frac{1+\cos t}{\sin t}.\end{aligned}$$

37. Using the trigonometric identity $\sin(2\theta) = 2\sin\theta\cos\theta$, we have

$$\begin{aligned}\sin(2\theta) - \cos\theta &= 0\\ 2\sin\theta\cos\theta - \cos\theta &= 0\\ \cos\theta(2\sin\theta - 1) &= 0\\ \cos\theta = 0 \quad \text{or} \quad \sin\theta &= \frac{1}{2}.\end{aligned}$$

If $\cos\theta = 0$, then we have two solutions: $\theta = \pi/2$ and $\theta = 3\pi/2$. On the other hand, if $\sin\theta = 1/2$, we first calculate the associated reference angle, which is $\sin^{-1}(1/2) = \pi/6$. Using a graph of the sine function on the interval $0 \le \theta \le 2\pi$, we see that the two solutions to $\sin\theta = 1/2$ are given by $\theta = \pi/6$ and $\theta = \pi - \pi/6 = 5\pi/6$. Combining the above observations, we see that there are four solutions to the original equation: $\pi/2, 3\pi/2, \pi/6$ and $5\pi/6$.

41. Not an identity. False for $x = 2$.

45. If we let $x = 1$, then we have

$$\sin(x^2) = \sin(1^2) = 0.841 \neq 1.683 = 2\sin 1 = 2\sin x.$$

Therefore, since the equation is not true for $x = 1$, it is not an identity.

49. Not an identity. False for $x = 0$.

53. Identity. $\dfrac{2\tan x}{1+\tan^2 x}\cdot\dfrac{\cos^2 x}{\cos^2 x} = \dfrac{2\sin x\cos x}{\cos^2 x+\sin^2 x} = \dfrac{\sin 2x}{1} = \sin 2x$.

57. Note the hypotenuse of the triangle is $\sqrt{1+y^2}$.

(a) $y = \dfrac{y}{1} = \tan\theta$.

(b) $\cos\phi = \sin(\pi/2-\phi) = \sin\theta$.

(c) Since $\cos\theta = \dfrac{1}{\sqrt{1+y^2}}$, we have $\sqrt{1+y^2} = \dfrac{1}{\cos\theta}$, or $1+y^2 = \left(\dfrac{1}{\cos\theta}\right)^2$. (Alternatively, $1+y^2 = 1+\tan^2\theta$.)

(d) Triangle area $= \dfrac{1}{2}$(base)(height) $= \dfrac{1}{2}(1)(y)$. But $y = \tan\theta$, so the area is $\frac{1}{2}\tan\theta$.

61. **(a)** Let $\theta = \cos^{-1}x$, so $\cos\theta = x$. Then, since $0 \le \theta \le \pi$, $\sin\theta = \sqrt{1-x^2}$ and $\tan\theta = \frac{\sqrt{1-x^2}}{x}$, and $\tan(2\cos^{-1}x) = \tan 2\theta = \frac{2\tan\theta}{1-\tan^2\theta}$. Now $1-\tan^2\theta = 1-\frac{1-x^2}{x^2} = \frac{2x^2-1}{x^2}$, so $\tan 2\theta = \frac{2\sqrt{1-x^2}}{x}\cdot\frac{x^2}{2x^2-1} = \frac{2x\sqrt{1-x^2}}{2x^2-1}$.

(b) Let $\theta = \tan^{-1}x$, so $\tan\theta = x$. Then $\sin\theta = \frac{x}{\sqrt{1+x^2}}$ and $\cos\theta = \frac{1}{\sqrt{1+x^2}}$, so $\sin(2\tan^{-1}x) = \sin 2\theta = 2\sin\theta\cos\theta = 2\left(\frac{x}{\sqrt{1+x^2}}\right)\left(\frac{1}{\sqrt{1+x^2}}\right) = \frac{2x}{1+x^2}$.

65. **(a)** Since $-\pi \le t < 0$ we have $0 < -t \le \pi$ so the double-angle formula for sine can be used for the angle $\theta = -t$. Therefore $\sin 2\theta = 2\sin\theta\cos\theta$ tells us that

$$\sin(-2t) = 2\sin(-t)\cos(-t).$$

(b) Since sine is odd, we have $\sin(-2t) = -\sin 2t$. Since sine is odd and cosine is even, we have

$$2\sin(-t)\cos(-t) = 2(-\sin t)\cos t = -2\sin t\cos t.$$

Substitution of these results into the results of part (a) shows that

$$-\sin(2t) = -2\sin t\cos t.$$

Multiplication by -1 gives

$$\sin 2t = 2\sin t\cos t.$$

Solutions for Section 9.3

Skill Refresher

S1. Since $f(4+3) = f(7)$, we have

$$f(7) = 3(7^2) = 147.$$

S5. Since $f(60° + 30°) = f(90°)$, we have $f(90°) = 1$.

EXERCISES

1. Applying the sum-of-angles formula for sine, we have

$$\begin{aligned}\sin(A+B) &= \sin A\cos B + \cos A\sin B\\ &= (0.84)(0.39) + (0.54)(0.92) = 0.8244.\end{aligned}$$

5. Applying the sum-of-angles formula for sine, we have

$$\begin{aligned}\sin(S+T) &= \sin S\cos T + \cos S\sin T\\ &= \left(\frac{7}{25}\right)\left(\frac{8}{17}\right) + \left(\frac{24}{25}\right)\left(\frac{15}{17}\right) = \frac{416}{425}.\end{aligned}$$

9. Write $\sin 15° = \sin(45° - 30°)$, and then apply the appropriate trigonometric identity.

$$\begin{aligned}\sin 15° &= \sin(45° - 30°)\\ &= \sin 45° \cos 30° - \sin 30° \cos 45°\\ &= \frac{\sqrt{6}}{4} - \frac{\sqrt{2}}{4}\end{aligned}$$

Similarly, $\sin 75° = \sin(45° + 30°)$.

$$\begin{aligned}\sin 75° &= \sin(45° + 30°)\\ &= \sin 45° \cos 30° + \sin 30° \cos 45°\\ &= \frac{\sqrt{6}}{4} + \frac{\sqrt{2}}{4}\end{aligned}$$

Also, note that $\cos 75° = \sin(90° - 75°) = \sin 15°$, and $\cos 15° = \sin(90° - 15°) = \sin 75°$.

13. **(a)** We have $\sin(15° + 42°) = \sin 15° \cos 42° + \sin 42° \cos 15° = 0.839$.
(b) See Figure 9.10.

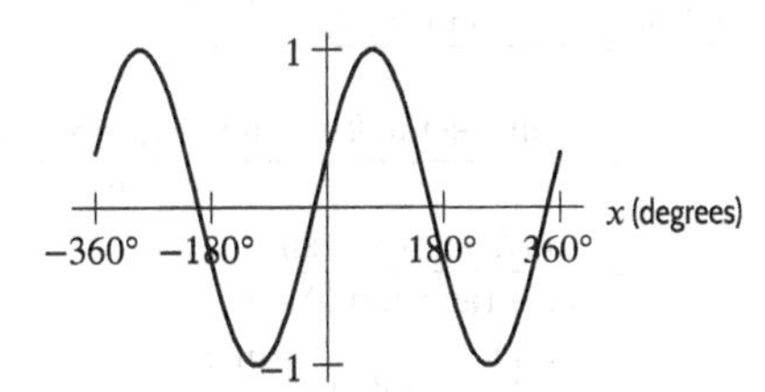

Figure 9.10

PROBLEMS

17. First, we note that the unlabeled side of the triangle has length $\sqrt{4-y^2}$.

(a) Using an angle difference formula, we have

$$\begin{aligned}\cos(\theta - \phi) &= \cos\theta\cos\phi + \sin\theta\sin\phi\\ &= \frac{\sqrt{4-y^2}}{2}\cdot\frac{y}{2} + \frac{y}{2}\cdot\frac{\sqrt{4-y^2}}{2}\\ &= \frac{y\sqrt{4-y^2}}{2}.\end{aligned}$$

(b) Using an angle difference formula, we have

$$\begin{aligned}\sin(\theta - \phi) &= \sin\theta\cos\phi - \cos\theta\sin\phi\\ &= \frac{y}{2}\cdot\frac{y}{2} - \frac{\sqrt{4-y^2}}{2}\cdot\frac{\sqrt{4-y^2}}{2}\\ &= \frac{y^2-(4-y^2)}{4}\\ &= \frac{y^2-2}{2}.\end{aligned}$$

(c) We have

$$\begin{aligned}\cos\theta - \cos\phi &= \frac{\sqrt{4-y^2}}{2} - \frac{y}{2}\\ &= \frac{\sqrt{4-y^2}-y}{2}.\end{aligned}$$

(d) Since $\theta + \phi = \pi/2$, we have

$$\sin(\theta + \phi) = \sin\left(\frac{\pi}{2}\right) = 1.$$

21. We have

$$\begin{aligned}\cos 3t &= \cos(2t+t)\\ &= \cos 2t\cos t - \sin 2t\sin t\\ &= (2\cos^2 t - 1)\cos t - (2\sin t\cos t)\sin t\\ &= \cos t((2\cos^2 t - 1) - 2\sin^2 t)\\ &= \cos t(2\cos^2 t - 1 - 2(1-\cos^2 t))\\ &= \cos t(2\cos^2 t - 1 - 2 + 2\cos^2 t)\\ &= \cos t(4\cos^2 t - 3)\\ &= 4\cos^3 t - 3\cos t,\end{aligned}$$

as required.

25. We manipulate the equation for the average rate of change as follows:

$$\begin{aligned}\frac{\tan(x+h)-\tan x}{h} &= \frac{\dfrac{\tan x+\tan h}{1-\tan x\tan h}-\tan x}{h}\\ &= \frac{(\tan x+\tan h-\tan x+\tan^2 x\tan h)/(1-\tan x\tan h)}{h}\\ &= \frac{\tan h+\tan^2 x\tan h}{(1-\tan x\tan h)\cdot h}\\ &= \frac{\dfrac{\sin h}{\cos h}+\tan^2 x\cdot\dfrac{\sin h}{\cos h}}{\left(1-\tan x\cdot\dfrac{\sin h}{\cos h}\right)\cdot h}\\ &= \frac{(1+\tan^2 x)\dfrac{\sin h}{\cos h}}{\left(1-\tan x\dfrac{\sin h}{\cos h}\right)\cdot h}\\ &= \frac{\left(\dfrac{1}{\cos^2 x}\right)\cdot\dfrac{\sin h}{\cos h}}{\left(1-\tan x\cdot\dfrac{\sin h}{\cos h}\right)\cdot h}\\ &= \frac{\dfrac{1}{\cos^2 x}\cdot\sin h}{(\cos h-\tan x\sin h)\cdot h}\\ &= \frac{\dfrac{1}{\cos^2 x}\cdot\sin h}{\cos h-\sin h\tan x}\cdot\left(\frac{1}{h}\right)\\ &= \frac{1}{\cos^2 x}\,\frac{\sin h}{h}\cdot\frac{1}{\cos h-\sin h\tan x}.\end{aligned}$$

29. **(a)** The coordinates of P_1 are $(\cos\theta,\sin\theta)$; for P_2 they are $(\cos(-\phi),\sin(-\phi)) = (\cos\phi,-\sin\phi)$; for P_3 they are $(\cos(\theta+\phi),\sin(\theta+\phi))$; and for P_4 they are $(1,0)$.

(b) The triangles P_1OP_2 and P_3OP_4 are congruent by the side-angle-side property because $\angle P_1OP_2 = \theta+\phi = \angle P_3OP_4$. Therefore their corresponding sides P_1P_2 and P_3P_4 are equal.

(c) We have

$$\begin{aligned}(P_1P_2)^2 &= (\cos\theta-\cos\phi)^2+(\sin\theta+\sin\phi)^2\\ &= \cos^2\theta-2\cos\theta\cos\phi+\cos^2\phi+\sin^2\theta+2\sin\theta\sin\phi+\sin^2\phi\\ &= \cos^2\theta+\sin^2\theta+\cos^2\phi+\sin^2\phi-2\cos\theta\cos\phi+2\sin\theta\sin\phi\\ &= 2-2(\cos\theta\cos\phi-\sin\theta\sin\phi)\end{aligned}$$

We also have

$$(P_3P_4)^2 = (\cos(\theta+\phi)-1)^2+(\sin(\theta+\phi)-0)^2$$

$$= \cos^2(\theta+\phi) - 2\cos(\theta+\phi) + 1 + \sin^2(\theta+\phi)$$
$$= 2 - 2\cos(\theta+\phi)$$

The distances P_1P_2 and P_3P_4 are the square roots of these expressions (but we will use the squares of the distances).

(d) $(P_3P_4)^2 = (P_1P_2)^2$ by part (b), so

$$2 - 2\cos(\theta+\phi) = 2 - 2(\cos\theta\cos\phi - \sin\theta\sin\phi)$$
$$\cos(\theta+\phi) = \cos\theta\cos\phi - \sin\theta\sin\phi.$$

Solutions for Section 9.4

Skill Refresher

S1. Since the point is 3 units to the right of the origin and then 4 units up, the distance from the origin is:

$$\text{Distance} = \sqrt{3^2+4^2} = \sqrt{9+16} = \sqrt{25} = 5.$$

Since the x and y coordinates are both positive, this point is in Quadrant I.

S5. Since the point is 5 units to the right of the origin and then 4 units down, the distance from the origin is:

$$\text{Distance} = \sqrt{5^2+4^2} = \sqrt{25+16} = \sqrt{41} = 6.403.$$

Since the x coordinate is positive and the y coordinate is negative, this point is in Quadrant IV.

EXERCISES

1. Quadrant IV.

5. Since $3.2\pi = 2\pi + 1.2\pi$, such a point is in Quadrant III.

9. Since -7 is an angle of -7 radians, corresponding to a rotation of just over 2π, or one full revolution, in the clockwise direction, such a point is in Quadrant IV.

13. We have $180° < \theta < 270°$. See Figure 9.11.

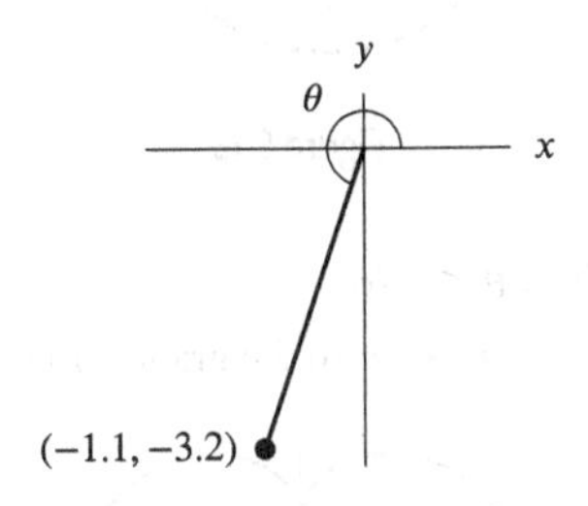

Figure 9.11

17. With $x = -\sqrt{3}$ and $y = 1$, find $r = \sqrt{(-\sqrt{3})^2 + 1^2} = \sqrt{4} = 2$. Find θ from $\tan\theta = y/x = 1/(-\sqrt{3})$. Thus, $\theta = \tan^{-1}(-1/\sqrt{3}) = -\pi/6$. Since $(-\sqrt{3}, 1)$ is in the second quadrant, $\theta = -\pi/6 + \pi = 5\pi/6$. The polar coordinates are $(2, 5\pi/6)$.

21. With $r = \sqrt{3}$ and $\theta = -3\pi/4$, we find $x = r\cos\theta = \sqrt{3}\cos(-3\pi/4) = \sqrt{3}(-\sqrt{2}/2) = -\sqrt{6}/2$ and $y = r\sin\theta = \sqrt{3}\sin(-3\pi/4) = \sqrt{3}(-\sqrt{2}/2) = -\sqrt{6}/2$.

The rectangular coordinates are $(-\sqrt{6}/2, -\sqrt{6}/2)$.

PROBLEMS

25. Multiply both sides by r to transform the right side of the equation into $6r\cos\theta$ so that we can substitute $x = r\cos\theta$. The left side is now $r^2 = x^2 + y^2$. In rectangular coordinates, the equation is $x^2 + y^2 = 6x$.

29. By substituting $r^2 = x^2 + y^2$, we have $r^2 = 5$. Since r is positive, this could also be written as $r = \sqrt{5}$.

33. Figure 9.12 shows that at 3 pm, we have:
In Cartesian coordinates, $H = (3, 0)$. In polar coordinates, $H = (3, 0)$; that is $r = 3, \theta = 0$. In Cartesian coordinates, $M = (0, 4)$. In polar coordinates, $M = (4, \pi/2)$, that is $r = 4, \theta = \pi/2$.

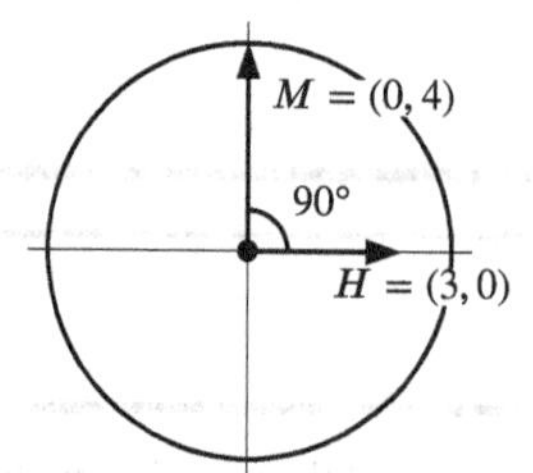

Figure 9.12

37. Figure 9.13 shows that at 7 am the polar coordinates of the point H are $r = 3$ and $\theta = 60° + 180° = 240° = 4\pi/3$. Thus, the Cartesian coordinates of H are given by

$$x = 3\cos\left(\frac{4\pi}{3}\right) = -\frac{3}{2} = -1.5, \quad y = 3\sin\left(\frac{4\pi}{3}\right) = -\frac{3\sqrt{3}}{2} \approx -2.598.$$

Thus, in Cartesian coordinates, $H = (-1.5, -2.598)$. In polar coordinates, $H = (3, 4\pi/3)$. In Cartesian coordinates, $M = (0, 4)$. In polar coordinates, $M = (4, \pi/2)$.

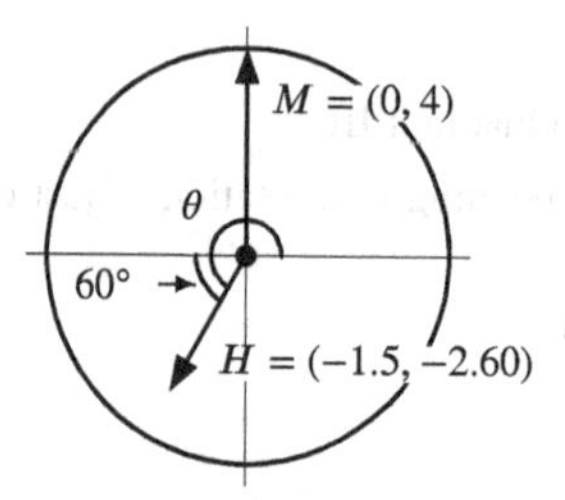

Figure 9.13

41. The region is given by $0 \le r \le 2$ and $-\pi/6 \le \theta \le \pi/6$.

45. The graph will begin to draw over itself for any $\theta \ge 2\pi$ so the graph will look the same in all three cases. See Figure 9.14.

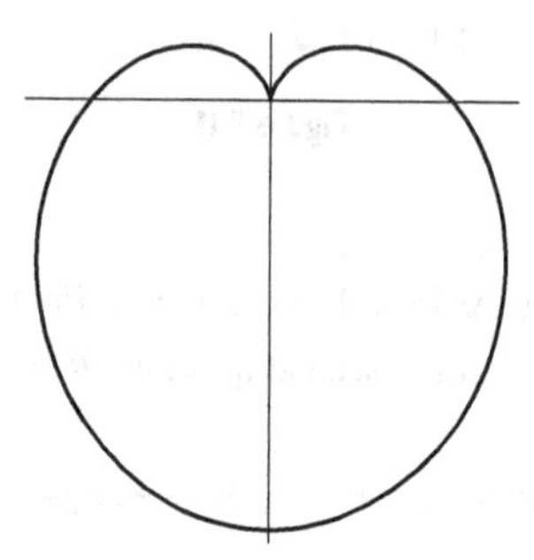

Figure 9.14

49. A loop starts and ends at the origin, that is, when $r = 0$. This happens first when $\theta = \pi/4$ and next when $\theta = 5\pi/4$. This can also be seen by using a trace mode on a calculator. Thus restricting θ so that $\pi/4 \leq \theta \leq 5\pi/4$ will graph the upper loop only. See Figure 9.15. To show only the other loop use $0 \leq \theta \leq \pi/4$ and $5\pi/4 \leq \theta \leq 2\pi$. See Figure 9.16.

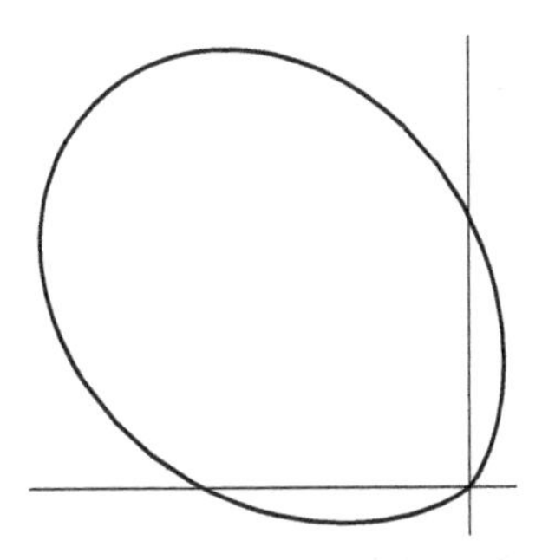

Figure 9.15: $\pi/4 \leq \theta \leq 5\pi/4$

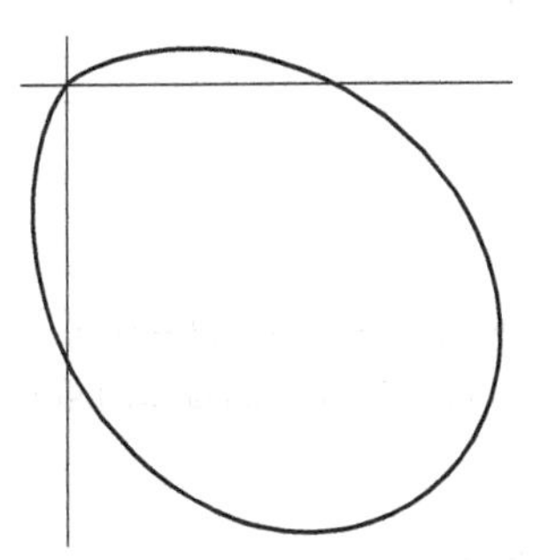

Figure 9.16: $0 \leq \theta \leq \pi/4$ and $5\pi/4 \leq \theta \leq 2\pi$

Solutions for Section 9.5

Skill Refresher

S1. We have

$$(1+x)(3+2x) = 3+2x+3x+2x^2 = 3+5x+2x^2.$$

S5. We have

$$\left(\frac{1+x}{2-x}\right)\left(\frac{x}{2+x}\right) = \frac{(1+x)x}{(2-x)(2+x)} = \frac{x+x^2}{4-x^2}.$$

EXERCISES

1. We have

$$4\sqrt{-50} = 4\sqrt{-1\cdot 50} = 4\sqrt{-1}\sqrt{50} = 4i\sqrt{25}\sqrt{2} = 20i\sqrt{2}.$$

5. Distributing and combining like terms gives

$$(-4+5i)(3-2i) = -12+8i+15i-10i^2 = -12+23i-10(-1) = -2+23i.$$

9. $5e^{i\pi}$

13. We have $(-3)^2+(-4)^2 = 25$, and $\arctan(4/3) \approx 4.069$. So the number is $5e^{i4.069}$.

17. $-3-4i$

21. We have $\sqrt{e^{i\pi/3}} = e^{(i\pi/3)/2} = e^{i\pi/6}$, thus $\cos\frac{\pi}{6} + i\sin\frac{\pi}{6} = \frac{\sqrt{3}}{2} + \frac{i}{2}$.

PROBLEMS

25. One value of $\sqrt[3]{i}$ is $\sqrt[3]{e^{i\frac{\pi}{2}}} = (e^{i\frac{\pi}{2}})^{\frac{1}{3}} = e^{i\frac{\pi}{6}} = \cos\frac{\pi}{6} + i\sin\frac{\pi}{6} = \frac{\sqrt{3}}{2} + \frac{i}{2}$

29. One value of $(\sqrt{3}+i)^{1/2}$ is
$(2e^{i\frac{\pi}{6}})^{1/2} = \sqrt{2}e^{i\frac{\pi}{12}} = \sqrt{2}\cos\frac{\pi}{12} + i\sqrt{2}\sin\frac{\pi}{12} \approx 1.366 + 0.366i$

33. Substituting $A_2 = i - A_1$ into the second equation gives

$$iA_1 - (i - A_1) = 3,$$

so

$$\begin{aligned} iA_1 + A_1 &= 3 + i \\ A_1 &= \frac{3+i}{1+i} = \frac{3+i}{1+i} \cdot \frac{1-i}{1-i} = \frac{3 - 3i + i - i^2}{2} \\ &= 2 - i \end{aligned}$$

Therefore $A_2 = i - (2 - i) = -2 + 2i$.

37. Using Euler's formula, we have:

$$e^{i(2\theta)} = \cos 2\theta + i \sin 2\theta$$

On the other hand,

$$e^{i(2\theta)} = (e^{i\theta})^2 = (\cos\theta + i\sin\theta)^2 = (\cos^2\theta - \sin^2\theta) + i(2\cos\theta\sin\theta)$$

Equating real parts, we find

$$\cos 2\theta = \cos^2\theta - \sin^2\theta.$$

41. One polar form for $z = -8$ is $z = 8e^{i\pi}$ with $(r, \theta) = (8, \pi)$. Two more sets of polar coordinates for z are $(8, 3\pi)$, and $(8, 5\pi)$. Three cube roots of z are given by

$$\begin{aligned} \left(8e^{\pi i}\right)^{1/3} &= 8^{1/3}e^{1/3\cdot\pi i} = 2e^{\pi i/3} = 2\cos(\pi/3) + i2\sin(\pi/3) \\ &= 1 + 1.732i. \\ \left(8e^{3\pi i}\right)^{1/3} &= 8^{1/3}e^{1/3\cdot 3\pi i} = 2e^{\pi i} = 2\cos\pi + i2\sin\pi \\ &= -2. \\ \left(8e^{5\pi i}\right)^{1/3} &= 8^{1/3}e^{1/3\cdot 5\pi i} = 2e^{5\pi i/3} = 2\cos(5\pi/3) + i2\sin(5\pi/3) \\ &= 1 - 1.732i. \end{aligned}$$

45. By de Moivre's formula we have

$$(\cos 2\pi/3 + i\sin 2\pi/3)^3 = \cos(3\cdot 2\pi/3) + i\sin(3\cdot 2\pi/3) = 1 + i0 = 1.$$

49. Using the exponent rules, we see from Euler's formula that

$$\begin{aligned} e^{i(\theta+\phi)} &= e^{i\theta}\cdot e^{i\phi} \\ &= (\cos\theta + i\sin\theta)(\cos\phi + i\sin\phi) \\ &= \cos\theta\cos\phi + \underbrace{i\cos\theta\sin\phi + i\sin\theta\cos\phi}_{i(\cos\theta\sin\phi+\sin\theta\cos\phi)} + \underbrace{i^2\sin\theta\sin\phi}_{-\sin\theta\sin\phi} \\ &= \underbrace{\cos\theta\cos\phi - \sin\theta\sin\phi}_{\text{Real part}} + i\underbrace{(\sin\theta\cos\phi + \cos\theta\sin\phi)}_{\text{Imaginary part}}. \end{aligned}$$

But Euler's formula also gives

$$e^{i(\theta+\phi)} = \underbrace{\cos(\theta+\phi)}_{\text{Real part}} + i\underbrace{\sin(\theta+\phi)}_{\text{Imaginary part}}.$$

Two complex numbers are equal only if their real and imaginary parts are equal. Setting real parts equal gives

$$\cos(\theta+\phi) = \cos\theta\cos\phi - \sin\theta\sin\phi.$$

Setting imaginary parts equal gives

$$\sin(\theta+\phi) = \sin\theta\cos\phi + \sin\phi\cos\theta.$$

STRENGTHEN YOUR UNDERSTANDING

1. True, because the factor e^{-t} decreases the oscillations of $\cos t$ as t grows.

5. True. This is an identity. Substitute using $\tan^2\theta = \sin^2\theta/\cos^2\theta$ and simplify to obtain $2 = 2\sin^2\theta + 2\cos^2\theta$. Divide by 2 to reach the Pythagorean identity.

9. True. There are many ways to prove this identity. We use the identity $\cos 2\theta = \cos^2\theta - \sin^2\theta$ to substitute in the right side of the equation. This becomes $\frac{1}{2}(1-(\cos^2\theta - \sin^2\theta))$. Now substitute using $1-\cos^2\theta = \sin^2\theta$ (a form of the Pythagorean identity.) The right side then simplifies to $\sin^2\theta$, which is the left side.

13. True. Start with the sine sum-of-angle identity:

$$\sin(\theta+\phi) = \sin\theta\cos\phi + \sin\phi\cos\theta$$

and let $\phi = \pi/2$, so

$$\sin(\theta+\pi/2) = \sin\theta\cos(\pi/2) + \sin(\pi/2)\cos\theta.$$

Simplify to

$$\sin(\theta+\pi/2) = \sin\theta\cdot 0 + 1\cdot\cos\theta = \cos\theta.$$

17. True. Since $A\cos(Bt) = A\sin(Bt+\pi/2) = A\sin(B(t+\pi/(2B)))$, the graph of $A\cos(Bt)$ is a shift of $A\sin(Bt)$ to the left by $\pi/(2B)$.

21. True. We use the assumption that a_1 and a_2 are nonzero. The amplitude of the single sine function is $A = \sqrt{a_1^2 + a_2^2}$. Thus, A is greater than either a_1 or a_2.

25. True. Hertz is a measure of cycles per second and so a single cycle will take 1/60th of a second.

29. True. Since $0 \le \cos^{-1}(x) \le \pi$, we have $0 \le \sin(\cos^{-1}x) \le 1$. Thus, $\cos^{-1}(\sin(\cos^{-1}x))$ is an angle θ whose cosine is between 0 and 1. In addition, we have $0 \le \theta \le \pi$, as this is part of the definition of $\cos^{-1}x$. Hence $0 \le \theta \le \pi/2$.

33. False. The point $(3,\pi)$ in polar coordinates is $(-3,0)$ in Cartesian coordinates.

37. True, since $\sqrt{a}$ is real for all $a \ge 0$.

41. True. We can write any nonzero complex number z as $re^{i\beta}$, where r and β are real numbers with $r > 0$. Since $r > 0$, we can write $r = e^c$ for $c = \ln r$. Therefore, $z = re^{i\beta} = e^c e^{i\beta} = e^{c+i\beta} = e^w$ where $w = c + i\beta$ is a complex number.

45. True. Since $i^4 = 1$, we have $i^{101} = (i^4)^{25} i = 1\cdot i = i$.

CHAPTER TEN

Solutions for Section 10.1

Skill Refresher

S1. Substituting the value into the function gives

$$f(1) = (1-2)^2 = (-1)^2 = 1.$$

S5. Substituting the value into the function gives

$$k(0) = \ln(e^0 + 2) = \ln(1+2) = \ln 3.$$

S9. Substituting the value into the function gives

$$f(0) = \cos\left(0 + \frac{\pi}{2}\right) = \cos\left(\frac{\pi}{2}\right) = 0.$$

S13. Substituting the value into the function gives

$$g\left(-\frac{\pi}{2}\right) = \ln e^{-\frac{\pi}{2}+\pi} = \ln e^{\frac{\pi}{2}} = \frac{\pi}{2}.$$

S17. Substituting the value into the function gives

$$f\left(-\frac{\pi}{2}\right) + \frac{3\pi}{2} = \cos\left(-\frac{\pi}{2} + \frac{\pi}{2}\right) + \frac{3\pi}{2} = \cos 0 + \frac{3\pi}{2} = 1 + \frac{3\pi}{2}.$$

S21. Substituting the given value into $g(x) = x^3$, we have:

(a)

$$g(a) = a^3$$

(b)

$$g(-a) = (-a)^3 = -a^3$$

(c)

$$g(2a) = (2a)^3 = 8a^3$$

(d)

$$g\left(\frac{a}{2}\right) = \left(\frac{a}{2}\right)^3 = \frac{a^3}{2^3} = \frac{a^3}{8}$$

(e)

$$g(a+h) = (a+h)^3 = (a+h)(a+h)^2 = (a+h)(a^2 + 2ah + h^2) = a^3 + 3a^2h + 3ah^2 + h^3$$

(f)

$$g(a) + h = a^3 + h$$

EXERCISES

1. **(a)** We have $g(-1) = -1$, so $f(g(-1)) = f(-1) = 2$.
(b) We have $f(2) = 1$, so $g(f(2)) = g(1) = 1$.
(c) We have $g(-2) = -2$, so $f(g(-2)) = f(-2) = -1$.
(d) We have $f(0) = 0$, so $f(g(0)) = g(0) = 2$
(e) We have $f(1) = -2$, so $f(f(1)) = f(-2) = -1$
(f) Since $g(-1) = -1$, we have $g(g(g(-1))) = g(g(-1)) = g(-1) = -1$.

5. We have

$$g(f(x)) = \frac{2^x}{2^x + 1}.$$

9. Since $g(x) = 9x - 2$, we substitute $9x - 2$ for x in $r(x)$, giving us $r(g(x)) = \sqrt{3(9x-2)}$, which simplifies to $r(g(x)) = \sqrt{27x - 6}$.

13. Since $f(x) = 3x^2$, we substitute $3x^2$ for x in $m(x)$, giving us $m(f(x)) = 4(3x^2)$, which simplifies to $m(f(x)) = 12x^2$, which we then substitute for x in $g(x)$, giving $g(m(f(x))) = 9(12x^2) - 2$, which simplifies to $g(m(f(x))) = 108x^2 - 2$.

17. Substituting and simplifying gives

$$\frac{f(2+h) - f(2)}{h} = \frac{(2+h)^2 - 2^2}{h} = \frac{4 + 4h + h^2 - 4}{h} = \frac{4h + h^2}{h} = 4 + h.$$

21. The inside function is $f(x) = \sin x$.

25. The inside function is $f(x) = 5 + 1/x$.

PROBLEMS

29. The function $t(f(H))$ gives the time of the trip as a function of temperature, H.

33. It is easiest to find values of h, because we can use the fact that $h(x) = g(f(x))$:

$$\begin{aligned} h(0) &= g(f(0)) \\ &= g(2) \qquad \text{because } f(0) = 2 \\ &= 3. \end{aligned}$$

Next, we will find values of f. To find $f(1)$, we know the output of $g(f(1))$ must be the same as $h(1)$, or 0. Since 0 is the output of g, we see from the table that its input must be 1. This means the value of $f(1)$ must be 1:

$$\begin{aligned} g(f(1)) = h(1) &= 0 \qquad \text{because } h(1) = 0 \\ g(\underbrace{f(1)}_{1}) &= 0 \qquad \text{because } g(1) = 0 \\ \text{so} \quad f(1) &= 1. \end{aligned}$$

Likewise, for $f(2)$, we see that

$$\begin{aligned} g(f(2)) = h(2) &= 2 \qquad \text{because } h(2) = 2 \\ g(\underbrace{f(2)}_{4}) &= 2 \qquad \text{because } g(4) = 2 \\ \text{so} \quad f(2) &= 4. \end{aligned}$$

Finally, we will find the values of g. To find $g(0)$, we know that the input of g is 0, so the output of f must be zero. This means that $x = 3$, because $f(3) = 0$. Thus, $g(0)$ is the same as $g(f(3))$, which equals $h(3)$ or 1:

$$\begin{aligned} g(0) &= g(\underbrace{f(3)}_{0}) \qquad \text{because } f(3) = 0 \\ &= h(3) \qquad \text{because } h(3) = g(f(3)) \\ &= 1. \end{aligned}$$

Likewise, to find $g(3)$, we have

$$g(3) = g(\underbrace{f(4)}_{3}) \qquad \text{because } f(4) = 3$$

$$= h(4) \qquad \text{because } g(f(4)) = h(4)$$
$$= 4.$$

See Table 10.1.

Table 10.1

x	$f(x)$	$g(x)$	$h(x)$
0	2	1	3
1	1	0	0
2	4	3	2
3	0	4	1
4	3	2	4

37. **(a)** $f(q(t))$ gives the total force on the dam, in kilonewtons, after t days of rain.

(b) (i) $q(1)$ is the height of the water in the reservoir after 1 day of rain. From the table, we see that $q(1) = 10.5$ meters. Therefore,
$$f(q(1)) = f(10.5) = 2500(10.5)^2 = 275{,}625 \text{ kilonewtons}$$
is the total force on the dam after 1 day of rain.

(ii) $q(2)$ is the height of the water in the reservoir after 2 days of rain. From the table, we see that $q(2) = 11.3$ meters. Therefore,
$$f(q(2)) = f(11.3) = 2500(11.3)^2 = 319{,}225 \text{ kilonewtons}$$
is the total force on the dam after 2 days of rain.

(c) The average rate of change of $f(q(t))$ tells us how fast the force on the dam is increasing with time. Between $t = 0$ and $t = 3$, we find that the average rate of change is
$$\frac{f(q(3)) - f(q(0))}{3-0} = \frac{f(11.8) - f(10)}{3} = \frac{2500(11.8)^2 - 2500(10)^2}{3} = 32{,}700 \text{ kilonewtons per day.}$$

41. First we find
$$f(x+h) - f(x) = (x+h)^2 + x + h - (x^2 + x) = (x^2 + 2xh + h^2 + x + h) - x^2 - x = 2xh + h^2 + h.$$
Then
$$\frac{f(x+h) - f(x)}{h} = \frac{2xh + h^2 + h}{h} = \frac{(2x+h+1)h}{h} = 2x + h + 1.$$

45. Since $k(f(x)) = e^{f(x)} = e^{2x}$, we can let $f(x) = 2x$.

49. $g(x) = x^2$ and $h(x) = x + 3$

53. **(a)** Since $v(x) = x^2$ and y can be written $\frac{1+(x^2)}{2+(x^2)}$, we take
$$u(x) = \frac{1+x}{2+x}.$$

(b) Since $v(x) = x^2 + 1$ and y can be written $\frac{1+x^2}{1+1+x^2}$, we take
$$u(x) = \frac{x}{1+x}.$$

57. **(a)** Since $u(x) = x^2$ and y can be written as $y = (\sin x)^2$, we take $v(x) = \sin x$.

(b) Since $v(x) = x^2$ and y can be written $y = \sin^2(\sqrt{x})^2$, we take $u(x) = \sin^2(\sqrt{x})$.

61. **(a)** From the graph of $f(x)$, we see that $f(g(x)) = 0$ when $g(x) = 0$ and when $g(x) = 4$. Since the solution to $g(x) = 0$ is $x = 4$, and the solution to $g(x) = 4$ is $x = 0$, we see that $x = 0$ and $x = 4$ are the only solutions to the equation $f(g(x)) = 0$.

(b) From the graph of $g(x)$, we see that $g(f(x)) = 0$ only when $f(x) = 4$, which occurs only when $x = 2$. Thus, $x = 2$ is the only solution to $g(f(x)) = 0$.

65. Reading values of the graph, we make an approximate table of values; we use these values to sketch Figure 10.1.

x	−2	−1	0	1	2
$f(x)$	−2	−0.3	0.7	1	0.7
$f(f(x))$	−2	0.4	1	1	1

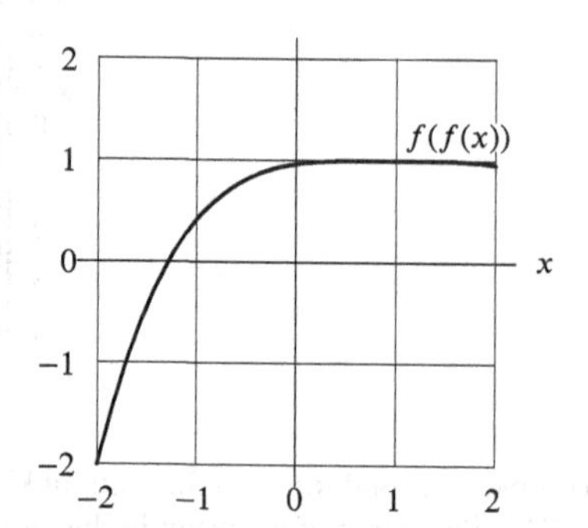

Figure 10.1: Graph of $f(f(x))$

69. **(a)** For $f(g(x)) = [g(x)]^2$, the graph always lies on or above the x-axis. It touches the x-axis where $g(x) = 0$, which is the x-intercept of the graph of $g(x)$ and is increasing to the right and decreasing from the left. The graph is shown in Figure 10.2.

(b) For $g(f(x)) = g(x^2)$, we only need to consider the graph of $g(x)$ for positive input values, that is to the right of the y-axis. We see then that the graph of $g(f(x))$ is increasing to the right of the y-intercept and decreasing from the left. The graph is shown in Figure 10.3.

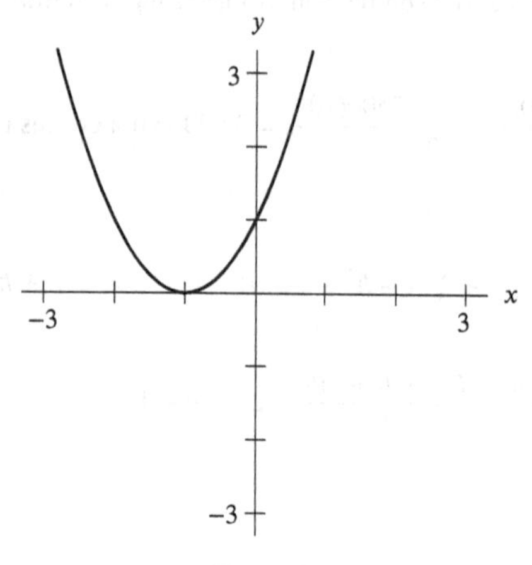

Figure 10.2

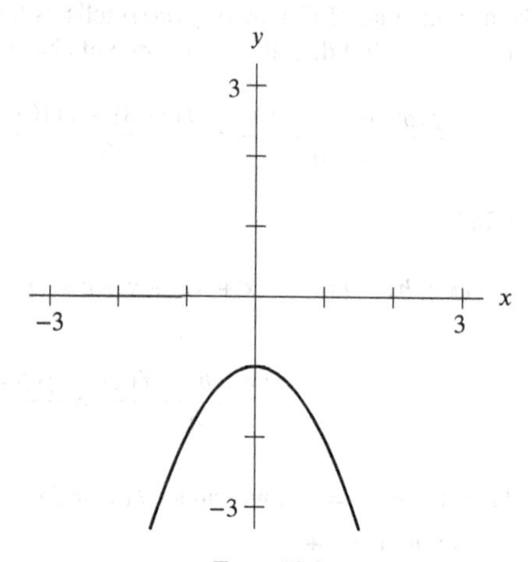

Figure 10.3

73. If $s(x) = 5 + \dfrac{1}{x+5} + x = x + 5 + \dfrac{1}{x+5}$ and $k(x) = x + 5$, then

$$s(x) = k(x) + \frac{1}{k(x)}.$$

However, $s(x) = v(k(x))$, so

$$v(k(x)) = k(x) + \frac{1}{k(x)}.$$

This is possible if

$$v(x) = x + \frac{1}{x}.$$

77. We have

$$\begin{aligned}
q(r(x)) &= \frac{8^{x^3}}{16^{x^2}} \\
&= 8^{x^3} \cdot 16^{-x^2} && \text{exponent rule} \\
&= (2^3)^{x^3} \cdot (2^4)^{-x^2} \\
&= 2^{3x^3} \cdot 2^{-4x^2} && \text{exponent rule} \\
&= 2^{3x^3-4x^2} && \text{exponent rule} \\
&= 2^{r(x)},
\end{aligned}$$

so $q(x) = 2^x$.

81. We have

$$\begin{aligned}
f(b) &= 1000\left(1 - 2^{-10/5}\right) && \text{because } b = 10 \\
&= 1000\left(1 - 2^{-2}\right) \\
&= 1000\left(1 - \frac{1}{4}\right) \\
&= 1000(3/4) \\
&= 750.
\end{aligned}$$

This tells us that after 10 weeks, 750 out of every 1000 cars have been repaired. Likewise,

$$\begin{aligned}
f(a) &= 1000\left(1 - 2^{-5/5}\right) && \text{because } a = 5 \\
&= 1000\left(1 - 2^{-1}\right) \\
&= 1000\left(1 - \frac{1}{2}\right) \\
&= 1000(1/2) \\
&= 500,
\end{aligned}$$

telling us that after 5 weeks, 500 out of every 1000 cars have been repaired. Thus,

$$\begin{aligned}
\frac{f(b) - f(a)}{b - a} &= \frac{750 - 500}{10 - 5} \\
&= \frac{250 \text{ cars}}{5 \text{ weeks}} \\
&= 50 \text{ cars/week},
\end{aligned}$$

telling us that an average of 50 cars per week are repaired between weeks 5 and 10.

85. **(a)** The function $y = f(x)$ has a y-intercept of 0, and a slope of $\frac{-10-0}{5-0} = -2$. So $f(x) = -2x$.

(b) The function is defined for the domain $0 \le x \le 5$, and it takes values in the range $-10 \le y \le 0$.

(c) The function $y = g(x)$ has a y-intercept of 1, and slope of $\frac{4-1}{1-0} = 3$. So $g(x) = 3x + 1$.

(d) The function $g(x)$ is defined on the domain $0 \le x \le 1$, and takes values in the range $1 \le y \le 4$.

(e) Since $f(x) = -2x$, and $g(x) = 3x + 1$,we know that $h(x) = f(g(x)) = -2(g(x)) = -2(3x + 1) = -6x - 2$.

(f) Since $g(x)$ is only defined for the domain $0 \le x \le 1$ and the range of $g(x)$ is contained in the domain of $f(x)$, $h(x)$ has the same domain as $g(x)$. Since $h(x)$ is a linear function, we can find its range by evaluating $h(x)$ for the extreme values of its domain, e.g. at $x = 0$ and at $x = 1$. We find that $h(0) = -6(0) - 2 = -2$ and $h(1) = -6(1) - 2 = -8$, so the range of $h(x)$ is $-8 \le y \le -2$.

(g) Since $h(x) = -6x - 2$, we know that its y-intercept is -2 and its slope is -6. Given a domain of $0 \le x \le 1$, $h(x)$ goes from $(0, -2)$ to $(1, -8)$.

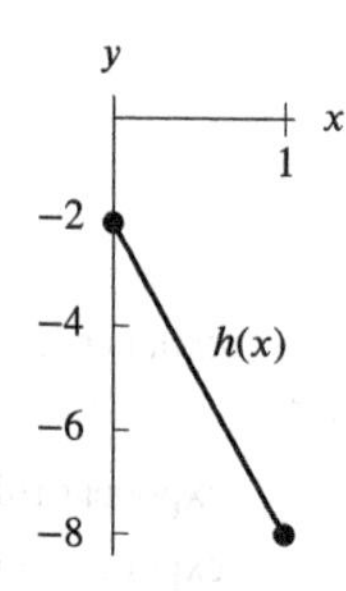

89. **(a)** We have $g(x) = g(-x)$, so $f(g(-x)) = f(g(x))$, making $f(g(x))$ even.

(b) We have $g(-x) = -g(x)$, so $f(g(-x)) = f(-g(x)) = f(g(x))$, making $f(g(x))$ even.

(c) We have $g(-x) = x$, so $f(g(-x)) = f(g(x))$, making $f(g(x))$ even.

(d) We have $g(-x) = -g(x)$, so

$$f(g(-x)) = f(-g(x)) = -f(g(x)),$$

which means $f(g(x))$ is odd.

(e) We have $g(-x) = g(x)$, so $f(g(-x)) = f(g(x))$, so $f(g(x))$ is even.

(f) Suppose that $g(x) = 5x$, which is odd, and $f(x) = 2x+1$, which is neither odd nor even. Then $f(g(x)) = 2(5x)+1 = 10x + 1$, which is neither even nor odd.

(g) Suppose that $f(x) = x^2$, which is even, and $g(x) = x + 1$, which is neither odd nor even. Then $f(g(x)) = (x + 1)^2$, which is neither odd nor even; for example $f(g(1)) = 4$ and $f(g(-1)) = 0$.

Solutions for Section 10.2

Skill Refresher

S1. Substituting $g(x)$ into the function $f(x)$ gives

$$f(g(x)) = f(x^{1/2}) = (x^{1/2})^2 = x.$$

S5. We have

$$\begin{aligned} y &= 4(x-5) \\ \frac{y}{4} &= x-5 \\ 5+\frac{y}{4} &= x, \end{aligned}$$

so

$$x = 5+\frac{y}{4} = \frac{20}{4}+\frac{y}{4} = \frac{20+y}{4}.$$

S9. We have

$$\begin{aligned} y &= 4e^x \\ \frac{y}{4} &= e^x \\ \ln\left(\frac{y}{4}\right) &= x, \end{aligned}$$

so

$$x = \ln\left(\frac{y}{4}\right).$$

S13. **(a)** We have $f(2) = 2^3 = 8$.
(b) Solving for x, we have

$$\begin{aligned} f(x) &= 2 \\ x^3 &= 2 \\ x &= 2^{\frac{1}{3}} \end{aligned}$$

S17. **(a)** We have $g(2t) = 3 \cdot (2t)^{-3} = \dfrac{3}{(2t)^3} = \dfrac{3}{8t^3}$.
(b) Solving for y, we have

$$\begin{aligned} g(y) &= 2t \\ 3y^{-3} &= 2t \\ 3 &= 2ty^3 \\ \frac{3}{2t} &= y^3 \\ y &= \left(\frac{3}{2t}\right)^{\frac{1}{3}} \end{aligned}$$

EXERCISES

1. A solution is $x = \tan^{-1}(-5) \approx -1.373$. Since $y = \tan x$ is a periodic function, there are multiple x-values for each of its y-values, so there are multiple solutions.

5. Let $y = x + 5$. Then $x = y - 5$, so $f^{-1}(x) = x - 5$.

9. Start with $x = f(f^{-1}(x))$ and substitute $y = f^{-1}(x)$. We have

$$\begin{aligned} x &= f(y) \\ x &= 3y - 7 \\ x + 7 &= 3y \\ \frac{x+7}{3} &= y \end{aligned}$$

Therefore,

$$f^{-1}(x) = \frac{x+7}{3}.$$

13. Since $f(x)$ passes the horizontal line test it is invertible.

17. It is not invertible.

21. It is not invertible.

25. Let $y = 1/(1 + \frac{1}{x})$. Then

$$y = \frac{1}{\left(\frac{1+x}{x}\right)} = \frac{x}{x+1}.$$

This means that

$$\begin{aligned} y(x+1) &= x \\ yx + y &= x \\ x - yx &= y \\ x(1-y) &= y \\ x &= \frac{y}{1-y}. \end{aligned}$$

Thus, $o^{-1}(x) = x/(1 - x)$.

29. Start with $x = h(h^{-1}(x))$ and substitute $y = h^{-1}(x)$. We have

$$\begin{aligned}
x &= h(y) \\
x &= \log \frac{y+5}{y-4} \\
10^x &= \frac{y+5}{y-4} \\
10^x(y-4) &= y+5 \\
10^x y - 4 \cdot 10^x &= y+5 \\
10^x y - y &= 5 + 4 \cdot 10^x \\
y(10^x - 1) &= 5 + 4 \cdot 10^x \\
y &= \frac{5 + 4 \cdot 10^x}{10^x - 1} \\
h^{-1}(x) &= \frac{5 + 4 \cdot 10^x}{10^x - 1}.
\end{aligned}$$

PROBLEMS

33. The inverse function $f^{-1}(P)$ gives the time, t, in years at which the population is P thousand. Its units are years.

37. **(a)** We have

$$R = 150 + 5T.$$

Solving for T, we have

$$5T = R - 150,$$

so

$$T = f^{-1}(R) = \frac{1}{5}R - 30.$$

(b) Table 10.2 shows values of f, and Table 10.3 shows values of f^{-1}. Notice that the columns in the two tables are reversed. This is because input values of f are output values of f^{-1}, and input values of f^{-1} are output values of f.

Table 10.2 *Values of f*

T, temperature (°C)	$R = f(T)$, resistance (ohms)
−20	50
−10	100
0	150
10	200
20	250
30	300
40	350
50	400

Table 10.3 *Values of f^{-1}*

R, resistance (ohms)	$T = f^{-1}(R)$, temperature (°C)
50	−20
100	−10
150	0
200	10
250	20
300	30
350	40
400	50

41. **(a)** When $t = 0$, $P = 37.8(1.044)^0 = 37.8(1) = 37.8$. This tells us that the population of the town when $t = 0$ is 37,800. The growth factor, 1.044, tells us that the population is 104.4% of what it had been the previous year, or that the town grows by 4.4% each year.

(b) Since $f(t) = 37.8(1.044)^t$, then $f(50) = 37.8(1.044)^{50} = 325.474$ This tells us that there will be 325,474 people after 50 years.

(c) To find $f^{-1}(P)$, which is the inverse function of $f(t)$, we need to solve

$$P = 37.8(1.044)^t$$

for t. Begin by dividing both sides by 37.8:

$$\frac{P}{37.8} = 1.044^t$$

Then, take the log of both sides, using the property $\log a^b = b \cdot \log a$.

$$\log\left(\frac{P}{37.8}\right) = \log 1.044^t = t \log 1.044.$$

So solving for t,

$$t = \frac{\log\left(\frac{P}{37.8}\right)}{\log 1.044}.$$

We can make this formula look a little simpler by recalling that $\log \frac{a}{b} = \log a - \log b$. The formula for our inverse function is now:

$$t = f^{-1}(P) = \frac{\log P - \log 37.8}{\log 1.044}.$$

(d) $f^{-1}(50) = \frac{\log 50 - \log 37.8}{\log 1.044} \approx 6.496$. It will take about 6.496 years for P to reach 50,000 people.

45. **(a)** $f^{-1}(-1)$ is the value of x with $f(x) = -1$. Reading off the graph, we see that $f(0) = -1$ and so $f^{-1}(-1) = 0$.

(b) $f^{-1}(0)$ is the value of x with $f(x) = 0$. Reading off the graph, we see that $f(1) = 0$ and so $f^{-1}(0) = 1$.

(c) $f^{-1}(3)$ is the value of x with $f(x) = 3$. Reading off the graph, we see that $f(2) = 3$ and so $f^{-1}(3) = 2$.

(d) $f^{-1}(-2)$ is the value of x with $f(x) = -2$. Reading off the graph, we see that $f(-2) = -2$ and so $f^{-1}(-2) = -2$.

(e) Since $f^{-1}(f(x)) = x$ for any x in the domain of f, we have $f^{-1}(f(\frac{1}{2})) = \frac{1}{2}$. Alternatively, we could read values from the graph.

(f) Since $f(f^{-1}(y)) = y$ for any y in the domain of f^{-1}, we have $f(f^{-1}(-0.05)) = -0.05$. Alternatively, we could read values from the graph.

49. **(a)** $f^{-1}(30)$ is the value of t with $f(t) = 30$. From the graph, we estimate that $f(4.9) = 30$, so $f^{-1}(30) = 4.9$. This means that 4.9 sec is the time when the charge remaining in the capacitor is 30 microcoulombs.

(b) $f^{-1}(20)$ is the value of t with $f(t) = 20$. From the graph, we estimate that $f(6.9) = 20$, so $f^{-1}(20) = 6.9$. Therefore,

$$f^{-1}(20) - 3 = 3.9 \text{ sec.}$$

This quantity represents the time 3 sec before the amount of charge in the capacitor is 20 microcoulombs.

(c) The equation $f^{-1}(q) = 2$ is equivalent to the equation $f(2) = q$. From the graph, we estimate that $f(2) = 54$, so $q = 54$ microcoulombs. This solution represents the amount of charge remaining in the capacitor after 2 sec.

53. **(a)** This represents the total force on the dam, in meganewtons, when the height of the water in the reservoir is 50 meters.

(b) This represents the height of water in the reservoir that produces a total force of 50 meganewtons on the dam.

(c) This represents 10 meganewtons more than the total force on the dam produced by a water height of 40 meters in the reservoir.

(d) This represents a water height 10 meters higher than the water height required to produce a total force of 40 meganewtons on the dam.

(e) This indicates that when the height of the water in the reservoir is 60 meters, the total force on the dam will be 7200 meganewtons.

(f) This indicates that when the total force on the dam is 60 meganewtons, the water height in the reservoir will be 5.5 meters.

57. **(a)** $f^{-1}(87.0)$ is the value of x with $f(x) = 87.0$. From the table, we see that $f(0.8) = 87.0$, so $f^{-1}(87.0) = 0.8$. This means that in order for the spring to store 87.0 joules of potential energy, it must be stretched 0.8 meters beyond its equilibrium position.

(b) $f^{-1}(348)$ is the value of x with $f(x) = 348$. From the table, we see that $f(1.6) = 348$, so $f^{-1}(348) = 1.6$. Therefore,

$$f^{-1}(348) + 0.2 = 1.6 + 0.2 = 1.8 \text{ meters.}$$

This quantity represents a distance 0.2 meters greater than the distance the spring must be stretched beyond its equilibrium position in order to store 348 joules of potential energy.

(c) From the table, we see that $f(0.4) = 21.7$, so $x = 0.4$. This means that 0.4 meters is the distance that the spring must be stretched beyond its equilibrium position in order to store 21.7 joules of potential energy.

(d) The equation $f^{-1}(U) = 0.8$ is equivalent to the equation $f(0.8) = U$. From the table, $f(0.8) = 87.0$, so $U = 87.0$. This means that 87.0 joules is the amount of potential energy stored by the spring when it is stretched 0.8 meters beyond its equilibrium position.

61. We have

$$
\begin{aligned}
y &= 0.5(x^2 + A^2)^{0.5} \\
2y &= (x^2 + A^2)^{0.5} \\
(2y)^2 &= x^2 + A^2 \\
x^2 &= 4y^2 - A^2 \\
x &= \left(4y^2 - A^2\right)^{0.5} \quad \text{because } x \geq 0 \\
\text{so} \quad f^{-1}(x) &= \left(4x^2 - A^2\right)^{0.5}.
\end{aligned}
$$

65. **(a)** If $f(t)$ is exponential, then $f(t) = AB^t$, and

$$
\begin{aligned}
f(12) &= AB^{12} = 20 \\
\text{and} \quad f(7) &= AB^7 = 13.
\end{aligned}
$$

Taking the ratios, we have

$$
\begin{aligned}
\frac{AB^{12}}{AB^7} &= \frac{20}{13} \\
B^5 &= \frac{20}{13} \\
B &= \left(\frac{20}{13}\right)^{\frac{1}{5}} \approx 1.08998.
\end{aligned}
$$

Substituting this value into $f(7)$, we get

$$
\begin{aligned}
f(7) &= A(1.08998)^7 = 13 \\
A &= \frac{13}{(1.08998)^7} \approx 7.112.
\end{aligned}
$$

Using these values of A and B, we have

$$f(t) = 7.112(1.08998)^t.$$

(b) Let $P = 7.112(1.08998)^t$, then solve for t:

$$
\begin{aligned}
P &= 7.112(1.08998)^t \\
P/7.112 &= 1.08998^t \\
\log 1.08998^t &= \log\left(\frac{P}{7.112}\right) \\
t \log 1.08998 &= \log\left(\frac{P}{7.112}\right) \\
t &= \frac{\log(P/7.112)}{\log 1.08998}.
\end{aligned}
$$

Thus

$$f^{-1}(P) = \frac{\log(P/7.112)}{\log 1.08998}.$$

(c)

$$f(25) = 7.112(1.08998)^{25} \approx 61.299.$$

This means that in year 25, the population is approximately 61,300 people.

$$f^{-1}(25) = \frac{\log(25/7.112)}{\log 1.08998} \approx 14.590.$$

This means that when the population is 25,000, the year is approximately 14.590.

69. Since f is assumed to be an increasing function, its inverse is well defined. This is an amount of caffeine: the amount giving a pulse 20 bpm higher than r_c, that is, 20 bpm higher than the pulse of a person having 1 serving of coffee.

73. Since f is assumed to be an increasing function, its inverse is well defined. This is an amount of caffeine. We know that $1.1f(q_c) = 1.1r_c$ is 10% higher than the pulse of a person who has had 1 serving of coffee. This makes $f^{-1}(1.1f(q_c))$ is the amount of caffeine that will lead to a pulse 10% higher than will a serving of coffee.

Solutions for Section 10.3

Skill Refresher

S1. Since the logarithm function is only defined for positive numbers, the domain is $x > 0$.

S5. Since an exponential function is defined for all real numbers, the function $f(x) = 2^{x+1} = 2 \cdot 2^x$ is also defined for all real numbers since it is a constant multiple of an exponential function.

S9. The function $h(x)$ is a vertical stretch of $y = \sin x$ by a factor of 2, so the maximum of $h(x)$ is 2, and the minimum is -2. The range is $-2 \le h(x) \le 2$.

EXERCISES

1. The graph of the inverse function $f^{-1}(x)$ of a function $f(x)$ is the reflection of the graph of $f(x)$ across the line $y = x$. Thus we see that (a) and (e) are inverse functions, (b) and (c) are inverse functions and (d) and (f) are inverse functions.

5. Since $f(x) = (x/4) - (3/2)$ and $g(t) = 4(t + 3/2)$, we have

$$f(g(t)) = \frac{4(t + \frac{3}{2})}{4} - \frac{3}{2} = t + \frac{3}{2} - \frac{3}{2} = t$$
$$g(f(x)) = 4\left(\frac{x}{4} - \frac{3}{2} + \frac{3}{2}\right) = 4\frac{x}{4} = x$$

9. Check using the two compositions

$$f(f^{-1}(x)) = e^{f^{-1}(x)+1} = e^{(\ln x - 1)+1} = e^{\ln x} = x$$

and

$$f^{-1}(f(x)) = \ln f(x) - 1 = \ln(e^{x+1}) - 1 = (x+1) - 1 = x.$$

They are inverses of one another.

PROBLEMS

13. Estimating from the graph of f, we see that $f(-4) = -3$, which means that $f^{-1}(-3) = -4$. Using similar logic, we estimate that $f^{-1}(-2) = -0.7, f^{-1}(-1) = 0.3, f^{-1}(0) = 0.9, f^{-1}(1) = 1.4, f^{-1}(2) = 1.9$, and $f^{-1}(3) = 2.5$. Using this data, we construct the graph for f^{-1} shown in Figure 10.4.

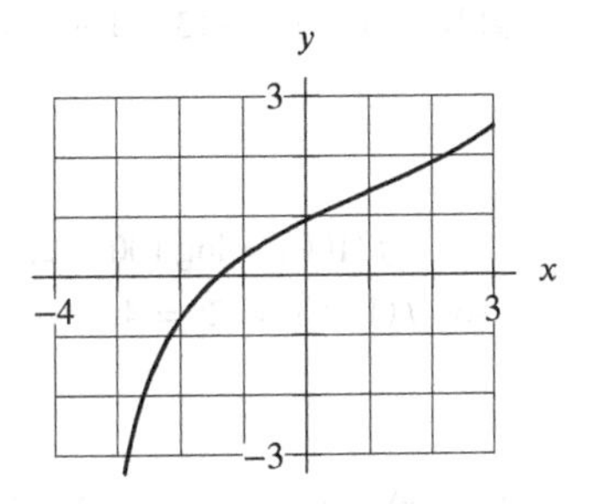

Figure 10.4

17. (a) Her maximum height is approximately 36 m.
(b) She lands on the trampoline approximately 6 seconds later.
(c) The graph is a parabola, and hence is symmetric about the vertical line $x = 3$ through its vertex. We choose the right half of the parabola, whose domain is the interval $3 \le t \le 6$. See Figure 10.5.
(d) The gymnast is part of a complicated stunt involving several phases. At 3 seconds into the stunt, she steps off the platform for the high wire, 36 meters in the air. Three seconds later (at $t = 6$) she lands on a trampoline at ground level.

(e) See Figure 10.6. As the gymnast falls, her height decreases steadily from 36 m to 0 m. In other words, she occupies each height for one moment of time only, and does not return to that height at any other time. This means that each value of t corresponds to a single height, and therefore time is a function of height.

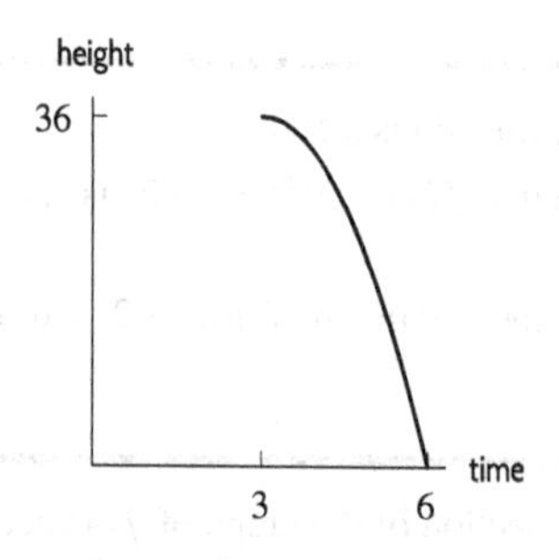

Figure 10.5

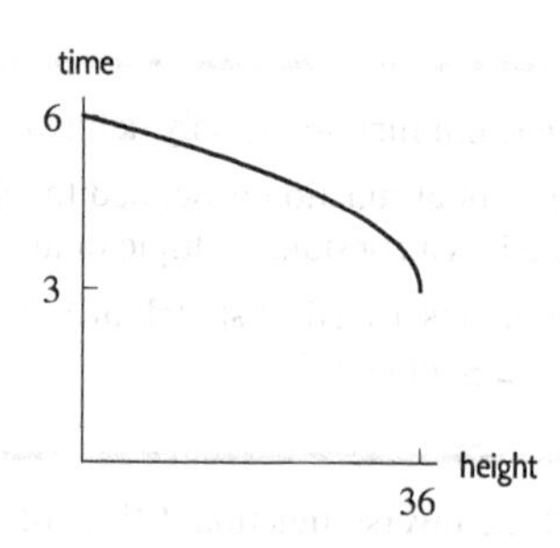

Figure 10.6

Solutions for Section 10.4

Skill Refresher

S1. Evaluating each function gives

$$\begin{aligned} f(-2) &= -(-2)^2 + 3 = -4 + 3 = -1, \\ \text{and } g(3) &= 4 \cdot 3 - 1 = 11, \\ \text{so, } f(-2) + g(3) &= -1 + 11 = 10. \end{aligned}$$

S5. Evaluating each function gives

$$\begin{aligned} f(4) &= -4^2 + 3 = -16 + 3 = -13, \\ g(3) &= 4 \cdot 3 - 1 = 11, \\ \text{and } f(-1) &= -(-1)^2 + 3 = 2. \\ \text{Thus, } f(4) \cdot g(3) - f(-1) &= -13 \cdot 11 - 2 = -145. \end{aligned}$$

S9. Evaluating gives

$$\begin{aligned} f(100) &= \log 100 = 2, \\ \text{so } (f(100))^2 &= 2^2 = 4. \end{aligned}$$

S13. We have

$$1 - (1 - x)(1 + x) = 1 - (1 - x^2) = 1 - 1 + x^2 = x^2$$

S17. We have

$$\frac{e^x(x^2 - 2x)}{2x} = \frac{x^2e^x - 2xe^x}{2x} = \frac{x^2e^x}{2x} - \frac{2xe^x}{2x} = \frac{xe^x}{2} - e^x.$$

EXERCISES

1. **(a)** We have $f(x) + g(x) = x + 5 + x - 5 = 2x$.
(b) We have $f(x) - g(x) = x + 5 - (x - 5) = 10$.
(c) We have $f(x)g(x) = (x + 5)(x - 5) = x^2 - 25$.
(d) We have $f(x)/g(x) = (x + 5)/(x - 5)$.

5. **(a)** We have $f(x) + g(x) = x^2 + 4 + x^2 + 2 = 2x^2 + 6$.
(b) We have $f(x) - g(x) = x^2 + 4 - (x^2 + 2) = 2$.
(c) We have $f(x)g(x) = (x^2 + 4)(x^2 + 2) = x^4 + 6x^2 + 8$.
(d) We have $f(x)/g(x) = (x^2 + 4)/(x^2 + 2)$.

9. $h(x) = 2(2x - 1) - 3(1 - x) = 4x - 2 - 3 + 3x = 7x - 5$.

13. We have $f(x) = e^x(2x + 1) = 2xe^x + e^x$.

17. Since $h(x) = f(x) + g(x)$, we know that $h(-1) = f(-1) + g(-1) = -4 + 4 = 0$. Similarly, $j(x) = 2f(x)$ tells us that $j(-1) = 2f(-1) = 2(-4) = -8$. Repeat this process for each entry in the table.

Table 10.4

x	$h(x)$	$j(x)$	$k(x)$	$m(x)$
-1	0	-8	16	-1
0	0	-2	1	-1
1	2	4	0	0
2	6	10	1	0.2
3	12	16	16	0.5
4	20	22	81	9/11

21. $\dfrac{f(x)}{g(x)} = \dfrac{\sin x}{x^2}$.

PROBLEMS

25. **(a)** A formula for $h(x)$ would be

$$h(x) = f(x) + g(x).$$

To evaluate $h(x)$ for $x = 3$, we use this equation:

$$h(3) = f(3) + g(3).$$

Since $f(x) = x + 1$, we know that

$$f(3) = 3 + 1 = 4.$$

Likewise, since $g(x) = x^2 - 1$, we know that

$$g(3) = 3^2 - 1 = 9 - 1 = 8.$$

Thus, we have

$$h(3) = 4 + 8 = 12.$$

To find a formula for $h(x)$ in terms of x, we substitute our formulas for $f(x)$ and $g(x)$ into the equation $h(x) = f(x) + g(x)$:

$$h(x) = \underbrace{f(x)}_{x+1} + \underbrace{g(x)}_{x^2-1}$$

$$h(x) = x + 1 + x^2 - 1 = x^2 + x.$$

To check this formula, we use it to evaluate $h(3)$, and see if it gives $h(3) = 12$, which is what we got before. The formula is $h(x) = x^2 + x$, so it gives

$$h(3) = 3^2 + 3 = 9 + 3 = 12.$$

This is the result that we expected.

(b) A formula for $j(x)$ would be

$$j(x) = g(x) - 2f(x).$$

To evaluate $j(x)$ for $x = 3$, we use this equation:

$$j(3) = g(3) - 2f(3).$$

We already know that $g(3) = 8$ and $f(3) = 4$. Thus,

$$j(3) = 8 - 2 \cdot 4 = 8 - 8 = 0.$$

To find a formula for $j(x)$ in terms of x, we again use the formulas for $f(x)$ and $g(x)$:

$$\begin{aligned} j(x) &= \underbrace{g(x)}_{x^2-1} - 2\underbrace{f(x)}_{x+1} \\ &= (x^2 - 1) - 2(x + 1) \\ &= x^2 - 1 - 2x - 2 \\ &= x^2 - 2x - 3. \end{aligned}$$

We check this formula using the fact that we already know $j(3) = 0$. Since we have $j(x) = x^2 - 2x - 3$,

$$j(3) = 3^2 - 2 \cdot 3 - 3 = 9 - 6 - 3 = 0.$$

This is the result that we expected.

(c) A formula for $k(x)$ would be

$$k(x) = f(x)g(x).$$

Evaluating $k(3)$, we have

$$k(3) = f(3)g(3) = 4 \cdot 8 = 32.$$

A formula in terms of x for $k(x)$ would be

$$\begin{aligned} k(x) &= \underbrace{f(x)}_{x+1} \cdot \underbrace{g(x)}_{x^2-1} \\ &= (x + 1)(x^2 - 1) \\ &= x^3 - x + x^2 - 1 \\ &= x^3 + x^2 - x - 1. \end{aligned}$$

To check this formula,

$$k(3) = 3^3 + 3^2 - 3 - 1 = 27 + 9 - 3 - 1 = 32,$$

which agrees with what we already knew.

(d) A formula for $m(x)$ would be

$$m(x) = \frac{g(x)}{f(x)}.$$

Using this formula, we have

$$m(3) = \frac{g(3)}{f(3)} = \frac{8}{4} = 2.$$

To find a formula for $m(x)$ in terms of x, we write

$$\begin{aligned} m(x) = \frac{g(x)}{f(x)} &= \frac{x^2 - 1}{x + 1} \\ &= \frac{(x + 1)(x - 1)}{(x + 1)} \\ &= x - 1 \text{for } x \neq -1 \end{aligned}$$

We were able to simplify this formula by first factoring the numerator of the fraction $\frac{x^2 - 1}{x + 1}$. To check this formula,

$$m(3) = 3 - 1 = 2,$$

which is what we were expecting.

(e) We have

$$n(x) = (f(x))^2 - g(x).$$

This means that

$$\begin{aligned} n(3) &= (f(3))^2 - g(3) \\ &= (4)^2 - 8 \\ &= 16 - 8 \\ &= 8. \end{aligned}$$

A formula for $n(x)$ in terms of x would be

$$\begin{aligned} n(x) &= (f(x))^2 - g(x) \\ &= (x+1)^2 - (x^2 - 1) \\ &= x^2 + 2x + 1 - x^2 + 1 \\ &= 2x + 2. \end{aligned}$$

To check this formula,

$$n(3) = 2 \cdot 3 + 2 = 8,$$

which is what we were expecting.

29. We have:

$$\begin{aligned} q(0) &= w^{-1}(v(0)) = w^{-1}(4) = 3 && \text{because } v(0) = 4 \text{ and } w(3) = 4 \\ q(1) &= w^{-1}(v(1)) = w^{-1}(3) = 2 && \text{because } v(1) = 3 \text{ and } w(2) = 3 \\ q(2) &= w^{-1}(v(2)) = w^{-1}(3) = 2 && \text{because } v(2) = 3 \text{ and } w(2) = 3 \\ q(3) &= w^{-1}(v(3)) = w^{-1}(5) = 5 && \text{because } v(3) = 5 \text{ and } w(5) = 5 \\ q(4) &= w^{-1}(v(4)) = w^{-1}(4) = 3 && \text{because } v(4) = 4 \text{ and } w(3) = 4 \\ q(5) &= w^{-1}(v(5)) = w^{-1}(4) = 3. && \text{because } v(5) = 4 \text{ and } w(3) = 4 \end{aligned}$$

See Table 10.5.

Table 10.5

x	0	1	2	3	4	5
$q(x)$	3	2	2	5	3	3

33. Where $g(x) = f(x)$, we see that $h(x) = g(x) - f(x) = 0$. Thus, the graph of h has an x-intercept wherever the graphs of f and g cross. See Figure 10.7.

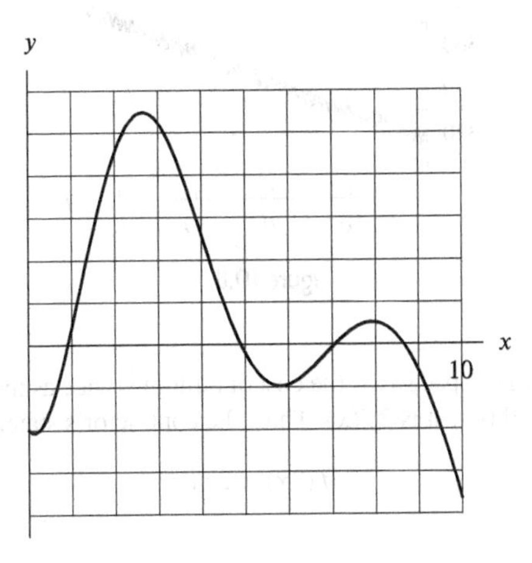

Figure 10.7

37. We can find the revenue function as a product:

$$\text{Revenue} = (\#\text{ of customers}) \cdot (\text{price per customer}).$$

At the current price, 50,000 people attend every day. Since 2500 customers will be lost for each \$1 increase in price, the function $n(i)$ giving the number of customers who will attend given i one-dollar price increases, is given by $n(i) = 50{,}000 - 2500i$. The price function $p(i)$ giving the price after i one-dollar price increases is given by $p(i) = 15 + i$. The revenue function $r(i)$ is given by

$$\begin{aligned} r(i) &= n(i)p(i) \\ &= (50{,}000 - 2500i)(15 + i) \\ &= -2500i^2 + 12{,}500i + 750{,}000 \\ &= -2500(i - 20)(i + 15). \end{aligned}$$

The graph $r(i)$ is a downward-facing parabola with zeros at $i = -15$ and $i = 20$, so the maximum revenue occurs at $i = 2.5$, which is halfway between the zeros. Thus, to maximize profits the ideal price is \$15+2.5(\$1.00) = \$15+\$2.50 = \$17.50.

41. We have

$$\begin{aligned} H(x) &= F(G(x)) \\ &= F\left(\sqrt{x}\right) \\ &= \cos\left(\sqrt{x}\right) \\ h(x) &= f(G(x)) \cdot g(x) \\ &= f\left(\sqrt{x}\right) \cdot \frac{1}{2\sqrt{x}} \\ &= -\sin\left(\sqrt{x}\right) \cdot \frac{1}{2\sqrt{x}} \\ &= -\frac{\sin\left(\sqrt{x}\right)}{2\sqrt{x}}. \end{aligned}$$

45. **(a)** Since the initial amount was 316.75 and the growth factor is 1.004, we have $A(t) = 316.75(1.004)^t$.
(b) Since the CO_2 level oscillates once per year, the period is 1 year. The amplitude is 3.25 ppm, so one possible answer is $V(t) = 3.25\sin(2\pi t)$. Any sinusoidal function with the same amplitude and period could describe the variation.
(c) The graph of $y = 316.75(1.004)^t + 3.25\sin(2\pi t)$ is in Figure 10.8.

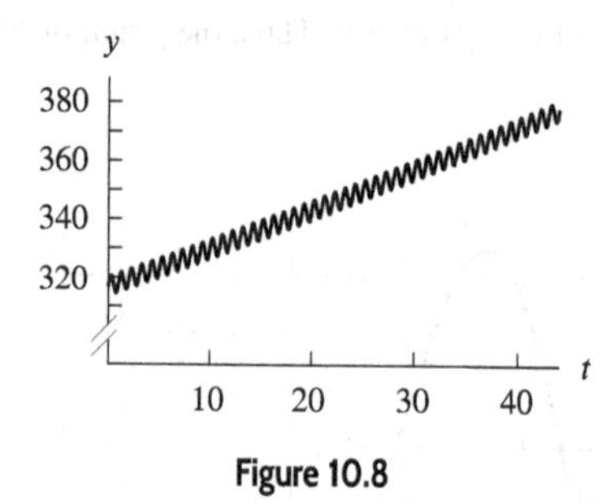

Figure 10.8

49. Since $2x$ represents twice as much office space as x, the cost of building twice as much space is $f(2x)$. The cost of building x amount of space is $f(x)$, so twice this cost is $2f(x)$. Thus, the contractor's statement is expressed

$$f(2x) < 2f(x).$$

53. The inequality $h(f(x)) < x$ tells us that Space can build fewer than x square feet of office space with the money Ace needs to build x square feet. You get more for your money with Ace.

STRENGTHEN YOUR UNDERSTANDING

1. False, since $f(4) + g(4) = \frac{1}{4} + \sqrt{4}$ but $(f + g)(8) = \frac{1}{8} + \sqrt{8}$.

5. True, since $g(f(x)) = g\left(\frac{1}{x}\right) = \sqrt{\frac{1}{x}}$.

9. True. Evaluate $\dfrac{f(3) + g(3)}{h(3)} = \dfrac{\frac{1}{3} + \sqrt{3}}{3 - 5}$ and simplify.

13. False. As a counterexample, let $f(x) = x^2$ and $g(x) = x+1$. Then $f(g(x)) = (x+1)^2 = x^2+2x+1$, but $g(f(x)) = x^2+1$.

17. False. $f(x + h) = \dfrac{1}{x + h} \neq \dfrac{1}{x} + \dfrac{1}{h}$.

21. False. If $f(x) = ax^2 + bx + c$ and $g(x) = px^2 + qx + r$, then

$$f(g(x)) = f(px^2 + qx + r) = a(px^2 + qx + r)^2 + b(px^2 + qx + r) + c.$$

Expanding shows that $f(g(x))$ has an x^4 term.

25. True, since $g(f(2)) = g(1) = 3$ and $f(g(3)) = f(1) = 3$.

29. True. The function g is not invertible if two different points in the domain have the same function value.

33. True. Each x value has only one y value.

37. True. The inverse of a function reverses the action of the function and returns the original value of the independent variable x.

CHAPTER ELEVEN

Solutions for Section 11.1

Skill Refresher

S1. $\sqrt{36t^2} = (36t^2)^{1/2} = 36^{1/2} \cdot (t^2)^{1/2} = 6|t^1| = 6|t|$

S5. We have

$$\begin{aligned} 10x^{5-2} &= 2 \\ 10x^3 &= 2 \\ x^3 &= 0.2 \\ x &= (0.2)^{1/3} = 0.585. \end{aligned}$$

S9. False

EXERCISES

1. This is a power function in the form $y = ax^p$:

$$y = \frac{48}{30625} \cdot x^{-2}, \qquad a = \frac{48}{30625}, \qquad p = -2.$$

We have

$$\begin{aligned} y &= 3\left(\frac{2}{5\sqrt{7x}}\right)^4 \\ &= 3 \cdot \frac{2^4}{5^4\left(\sqrt{7}\sqrt{x}\right)^4} \\ &= 3 \cdot \frac{2^4}{5^4\left(7^{\frac{1}{2}}x^{\frac{1}{2}}\right)^4} \\ &= \frac{48}{625 \cdot 49x^2} \\ &= \frac{48}{30625} \cdot x^{-2}. \end{aligned}$$

5. Although y is a power function of $(x+7)$, it is not a power function of x and cannot be written in the form $f(x) = kx^p$. In particular, expanding we see that $4(x+7)^2 = 4(x^2 + 14x + 49)$, so the expression involves more than one power of x.

9. Since the graph is symmetric about the y-axis, the power function is even.

13. Since $f(1) = k \cdot 1^p = k$, we know $k = f(1) = \frac{3}{2}$.

Since $f(2) = k \cdot 2^p = \frac{3}{8}$, and since $k = \frac{3}{2}$, we know

$$\left(\frac{3}{2}\right) \cdot 2^p = \frac{3}{8}$$

which implies

$$2^p = \frac{3}{8} \cdot \frac{2}{3} = \frac{1}{4}.$$

Thus $p = -2$, and $f(x) = \frac{3}{2} \cdot x^{-2}$.

17. We need to solve $j(x) = kx^p$ for p and k. We know that $j(x) = 2$ when $x = 1$. Since $j(1) = k \cdot 1^p = k$, we have $k = 2$. To solve for p, use the fact that $j(2) = 16$ and also $j(2) = 2 \cdot 2^p$, so

$$2 \cdot 2^p = 16,$$

giving $2^p = 8$, so $p = 3$. Thus, $j(x) = 2x^3$.

21. Substituting into the general formula $c = kd^2$, we have $45 = k(3)^2$ or $k = 45/9 = 5$. So the formula for c is

$$c = 5d^2.$$

When $d = 5$, we get $c = 5(5)^2 = 125$.

25. **(a)** $\lim_{x\to\infty} x^{-4} = \lim_{x\to\infty}(1/x^4) = 0.$

(b) $\lim_{x\to-\infty} 2x^{-1} = \lim_{x\to-\infty}(2/x) = 0.$

PROBLEMS

29. **(a)** For $f(x) = x^{1/2}$, between $x = 0$ and $x = 2$, we have

$$\text{Average rate of change} = \frac{f(2) - f(0)}{2 - 0} = \frac{\sqrt{2} - \sqrt{0}}{2 - 0} = 0.707.$$

Similar calculations show the rate of change of $f(x)$ between $x = 2$ and $x = 4$ is 0.293 and give all the values in Table 11.1.

Table 11.1

Interval	$0-2$	$2-4$	$4-6$	$6-8$
Rate of change of $f(x) = x^{1/2}$	0.707	0.293	0.225	0.189
Rate of change of $g(x) = x^2$	2	6	10	14

(b) For $f(x) = x^{1/2}$, as x increases, the rate of change decreases (from 0.707 to 0.293 to 0.225 to 0.189). This reflects the fact that the graph of $f(x) = x^{1/2}$ is concave down.

For $g(x) = x^2$, as x increases, the rate of change increases. This reflects the fact that the graph of $g(x) = x^2$ is concave up.

33. $c(t) = \frac{1}{t}$ is indeed one possible formula. It is not, however, the only one. Because the vertical and horizontal axes are asymptotes for this function, we know that the power p is a negative number and

$$c(t) = kt^p.$$

If $p = -3$ then $c(t) = kt^{-3}$. Since $(2, \frac{1}{2})$ lies on the curve, $\frac{1}{2} = k(2)^{-3}$ or $k = 4$. So, $c(t) = 4t^{-3}$ could describe this function. Similarly, so could $c(t) = 16x^{-5}$ or $c(t) = 64x^{-7}$...

37. **(a)** Since the cost of the fabric, $C(x)$, is directly proportional to the amount purchased, x, we know that the formula will be of the form

$$C(x) = kx.$$

(b) Since 3 yards cost \$28.50, we know that $C(3) = \$28.50$. Thus, we have

$$28.50 = 3k$$
$$k = 9.5$$

Our formula for the cost of x yards of fabric is

$$C(x) = 9.5x.$$

(c) Notice that the graph in Figure 11.1 goes through the origin.

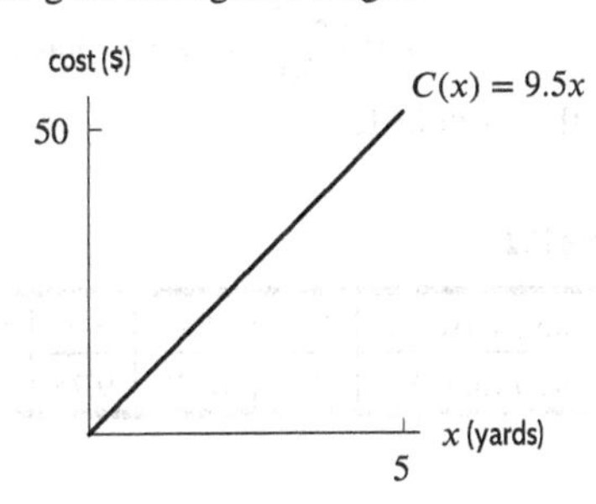

Figure 11.1

(d) To find the cost of 5.5 yards of fabric, we evaluate $C(x)$ for $x = 5.5$:

$$C(5.5) = 9.5(5.5) = \$52.25.$$

41. Since the volume is inversely proportional to the pressure, we know that

$$V = \frac{k}{P},$$

where k is constant. We also know that $P = 12$ atmospheres when $V = 10$ liters, so we have:

$$10 = \frac{k}{12}$$
$$k = 120.$$

Thus, when $V = 5$ liters, we have:

$$5 = \frac{120}{P}$$
$$P = \frac{120}{5} = 24 \text{ atmospheres.}$$

45. Since we know

$$\text{Cooling amount } = k(T - W),$$

we use the given information to find the value of k for this swamp cooler. The cooling amount is the drop in temperature, $40 - 23$, so

$$40 - 23 = k(40 - 20)$$
$$17 = 20k$$
$$0.85 = k$$

(a) Using $T = 48$ and $W = 22$, the swamp cooler can reduce the air temperature by $0.85(48 - 22) = 22.1°$ Celsius. Therefore, the air temperature can be cooled to $48 - 22.1 = 25.9°$ Celsius.
(b) Using $T = 32$ and $W = 26$, the swamp cooler can reduce the air temperature by $0.85(32 - 26) = 5.1°$ Celsius. Therefore, the air temperature can be cooled to $32 - 5.1 = 26.9°$ Celsius.

49. (a) Since cooking time, t, is inversely proportional to power level, w, we know that for some constant k

$$t = \frac{k}{w}.$$

We know that $t = 6.5$ minutes when $w = 750$ watts, so we solve for k:

$$6.5 = \frac{k}{750}$$
$$k = 6.5(750) = 4875.$$

Thus, the function is

$$t = f(w) = \frac{4875}{w}.$$

(b) When $w = 250$ watts
$$t = \frac{4875}{250} = 19.5 \text{ minutes.}$$
Continuing in the same way gives the result in Table 11.2.

Table 11.2

Power, w (watts)	250	300	500	650
Time, t (mins)	19.5	16.25	9.75	7.5

(c) The graph of $t = f(w) = 4875/w$ is shown in Figure 11.2.

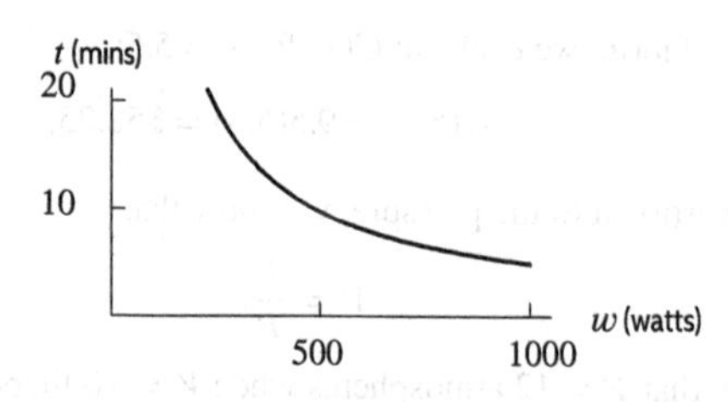

Figure 11.2

(d) For a new dish, there is a new value of k. Since $t = 2$ when $w = 250$,
$$2 = \frac{k}{250}$$
$$k = 500,$$
Calculating t when $w = 500$, we have
$$t = \frac{500}{500} = 1 \text{ minute.}$$

53. **(a)** Since F is inversely proportional to r, we know that $F = k/r$, where k is a constant. Therefore, since $F = 5$ millinewtons when $r = 10$ cm, we have:
$$5 = \frac{k}{10}$$
$$k = 50.$$
Thus, our formula is
$$F = \frac{50}{r}.$$

(b) The graph of F as a function of r is shown in Figure 11.3. We observe that the force on the ball decreases as the radius of its path increases.

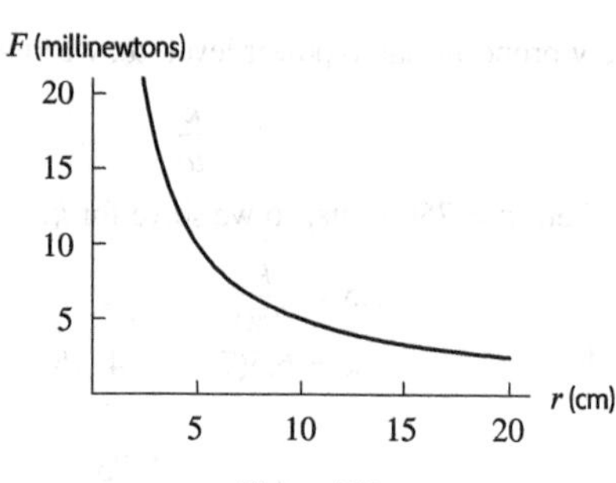

Figure 11.3

(c) Since F is proportional to v^2, we know that $F = kv^2$. Therefore, since $F = 10$ millinewtons when $v = 20$ cm/sec, we have:

$$10 = k(20)^2$$
$$k = \frac{10}{20^2} = \frac{1}{40}.$$

Thus, our formula is

$$F = \frac{1}{40}v^2.$$

(d) The graph of F as a function of v is shown in Figure 11.4. We observe that the force on the ball increases as its speed increases.

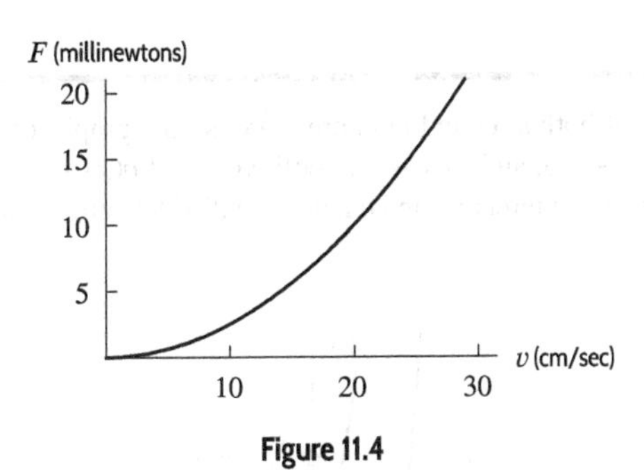

Figure 11.4

Solutions for Section 11.2

Skill Refresher

S1. Rewriting, we have

$$x^2 - 3x + 4 = -3x^3 + x^2 + 4x.$$

S5. **(a)** A positive number cubed is still positive and thus $f(100)$ is positive.
(b) A negative number to an odd power is negative and thus $f(-100)$ is negative.

S9. $x^2 + 2x - x - 2 = x^2 + x - 2$

S13. $\frac{1}{3}$

S17. The constant term of the polynomial is the product of the constant terms of each factor. We have

$$2(-6) = -12,$$

so the constant term is -12.

S21. The degree of the polynomial is determined by the product of the highest power term in each factor. We have

$$x^2 \cdot x \cdot 3x = 3x^4,$$

so the degree is 4.

EXERCISES

1. Since 5^x is not a power function, this is not a polynomial.

5. This is not a polynomial because $2e^x$ is not a power function.

9. Rewriting in standard form, we get

$$\begin{aligned} 5x^4 - 3x^2 - 2x^4 + 1 - 3x^4 &= 5x^4 - 2x^4 - 3x^4 - 3x^2 + 1 \\ &= -3x^2 + 1, \end{aligned}$$

so we see that the leading term is $-3x^2$.

13. Since n is a positive integer, we see that $n+3$ is larger than the other powers. This means the leading term is $4x^{n+3}$.

17. Here, the leading term is $x^2 \cdot 2x^3 = 2x^5$, and the contributing terms are $x^2, 2x^3$.

21. **(a)** $\lim_{x\to\infty}(3x^2 - 5x + 7) = \lim_{x\to\infty}(3x^2) = \infty$.

(b) $\lim_{x\to-\infty}(7x^2 - 9x^3) = \lim_{x\to-\infty}(-9x^3) = \infty$.

PROBLEMS

25. **(a)** Note that the leading terms of both $u(x)$ and $v(x)$ are $-\frac{1}{5}x^3$ so the graphs of u and v have the same end behavior. As $x \to -\infty$, both $u(x)$ and $v(x) \to \infty$, and as $x \to \infty$, both $u(x)$ and $v(x) \to -\infty$.

The graphs have different y-intercepts, and u has three distinct zeros. The function v has a multiple zero at $x = 0$. See Figure 11.5.

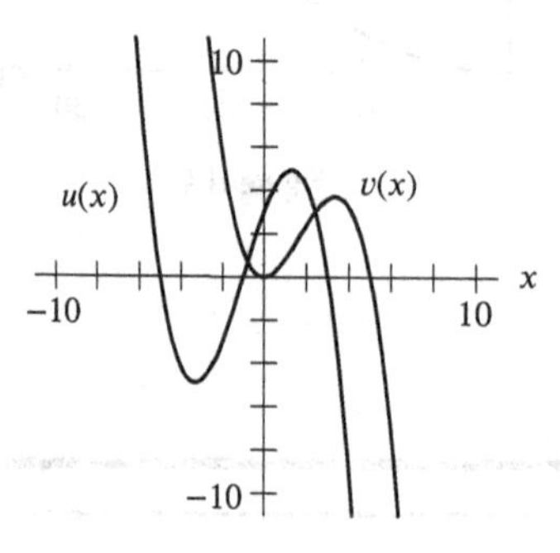

Figure 11.5

(b) On the window $-20 \le x \le 20$ by $-1600 \le y \le 1600$, the peaks and valleys of both functions are not distinguishable. Near the origin, the behavior of both functions looks the same. The functions are still distinguishable from one another on the ends.

On the window $-50 \le x \le 50$ by $-25{,}000 \le y \le 25{,}000$, the functions are still slightly distinct from one another on the ends—but barely.

On the last window the graphs of both functions appear identical. Both functions look like the function $y = -\frac{1}{5}x^3$.

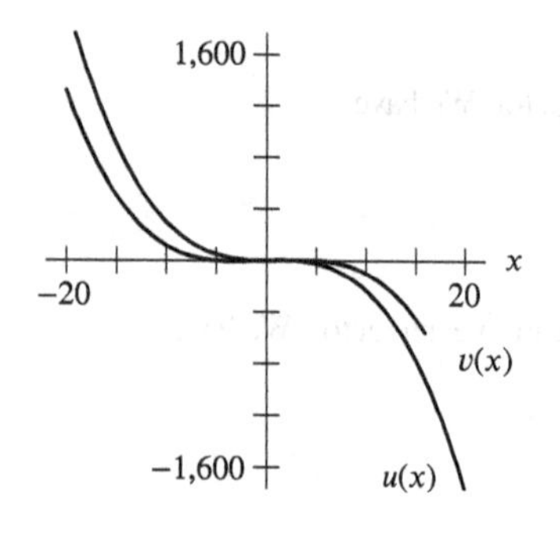

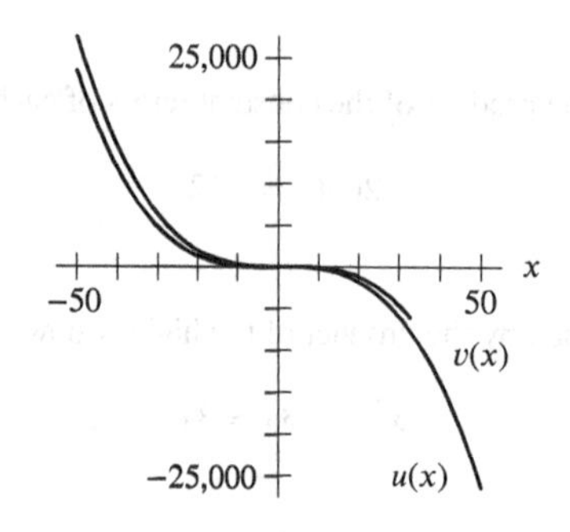

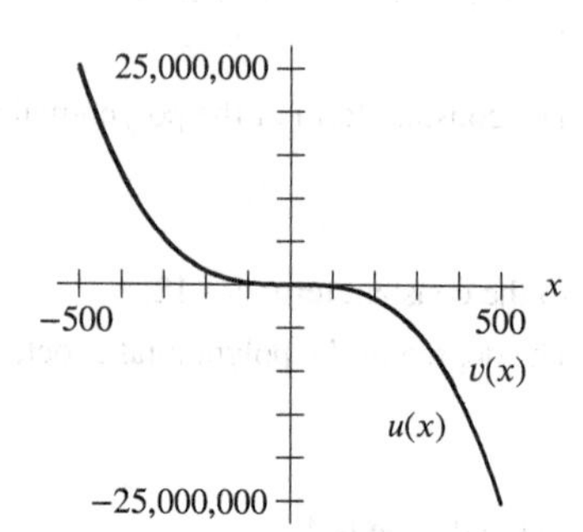

Figure 11.6

29. We have $-1.1 \le x \le -0.9$, $f(-1.1) \le y \le f(-0.9)$ or $-0.121 \le y \le 0.081$.

33. This is a fourth-degree polynomial whose leading coefficient is positive. Thus, in the long run its graph resembles Figure (II).

37. **(a)** Using a computer or a graphing calculator, we can get a picture of $f(x)$ like the one in Figure 11.7. On this window f appears to be invertible because it passes the horizontal line test.

(b) Substituting gives

$$f(0.5) = (0.5)^3 + 0.5 + 1 = 1.625.$$

To find $f^{-1}(0.5)$, we solve $f(x) = 0.5$. With a computer or graphing calculator, we trace along the graph of f in Figure 11.8 to find

$$f^{-1}(0.5) \approx -0.424.$$

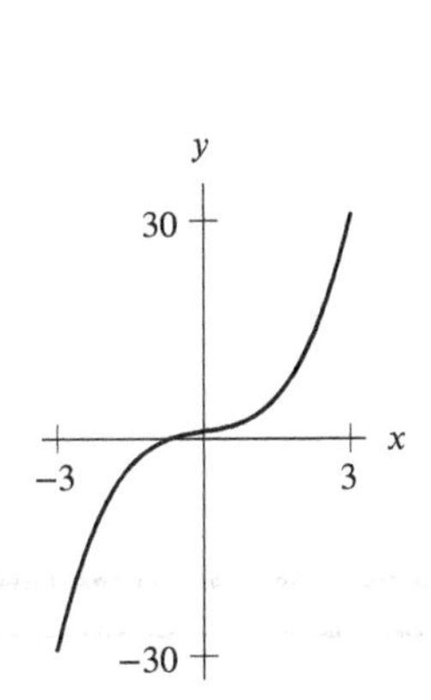

Figure 11.7: $f(x) = x^3 + x + 1$

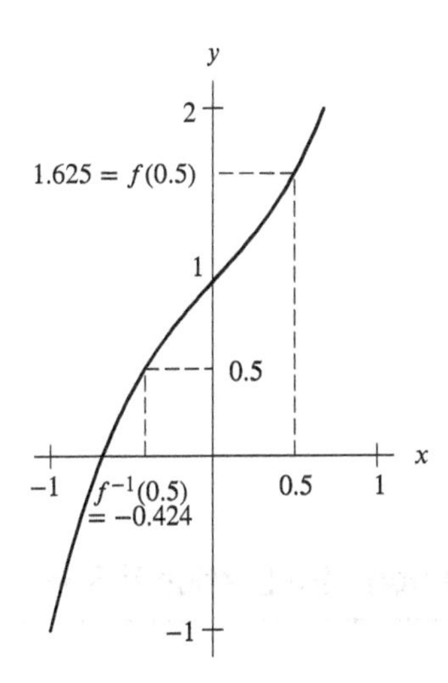

Figure 11.8

41. Yes. For the sake of illustration, suppose $f(x) = x^2 + x + 1$, a second-degree polynomial. Then

$$\begin{aligned} f(g(x)) &= (g(x))^2 + g(x) + 1 \\ &= g(x) \cdot g(x) + g(x) + 1. \end{aligned}$$

Since $f(g(x))$ is formed from products and sums involving the polynomial g, the composition $f(g(x))$ is also a polynomial. In general, $f(g(x))$ will be a sum of powers of $g(x)$, and thus $f(g(x))$ will be formed from sums and products involving the polynomial $g(x)$. A similar situation holds for $g(f(x))$, which will be formed from sums and products involving the polynomial $f(x)$. Thus, either expression will yield a polynomial.

45. **(a)** A graph of V is shown in Figure 11.9 for $0 \le t \le 5$, $0 \le V \le 1$.

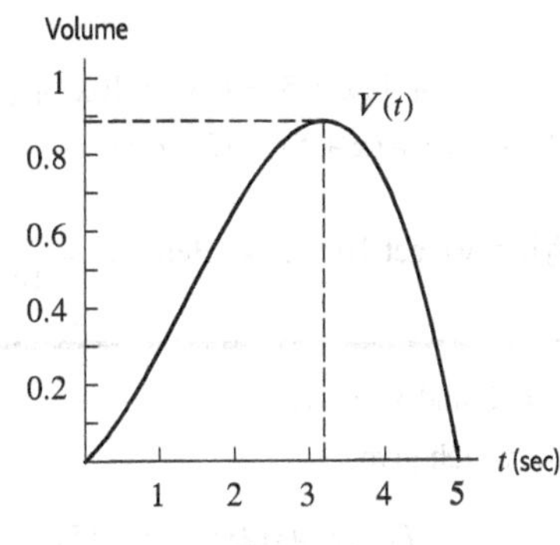

Figure 11.9

(b) The maximum value of V for $0 \le t \le 5$ occurs when $t \approx 3.195$, $V \approx .886$. Thus, at just over 3 seconds into the cycle, the lungs contain ≈ 0.88 liters of air.

(c) The volume is zero at $t = 0$ and again at $t \approx 5$. This indicates that at the beginning and end of the 5-second cycle the lungs are empty.

49. **(a)** Substituting $x = 0.5$ into p, we have

$$p(0.5) = 1 - 0.5 + 0.5^2 - 0.5^3 + 0.5^4 - 0.5^5 \approx 0.65625.$$

Since $f(0.5) = 2/3 = 0.6666....$, the approximation is accurate to 2 decimal places.

(b) We have $p(1) = 1 - 1 + 1 - 1 + 1 - 1 = 0$, but $f(1) = 0.5$. Thus $p(1)$ is a poor approximation to $f(1)$.

(c) See Figure 11.10. The two graphs are difficult to tell apart for $-0.5 \le x \le 0.5$, but for x outside this region the fit is not good.

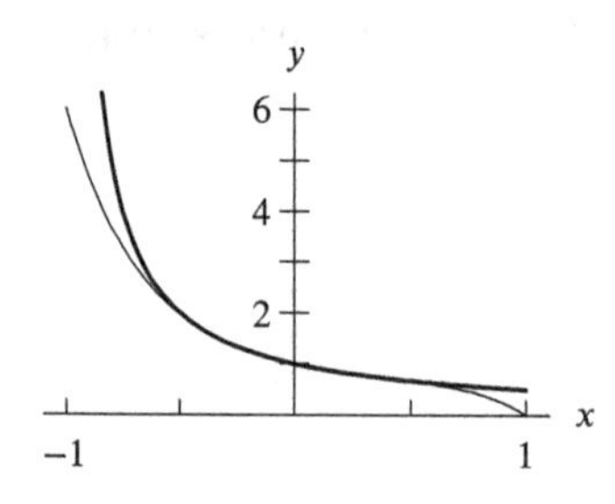

Figure 11.10

Solutions for Section 11.3

Skill Refresher

S1. Removing the common factor of x gives

$$x^2 + 10x = x(x + 10).$$

S5. Factoring out the common factor and then factoring the difference of two squares gives

$$27x - 3x^3 = 3x(9 - x^2) = 3x(3 + x)(3 - x).$$

S9. The leading term is $2x^2$, so the leading terms of the factors must have a product of $2x^2$. Rewriting, gives

$$2x^2 + 11x + 5 = (2x^2 + 10x) + (x + 5),$$
$$2x(x + 5) + (x + 5) = (2x + 1)(x + 5).$$

S13. We set $10 = k(3 - 1)(3 + 1)^2$. Simplifying, we get $10 = 32k$. Hence $k = \dfrac{5}{16}$.

EXERCISES

1. Zeros occur where $y = 0$, at $x = -3$, $x = 2$, and $x = -7$.

5. The function has a common factor of $4x$ which gives

$$f(x) = 4x(2x^2 - x - 15),$$

and the quadratic factor reduces further, giving

$$f(x) = 4x(2x + 5)(x - 3).$$

Thus, the zeros of f are $x = 0$, $x = \frac{-5}{2}$, and $x = 3$.

9. If $x = -2$ is a zero of f, then $bx - k$ is a factor with $\frac{k}{b} = -2$. Letting $k = -2$ and $b = 1$, one possible factor is $x + 2$. (Answer is not unique.)

13. This polynomial must be of fourth (or higher even-powered) degree, so either (but not both) of the zeros at $x = 2$ or $x = 5$ could be doubled. One possible formula is $y = k(x+2)(x-2)^2(x-5)$. Solving for k gives

$$\begin{aligned} k(0+2)(0-2)^2(0-5) &= 5 \\ -40k &= 5 \\ k &= -\frac{1}{8}, \end{aligned}$$

so

$$y = -\frac{1}{8}(x+2)(x-2)^2(x-5).$$

Another possible formula is $y = k(x+2)(x-2)(x-5)^2$. Solving for k gives

$$\begin{aligned} k(0+2)(0-2)(0-5)^2 &= 5 \\ -100k &= 5 \\ k &= -\frac{1}{20}, \end{aligned}$$

so

$$y = -\frac{1}{20}(x+2)(x-2)(x-5)^2.$$

There are other possible polynomials, but all are of degree higher than 4, so these are the simplest.

PROBLEMS

17. The graph represents a polynomial of even degree, at least fourth. Zeros are shown at $x = -2$, $x = -1$, $x = 2$, and $x = 3$. Since $f(x) \to -\infty$ as $x \to \infty$ or $x \to -\infty$, the leading coefficient must be negative. Thus, of the choices in the table, only C and E are possibilities. When $x = 0$, function C gives

$$y = -\frac{1}{2}(2)(1)(-2)(-3) = -\frac{1}{2}(12) = -6,$$

and function E gives

$$y = -(2)(1)(-2)(-3) = -12.$$

Since the y-intercept appears to be $(0, -6)$ rather than $(0, -12)$, function C best fits the polynomial shown.

21. The function f has zeros at $x = 0$ and 3. Thus, let $f(x) = kx(x-3)$. Use $f(-2) = 1$ to solve for k: $f(-2) = k(-2)(-2-3) = -10k$. Thus $-10k = 1$ and $k = -\frac{1}{10}$. This gives

$$f(x) = -\frac{1}{10}x(x-3).$$

25. To pass through the given points, the polynomial must be of at least degree 2. Thus, let f be of the form

$$f(x) = ax^2 + bx + c.$$

Then using $f(0) = 0$ gives

$$a(0)^2 + b(0) + c = 0,$$

so $c = 0$. Then, with $f(2) = 0$, we have

$$\begin{aligned} a(2)^2 + b(2) + 0 &= 0 \\ 4a + 2b &= 0 \\ \text{so} \quad b &= -2a. \end{aligned}$$

Using $f(3) = 3$ and $b = -2a$ gives

$$a(3)^2 + (-2a)(3) + 0 = 3$$

so

$$\begin{aligned} 9a - 6a &= 3 \\ 3a &= 3 \\ a &= 1. \end{aligned}$$

Thus, $b = -2a$ gives $b = -2$. The unique polynomial of degree ≤ 2 which satisfies the given conditions is $f(x) = x^2 - 2x$.

29. $y = 4x^2 - 1 = (2x - 1)(2x + 1)$, which implies that $y = 0$ for $x = \pm\frac{1}{2}$.

33. $y = 4x^2 + 1 = 0$ implies that $x^2 = -\frac{1}{4}$, which has no solutions. There are no real zeros.

37. To obtain the flattened effect of the graph near $x = 0$, let $x = 0$ be a multiple zero (of odd multiplicity). Thus, a possible choice would be $f(x) = kx^3(x + 1)(x - 2)$ for $k > 0$.

41. We see that h has zeros at $x = -2$, $x = -1$ (a double zero), and $x = 1$. Thus, $h(x) = k(x + 2)(x + 1)^2(x - 1)$. Then $h(0) = (2)(1)^2(-1)k = -2k$, and since $h(0) = -2$, $-2k = -2$ and $k = 1$. Thus,

$$h(x) = (x + 2)(x + 1)^2(x - 1)$$

is a possible formula for h.

45. **(a)** We could think of $f(x) = (x - 2)^3 + 4$ as $y = x^3$ shifted right 2 units and up 4. Thus, since $y = x^3$ is invertible, f should also be. Algebraically, we let

$$y = f(x) = (x - 2)^3 + 4.$$

Thus,

$$\begin{aligned} y - 4 &= (x - 2)^3 \\ \sqrt[3]{y - 4} &= x - 2 \\ \sqrt[3]{y - 4} + 2 &= x \end{aligned}$$

So $f(x)$ is invertible with an inverse

$$f^{-1}(x) = \sqrt[3]{x - 4} + 2.$$

(b) Since g is not so obvious, we might begin by graphing $y = g(x)$. Figure 11.11 shows that the function $g(x) = x^3 - 4x^2 + 2$ does not satisfy the horizontal line test, so g is not invertible.

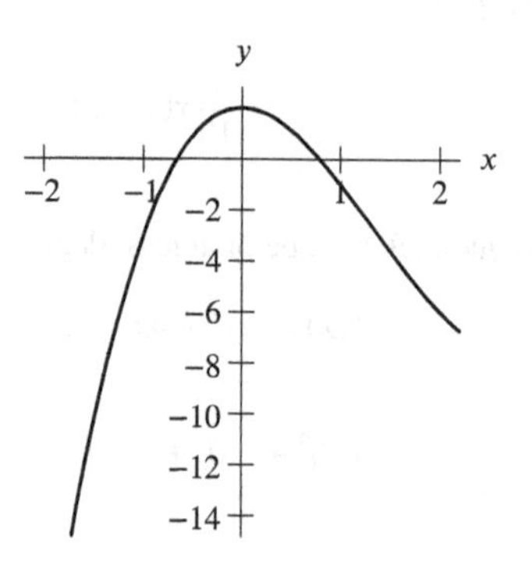

Figure 11.11

49. **(a)** $V(x) = x(6 - 2x)(8 - 2x)$

(b) Values of x for which $V(x)$ makes sense are $0 < x < 3$, since if $x < 0$ or $x > 3$ the volume is negative.

(c) See Figure 11.12.

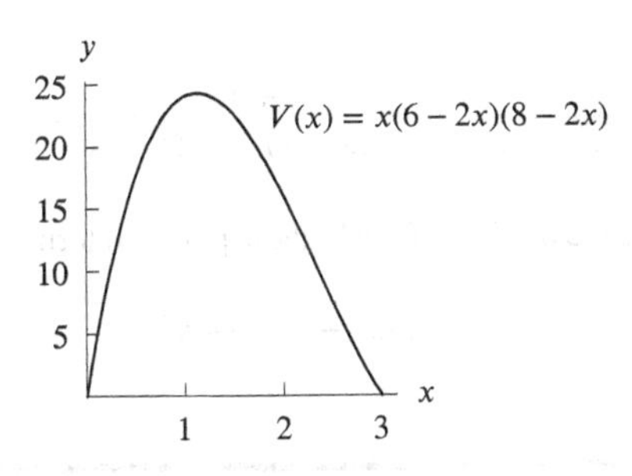

Figure 11.12

(d) Using a graphing calculator, we find the peak between $x = 0$ and $x = 3$ to occur at $x \approx 1.13$. The maximum volume is ≈ 24.26 in^3.

53. The domain is $x \geq c$ and $a \leq x \leq b$. Taking the hint, we see that the function $y = (x-a)(x-b)(x-c)$ has zeros at a, b, c and long-run behavior like $y = x^3$. Thus, the graph of $y = (x-a)(x-b)(x-c)$ looks something like the graph in the figure (though of course the zeros may be spaced differently). We see that $y < 0$ on the intervals $x < a$ and $b < x < c$. Thus, the function $y = \sqrt{(x-a)(x-b)(x-c)}$ is undefined on these intervals, but is defined everywhere else.

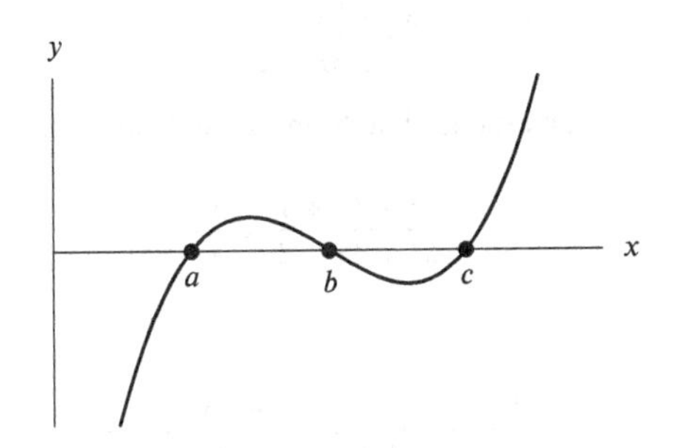

Figure 11.13

Solutions for Section 11.4

Skill Refresher

S1. Simplifying x^4/x gives

$$\frac{7x^4}{3x} = \frac{7x^3}{3}.$$

S5. Finding a common denominator gives

$$5 + \frac{3}{x+2} = \frac{5(x+2)}{x+2} + \frac{3}{x+2} = \frac{5x+10}{x+2} + \frac{3}{x+2} = \frac{5x+13}{x+2}.$$

S9.

$$\begin{aligned}\frac{5}{(x-2)^2(x+1)} - \frac{18}{(x-2)} &= \frac{5 - 18(x-2)(x+1)}{(x-2)^2(x+1)}\\ &= \frac{5 - 18x^2 + 18x + 36}{(x-2)^2(x+1)0}\\ &= \frac{-18x^2 + 18x + 41}{(x-2)^2(x+1)}\end{aligned}$$

S13. $\dfrac{x^{-1}+x^{-2}}{1-x^{-2}} = \dfrac{\frac{1}{x}+\frac{1}{x^2}}{1-\frac{1}{x^2}} = \dfrac{\frac{x+1}{x^2}}{\frac{x^2-1}{x^2}} = \dfrac{x+1}{x^2}\cdot\dfrac{x^2}{x^2-1} = \dfrac{x+1}{(x+1)(x-1)} = \dfrac{1}{x-1}.$

S17. The leading term of the polynomial is the product of the highest-powered terms in the factors of the polynomial. Thus, the leading term is

$$(4x)(-x) = -4x^2.$$

EXERCISES

1. This is a rational function, and it is already in the form of one polynomial divided by another.

5. This is not a rational function, as we cannot put it in the form of one polynomial divided by another, since $\sqrt{x}+1$ is not a polynomial.

9. Rewriting this in the form $p(x)/q(x)$ where $p(x)$ and $q(x)$ are polynomials in standard form, we have:

$$\begin{aligned}\frac{1}{x+2}+\frac{x}{x+1} &= \frac{1}{x+2}\cdot\frac{x+1}{x+1}+\frac{x}{x+1}\cdot\frac{x+2}{x+2}\\ &= \frac{x+1}{(x+1)(x+2)}+\frac{x(x+2)}{(x+1)(x+2)}\\ &= \frac{x^2+3x+1}{x^2+3x+2}.\end{aligned}$$

The leading term in the numerator is x^2 and the leading term in the denominator is x^2, so for large enough values of x (either positive or negative),

$$\frac{x^2+3x+1}{x^2+3x+2} \approx \frac{x^2}{x^2}.$$

Since $\dfrac{x^2}{x^2} \to 1$ as $x \to \pm\infty$ we see that

$$\frac{1}{x+2}+\frac{x}{x+1} = \frac{x^2+3x+1}{x^2+3x+2} \to 1$$

as $x \to \pm\infty$.

13. As $x \to \infty$, we see that $y \to -\dfrac{3x^2}{x^2}$, so the long-run behavior is $y \to -3$. This matches graph (III).

17. We have

$$\lim_{x\to\infty}(2x^{-3}+4) = \lim_{x\to\infty}(2/x^3+4) = 0+4 = 4.$$

21. As $x \to \pm\infty$, $1/x \to 0$ and $x/(x+1) \to 1$, so $h(x)$ approaches $3-0+1=4$. Therefore $y=4$ is the horizontal asymptote.

PROBLEMS

25. Let $y = f(x)$. Then $x = f^{-1}(y)$. Solving for x,

$$\begin{aligned} y &= \frac{4-3x}{5x-4}\\ y(5x-4) &= 4-3x\\ 5xy-4y &= 4-3x\\ 5xy+3x &= 4y+4\\ x(5y+3) &= 4y+4 \quad \text{(factor out an } x)\\ x &= \frac{4y+4}{5y+3}\end{aligned}$$

Therefore,

$$f^{-1}(x) = \frac{4x+4}{5x+3}.$$

29. **(a)** The time to travel the first 10 miles is $\frac{10}{40} = 0.25$ hours. The time for the remaining 50 miles in $50/V$ hours so the total journey time is $T = 0.25 + 50/V$. Thus, the average speed is

$$\text{Average speed} = \frac{60}{T} = \frac{60}{\left(0.25 + \frac{50}{V}\right)} = \frac{240V}{V + 200}.$$

(b) If you want to average 60 mph for the trip then you need

$$\frac{240V}{V+200} = 60.$$

Solving this equation gives $V = 200/3$ mph, nearly 70 mph.

33. **(a)** We have

$$f(x) = \frac{\text{Amount of Alcohol}}{\text{Amount of Liquid}} = \frac{x}{x+5}.$$

(b) Substituting $x = 7$ gives

$$f(7) = \frac{7}{7+5} = \frac{7}{12} \approx 58.333\%.$$

The quantity $f(7)$ is the concentration of alcohol in a solution consisting of 5gallons of water and 7 gallons of alcohol.

(c) If $f(x) = 0$, then

$$\frac{x}{x+5} = 0 \quad \text{so} \quad x = 0.$$

The concentration of alcohol is 0% when there is no alcohol in the solution, that is, when $x = 0$.

(d) The horizontal asymptote is given by the ratio of the highest-power terms of the numerator and denominator:

$$y = \frac{x}{x} = 1 = 100\%$$

This means that as the amount of alcohol added, x, grows large, the concentration of alcohol in the solution approaches 100%.

37. Line l_1 has a smaller slope than line l_2. We know the slope of line l_1 represents the average cost of producing n_1 units, and the slope of l_2 represents the average cost of producing n_2 units. Thus, the average cost of producing n_2 units is more than that of producing n_1 units. For these goods, the average cost actually goes up between n_1 and n_2 units.

Solutions for Section 11.5

Skill Refresher

S1. Substituting $x = -4$, gives

$$\frac{3(-4)}{-4+2} = \frac{-12}{-2} = 6.$$

S5. Substituting $x = 6$, gives

$$\frac{(6-2)^2}{6^2 - 4(6) + 4} = \frac{4^2}{36 - 24 + 4} = \frac{16}{16} = 1.$$

S9. Since the numerator is in factored form, we set each factor equal to zero, and we have

$$x + 1 = 0 \text{ so } x = -1,$$
$$\text{and } x - 12 = 0 \text{ so } x = 12.$$

Thus, $x = -1$ and $x = 12$ make the numerator equal to zero.

S13. Setting the denominator equal to zero gives

$$\begin{aligned} 2x + 7 &= 0 \\ 2x &= -7, \\ x &= \frac{-7}{2}. \end{aligned}$$

S17. We have

$$\begin{aligned} f(0) &= 1 \\ k \cdot \frac{0+1}{0-1} &= 1 \\ k(-1) &= 1 \\ k &= -1. \end{aligned}$$

S21. We have

$$\begin{aligned} f(0) &= 1 \\ k \cdot \frac{(0-1)(0+2)}{3(0-2)^2} &= 1 \\ k \cdot \frac{-2}{12} &= 1 \\ \frac{-2k}{12} &= 1 \\ k &= \frac{12}{-2} = -6. \end{aligned}$$

EXERCISES

1. The zero of this function is at $x = 4$. It has vertical asymptotes at $x = \pm 3$. Its long-run behavior is: $y \to 0$ as $x \to \pm\infty$. See Figure 11.14.

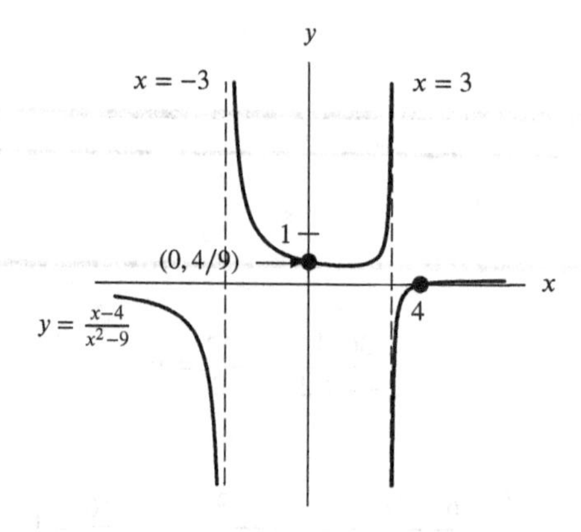

Figure 11.14

5. Since

$$g(x) = \frac{x^2 - 4}{x^3 + 4x^2} = \frac{(x-2)(x+2)}{x^2(x+4)},$$

the x-intercepts are $x = \pm 2$; there is no y-intercept; the horizontal asymptote is $y = 0$; the vertical asymptotes are $x = 0, x = -4$.

9. **(a)** $y = -\frac{1}{(x-5)^2} - 1$ has a vertical asymptote at $x = 5$, no x intercept, horizontal asymptote $y = -1$: (iii)

(b) $y = \frac{x-2}{(x+1)(x-3)}$ has vertical asymptotes at $x = -1, 3$, x intercept at 2, horizontal asymptote $y = 0$: (i)

(c) $y = \frac{2x+4}{x-1}$ has a vertical asymptote at $x = 1$, x intercept at $x = -2$, horizontal asymptote $y = 2$: (ii)

(d) $y = \frac{x-3+x+1}{(x+1)(x-3)} = \frac{2x-2}{(x+1)(x-3)}$ has vertical asymptotes at $x = -1, 3$, x intercept at 1, horizontal asymptote at $y = 0$: (iv)

(e) $y = \frac{(1+x)(1-x)}{x-2}$ has vertical asymptote at $x = 2$, two x intercepts at ± 1: (vi)

(f) $y = \frac{1-4x}{2x+2}$ has a vertical asymptote at $x = -1$, x intercept at $x = \frac{1}{4}$, horizontal asymptote at $y = -2$: (v)

PROBLEMS

13. **(a)** To estimate

$$\lim_{x\to 2^+} \frac{5-x}{(x-2)^2},$$

we consider what happens to the function when x is slightly larger than 2. The numerator is positive and the denominator is positive and is approaching 0 as x approaches 2. We suspect that $\frac{5-x}{(x-2)^2}$ gets larger and larger as x approaches 2 from the right. We can also use either a graph or a table of values as in Table 11.3 to estimate this limit. We see that

$$\lim_{x\to 2^+} \frac{5-x}{(x-2)^2} = +\infty.$$

Table 11.3

x	2.1	2.01	2.001	2.0001
$f(x)$	290	29900	2999000	299990000

(b) To estimate

$$\lim_{x\to 2^-} \frac{5-x}{(x-2)^2},$$

we consider what happens to the function when x is slightly smaller than 2. The numerator is positive and the denominator is positive and is approaching 0 as x approaches 2. We suspect that $\frac{5-x}{(x-2)^2}$ gets larger and larger as x approaches 2 from the left. We can also use either a graph or a table of values to estimate this limit. We see that

$$\lim_{x\to 2^-} \frac{5-x}{(x-2)^2} = +\infty.$$

17. **(a)** If $f(n)$ is large, then $\frac{1}{f(n)}$ is small.

(b) If $f(n)$ is small, then $\frac{1}{f(n)}$ is large.

(c) If $f(n) = 0$, then $\frac{1}{f(n)}$ is undefined.

(d) If $f(n)$ is positive, then $\frac{1}{f(n)}$ is also positive.

(e) If $f(n)$ is negative, then $\frac{1}{f(n)}$ is negative.

21. **(a)** The graph shows $y = 1/x$ flipped across the x-axis and shifted left 2 units. Therefore

$$y = -\frac{1}{x+2}$$

is a choice for a formula.

(b) The formula $y = -1/(x+2)$ is already written as a ratio of two linear functions.

(c) The graph has a y-intercept if $x = 0$. Thus, $y = -\frac{1}{2}$. Since y cannot be zero if $y = -1/(x+2)$, there is no x-intercept. The only intercept is $(0, -\frac{1}{2})$.

25. **(a)** The graph appears to be $y = 1/x^2$ shifted 3 units to the right and flipped across the x-axis. Thus,

$$y = -\frac{1}{(x-3)^2}$$

is a possible formula for it.

(b) The equation $y = -1/(x-3)^2$ can be written as

$$y = \frac{-1}{x^2 - 6x + 9}.$$

(c) Since y can not equal zero if $y = -1/(x^2 - 6x + 9)$, the graph has no x-intercept. The y-intercept occurs when $x = 0$, so $y = \frac{-1}{(-3)^2} = -\frac{1}{9}$. The y-intercept is at $(0, -\frac{1}{9})$.

29. **(a)** The table indicates symmetry about the y-axis. The fact that the function values have the same sign on both sides of the vertical asymptotes indicates a transformation of $y = 1/x^2$ rather than $y = 1/x$.

(b) As x takes on large positive or negative values, $y \to 1$. Thus, we try

$$y = \frac{1}{x^2} + 1.$$

This formula works and can be expressed as

$$y = \frac{1 + x^2}{x^2}.$$

33. The graph is the graph of $y = 1/x$ moved up by 2. See Figure 11.15.

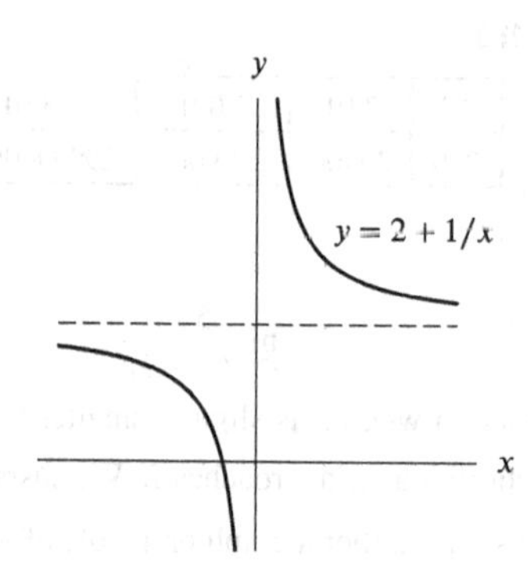

Figure 11.15

37. A guess of

$$y = \frac{(x-3)(x+2)}{(x+1)(x-2)}$$

fits the zeros and vertical asymptote of the graph. However, in order to satisfy the y-intercept at $(0, -3)$ and end behavior of $y \to -1$ as $x \to \pm\infty$, the graph should be "flipped" across the x-axis. Thus, we have

$$y = -\frac{(x-3)(x+2)}{(x+1)(x-2)}.$$

41.

$$f(x) = \frac{p(x)}{q(x)} = \frac{-3(x-2)(x-3)}{(x-5)^2}$$

We need the factor of -3 in the numerator and the exponent of 2 in the denominator, because we have a horizontal asymptote of $y = -3$. The ratio of highest term of $p(x)$ to highest term of $q(x)$ will be $\frac{-3x^2}{x^2} = -3$.

45. **(a)** Both graphs approach zero as $x \to \infty$ or $x \to -\infty$. Both have two vertical asymptotes. Both graphs are positive, increasing, and concave up to the left of their leftmost asymptote, both are negative and ∩-shaped between their vertical asymptotes, and both are positive, decreasing and concave up to the right of their rightmost asymptote. Except for the location of the vertical asymptotes, the graphs are quite similar.

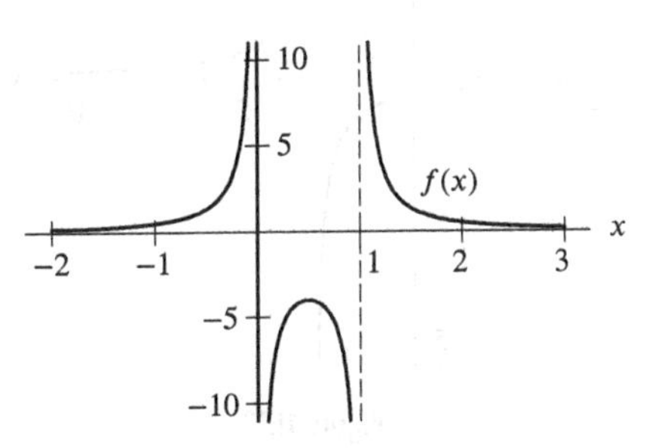

Figure 11.16

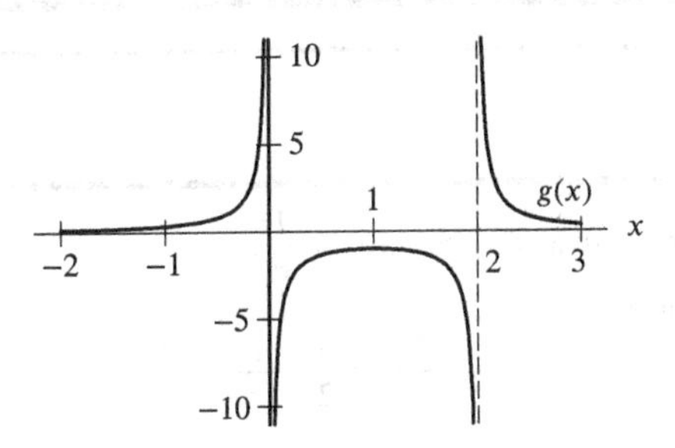

Figure 11.17

(b) We have

$$
\begin{aligned}
4g(2x) &= 4\frac{1}{(2x)(2x-2)} \\
&= \frac{4}{4x(x-1)} \\
&= \frac{1}{x(x-1)} = f(x).
\end{aligned}
$$

(c) To transform the graph of $g(x)$ into the graph of $g(2x)$, compress the horizontal scale by a factor of 2. Then to transform the graph of $g(2x)$ into the graph of $4g(2x) = f(x)$, stretch the vertical scale by a factor of 4.

49. Writing

$$
\begin{aligned}
y &= \frac{x-3}{x^2-6x+9} \\
&= \frac{x-3}{(x-3)^2} \\
&= \frac{1}{x-3} \qquad \text{for } x \neq 3,
\end{aligned}
$$

we see that y is undefined at $x = 3$. However, the behavior near $x = 3$ is like $y = 1/(x-3)$, so the graph has a vertical asymptote at $x = 3$, not a hole. See Figure 11.18.

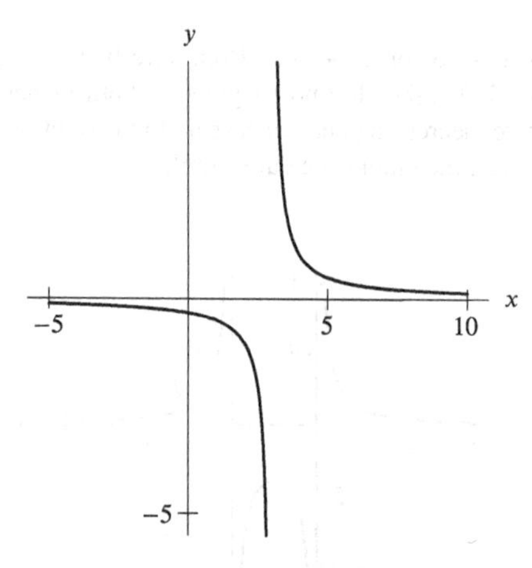

Figure 11.18

Solutions for Section 11.6

Skill Refresher

S1. By definition these expressions are equivalent, since $x^{-a} = \frac{1}{x^a}$.

S5. The expressions are equivalent since

$$\frac{3}{2x^{-4}} = \frac{3}{2} \cdot \frac{1}{x^{-4}} = \frac{3}{2}x^4.$$

S9. We have

$$\begin{aligned} x^2 &= 20x \\ x^2 - 20x &= 0 \\ x(x-20) &= 0, \end{aligned}$$

so our solutions are $x = 0$ and $x = 20$.

S13. We have

$$\begin{aligned} x^2 &= 10x^4 \\ 10x^4 - x^2 &= 0 \\ x^2(10x^2 - 1) &= 0, \end{aligned}$$

so either $x^2 = 0$ or $10x^2 - 1 = 0$. If $x^2 = 0$, then $x = 0$. On the other hand, if $10x^2 - 1 = 0$, then

$$\begin{aligned} 10x^2 &= 1 \\ x^2 &= \frac{1}{10} \\ x &= \pm\sqrt{\frac{1}{10}}. \end{aligned}$$

Therefore, our solutions are $x = 0$, $\sqrt{1/10}$, and $-\sqrt{1/10}$.

EXERCISES

1. Larger powers of x give smaller values for $0 < x < 1$.
A - (iii)
B - (ii)
C - (iv)
D - (i)

5. The function is exponential, because $p(x) = (5^x)^2 = 5^{2x} = (5^2)^x = 25^x$.

9. The function fits an exponential, because $r(x) = 2 \cdot 3^{-2x} = 2(3^{-2})^x = 2(\frac{1}{9})^x$.

13. Since 0.99^x is a decreasing exponential function, $y = 7 \cdot 0.99^x \to 0$ as $x \to \infty$, so $y = 6x^{35}$ dominates.

PROBLEMS

17. **(a)** We find the intersection points of the two graphs by setting the y-values equal:

$$x^3 = 100x^2$$
$$x^2(x - 100) = 0$$
$$x = 0 \quad \text{and} \quad x = 100.$$

To show the points of intersection, the viewing window must include $x = 0$ and $x = 100$. We choose a window with $0 \leq x \leq 120$. Since $x = 120$ gives output values $y = 120^3 = 1{,}728{,}000$ and $y = 100 \cdot 120^2 = 1{,}440{,}000$, we can choose the window with $0 \leq y \leq 1{,}500{,}000$. See Figure 11.19.

(b) Note that $x^3 = x \cdot x^2$. Therefore, for $x > 100$, we see that $x \cdot x^2 > 100x^2$, so x^3 dominates $100x^2$.

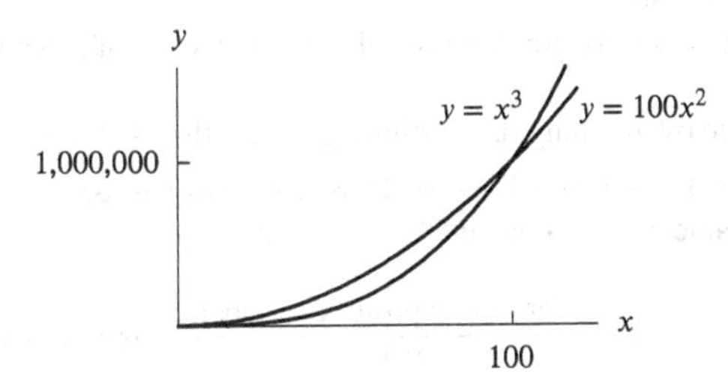

Figure 11.19: Dominance of $y = x^3$ over $y = 100x^2$

21. Since $s < r < 0$, we know that these are negative-powered power functions, so their graphs have horizontal asymptotes at $y = 0$ and vertical asymptotes at $x = 0$.

As $x \to \infty$, the graph of g approaches its horizontal asymptote more rapidly than the graph of f, because $s < r$; that is, s is "more negative" than r. Thus, the graph of g should lie below the graph of f for large x-values, as it does at $x = x_1$. Thus, $x_1 \geq x_0$.

25. Table 11.4 shows that 3^{-x} approaches zero faster than x^{-3} as $x \to \infty$.

Table 11.4

x	2	10	100
3^{-x}	1/9	0.000017	1.94×10^{-48}
x^{-3}	1/8	0.001	10^{-6}

29. **(a)** If f is linear,

$$m = \frac{128 - 16}{2 - 1} = 112,$$

and

$$16 = 112(1) + b, \qquad \text{so} \qquad b = -96.$$

Thus,

$$f(x) = 112x - 96.$$

(b) If f is exponential, then

$$\frac{128}{16} = \frac{a(b)^2}{a(b)} = b, \qquad \text{so} \qquad b = 8$$

and

$$16 = a(8), \qquad \text{so} \qquad a = 2.$$

Therefore

$$f(x) = 2(8)^x.$$

(c) If f is a power function, $f(x) = k(x)^p$. Then

$$\frac{f(2)}{f(1)} = \frac{k(2)^p}{k(1)^p} = (2)^p = \frac{128}{16} = 8,$$

so $p = 3$. Using $f(1) = 16$ to solve for k, we have

$$16 = k(1^3), \qquad \text{so} \qquad k = 16.$$

Thus,

$$f(x) = 16x^3.$$

33. **(a)** We are given $t(v) = v^{-2} = 1/v^2$ and $r(v) = 40v^{-3} = 40/v^3$; therefore

$$\frac{1}{v^2} = \frac{40}{v^3}.$$

Multiplying by v^3, we get $v = 40$.

(b) We found in part (a) that $v = 40$. By graphing or substituting values of x between 0 and 40, we see that $r(x) > t(x)$ for $0 < x < 40$.

(c) For values of $x > 40$ we see by graphing or substituting values that $t(x) > r(x)$.

37. For large positive t, the value of $3^{-t} \to 0$ and $4^t \to \infty$. Thus, $y \to 0$ as $t \to \infty$.

For large negative t, the value of $3^{-t} \to \infty$ and $4^t \to 0$. Thus,

$$y \to \frac{\text{Very large positive number}}{0+7} \qquad \text{as } t \to -\infty.$$

So $y \to \infty$ as $t \to -\infty$.

41. Since e^t and t^2 both dominate $\ln|t|$, we have $y \to \infty$ as $t \to \infty$.

For large negative t, the value of $e^t \to 0$, but t^2 is large and dominates $\ln|t|$. Thus, $y \to \infty$ as $t \to -\infty$,

45. Since e^{3t} dominates e^{2t}, the value of y is very small for large positive t. Thus, $y \to 0$ as $t \to \infty$.

For large negative t, the value of $e^{2t} \to 0$, so $y \to 0$ as $t \to -\infty$.

Solutions for Section 11.7

Skill Refresher

S1. We have

$$\begin{aligned} 2x^4 &= 8 \\ x^4 &= 4 \\ x &= \pm(4)^{1/4} = \pm 1.141. \end{aligned}$$

S5. We have

$$\begin{aligned} 3(2)^x &= 12 \\ (2)^x &= \frac{12}{3} = 4 \\ x &= 2. \end{aligned}$$

S9. We first use the points to find the slope m:

$$m = \frac{y_2 - y_1}{x_2 - x_1} = \frac{5-2}{3-0} = \frac{3}{3} = 1.$$

Next we use the equation:

$$y = b + mx.$$

Substituting 1 for m, we have

$$y = b + x.$$

Using the point $(0, 2)$, we have:

$$2 = b + (0)$$
$$2 = b$$

so,

$$y = 2 + x.$$

S13. Since this function is exponential, we know $y = ab^x$ with $b > 0$. We also know that $(0, 2)$ and $(2, 18)$ are on the graph of this function, so

$$2 = ab^0$$

and

$$18 = ab^2.$$

From the first equation, we see that $a = 2$. Substituting into the second equation, we get

$$18 = 2b^2$$
$$9 = b^2$$
$$3 = b$$

Thus, $y = 2(3)^x$.

EXERCISES

1. Judging from the figure, an exponential function might best model this data.
5. $f(x) = ax^p$ for some constants a and c. Since $f(1) = 1 = a(1)^p$, it follows that $a = 1$. Also, $f(2) = 2^p = c$. Solving for p, we have $p = \ln c / \ln 2$. Thus, $f(x) = x^{\ln c/\ln 2}$.
9. **(a)** (i) Calculator result: $y = 46.79t^{0.301}$. Answers may vary.

   (ii) Calculator result: $y = 0.822t^2 - 6.282t + 76.53$. Answers may vary.

   (b) For the time period 2002–2012, the quadratic function is the better fit. The power function is approximately the square root function, so it is concave down, but the catch values increase rapidly toward the end of this period. Notice that the power function goes through $(0, 0)$, meaning that the predicted value of the 2000 catch is zero, which is not a realistic prediction. The quadratic function is shifted and stretched, so is the better fit. See Figure 11.20. However, outside of the interval 2002–2012, there is no reason to suppose that either function is a good fit. Our results hold only for this time period.

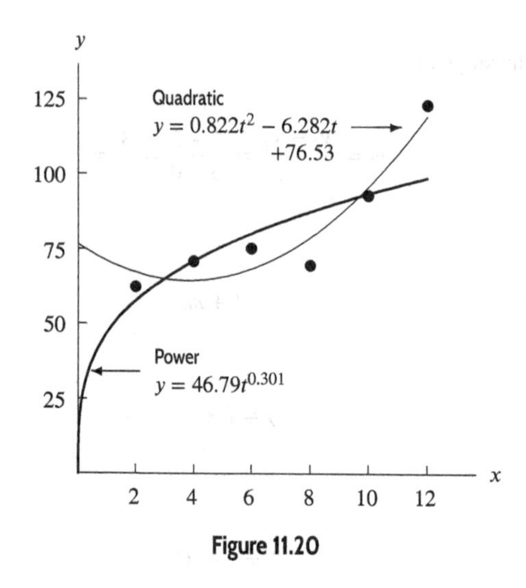

Figure 11.20

13. The slope of this line is $m = \frac{y_2-y_1}{x_2-x_1} = \frac{3}{2}$. The vertical intercept is 0, thus $y = \frac{3}{2}x$.

PROBLEMS

17. (a) The function $y = -83.039 + 61.514x$ gives a superb fit, with correlation coefficient $r = 0.99997$.
    (b) When the power function is plotted for $2 \leq x \leq 2.05$, it resembles a line. This is true for most of the functions we have studied. If you zoom in close enough on any given point, the function begins to resemble a line. However, for other values of x (say, $x = 3, 4, 5 \ldots$), the fit no longer holds.

21. (a) A computer or calculator gives

$$N = -14t^4 + 433t^3 - 2255t^2 + 5634t - 4397.$$

    (b) The graph of the data and the quartic in Figure 11.21 shows a good fit between 1980 and 1996.

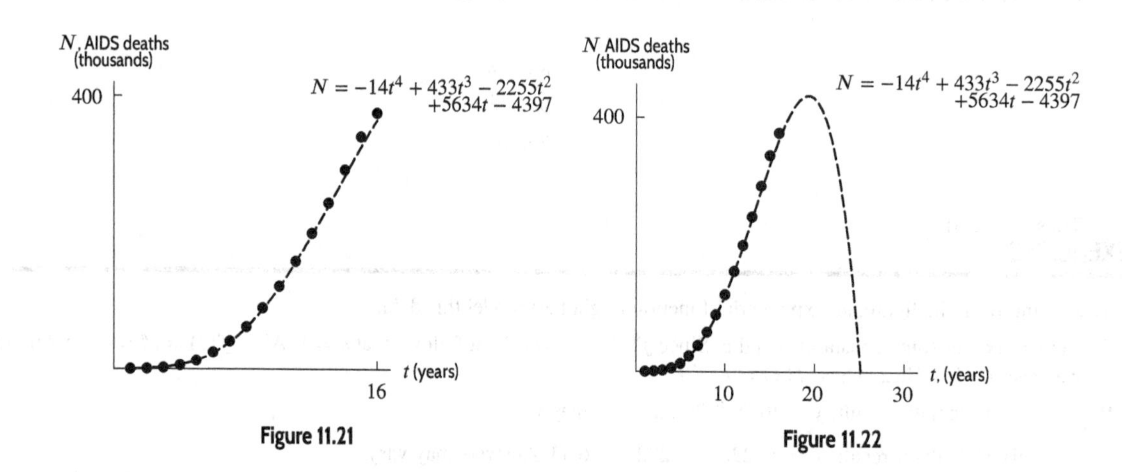

Figure 11.21

Figure 11.22

    (c) Figure 11.21 shows that the quartic model fits the 1980-1996 data well. However, this model predicts that in 2000, the number of deaths decreases. See Figure 11.22. Since N is the total number of deaths since 1980, this is impossible. Therefore the quartic is definitely not a good model for $t > 20$.

25. (a) See Figure 11.23.

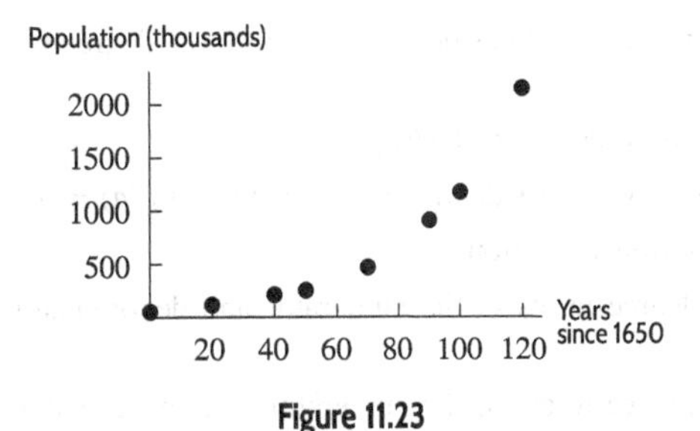

Figure 11.23

(b) Using a calculator or computer, we get $P(t) = 56.108(1.031)^t$. Answers may vary.

(c) The 56.108 represents a population of 56,108 people in 1650. Note that this is more than the actual population of 50,400. The growth factor of 1.031 means the rate of growth is approximately 3.1% per year.

(d) We find $P(100) = 1194.308$, which is slightly higher than the given data value of 1,170.8.

(e) The estimated population, $P(150) = 5510.118$, is higher than the given census population.

29. **(a)** The formula is $N = 1184.6e^{0.3625t}$. See Figure 11.24.

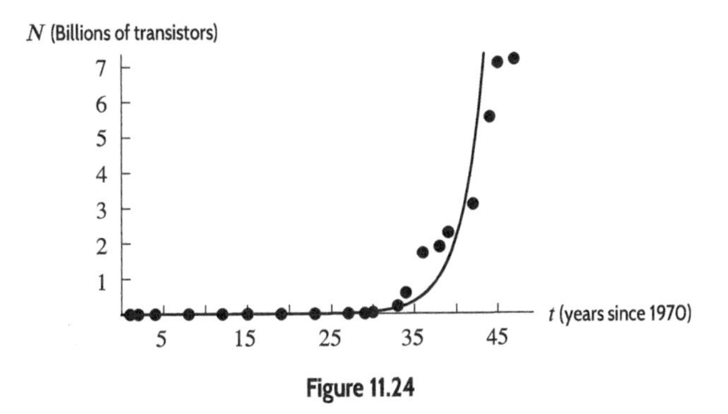

Figure 11.24

(b) The doubling time is given by $\ln 2/0.3625 \approx 1.912$. This is consistent with *Moore's Law*, which states that the number of transistors doubles about once every two years. Dr. Gordon E. Moore is Chairman Emeritus of Intel Corporation According to the Intel Corporation, "Gordon Moore made his famous observation in 1965, just four years after the first planar integrated circuit was discovered. The press called it 'Moore's Law'[1] and the name has stuck. In his original paper, Moore observed an exponential growth in the number of transistors per integrated circuit and predicted that this trend would continue."

33. **(a)** Quadratic is the only choice that increases and then decreases to match the data.

(b) Using ages of $x = 20, 30, \ldots, 80$, a quadratic function is $y = -38.2792x^2 + 3965.24x - 45{,}017.8$. Answers may vary.

(c) The value of the function at 37 is $y = -38.2792 \cdot 37^2 + 3965.24 \cdot 37 - 45{,}017.8 = \$49{,}292$.

(d) The value of the function for age 10 is $-38.2792 \cdot 10^2 + 3965.24 \cdot 10 - 45{,}017.8 = -\$9193.29$. Answers may vary. Not reasonable, as income is positive. In addition, 10-year-olds do not usually work.

STRENGTHEN YOUR UNDERSTANDING

1. False. The quadratic function $y = 3x^2 + 5$ is not of the form $y = kx^n$, so it is not a power function.

5. True. All positive even power functions have an upward opening U shape.

9. True. The x-axis is an asymptote for $f(x) = x^{-1}$, so the values approach zero.

13. False. As x grows very large the exponential decay function g approaches the x-axis faster than any power function with a negative power.

[1] The Intel Corporation, www.intel.com/museum/archives/history_docs/mooreslaw.htm, intel.com, accessed March 23, 2017.

17. False. For example, the polynomial $x^2 + x^3$ has degree 3 because the degree is the highest power, not the first power, in the formula for the polynomial.

21. True. The graph crosses the y-axis at the point $(0, p(0))$.

25. True. We can write $p(x) = (x - a) \cdot C(x)$. Evaluating at $x = a$, we get $p(a) = (a - a) \cdot C(a) = 0 \cdot C(a) = 0$.

29. True. This is the definition of a rational function.

33. True. The ratio of the highest-degree terms in the numerator and denominator is $2x/x^2 = 2/x$, so for large positive x-values, y approaches 0.

37. False. The ratio of the highest-degree terms in the numerator and denominator is $3x^4/x^2 = 3x^2$. So for large positive x-values, y behaves like $y = 3x^2$.

41. True. At $x = -4$, we have $f(-4) = (-4 + 4)/(-4 - 3) = 0/(-7) = 0$, so $x = -4$ is a zero.

45. False. If $p(x)$ has no zeros, then $r(x)$ has no zeros. For example, if $p(x)$ is a nonzero constant or $p(x) = x^2 + 1$, then $r(x)$ has no zeros.

CHAPTER TWELVE

Solutions for Section 12.1

Skill Refresher

S1. Since the sum of the interior angles of any triangle is $180°$, we have:

$$\begin{aligned} 50° + 60° + \theta &= 180° \\ \theta &= 180° - 50° - 60° = 70°. \end{aligned}$$

S5. Using the Pythagorean Theorem, we have:

$$\begin{aligned} x^2 &= 3^2 + 5^2 = 34 \\ x &= \sqrt{34}. \end{aligned}$$

EXERCISES

1. We can describe an elevation with one number, so this is a scalar.

5. Scalar.

9.

$$\vec{p} = 2\vec{w}, \quad \vec{q} = -\vec{u}, \quad \vec{r} = \vec{w} + \vec{u} = \vec{u} + \vec{w},$$
$$\vec{s} = \vec{p} + \vec{q} = 2\vec{w} - \vec{u}, \quad \vec{t} = \vec{u} - \vec{w}$$

PROBLEMS

13. See Figure 12.1.

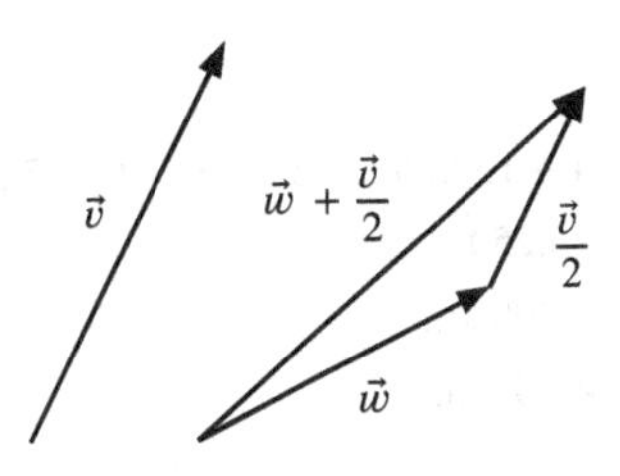

Figure 12.1

17. **(a)** From Figure 12.2,

$$\text{Distance from Oracle Road} = 5 \sin 20° = 1.710 \text{ miles}.$$

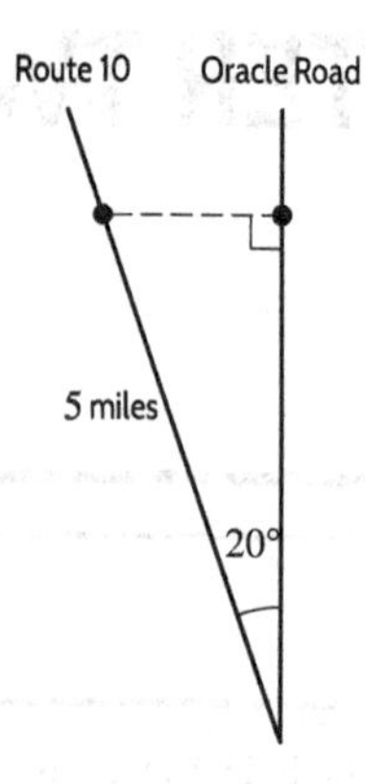

Figure 12.2

(b) If the distance along Route 10 is x miles, we have

$$x \sin 20° = 2 \text{ miles}$$
$$x = \frac{2}{\sin 20°} = 5.848 \text{ miles.}$$

21.

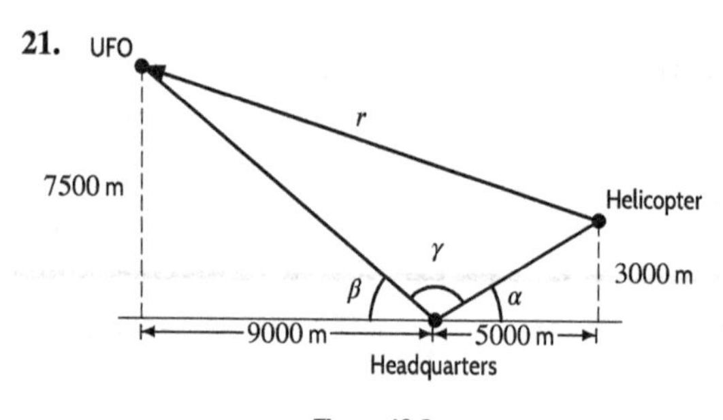

Figure 12.3

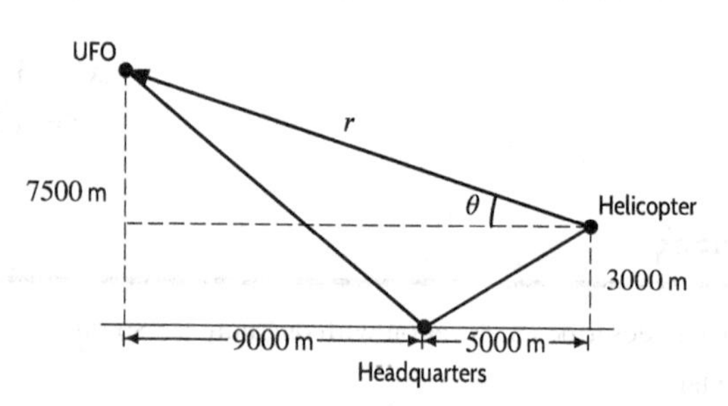

Figure 12.4

Figure 12.3 shows the headquarters at the origin, and a positive y-value as up, and a positive x-value as east. To solve for r, we must first find γ:

$$\begin{aligned}\gamma &= 180° - \alpha - \beta \\ &= 180° - \arctan\frac{3000}{5000} - \arctan\frac{7500}{9000} \\ &= 109.231°.\end{aligned}$$

We now can find r using the Law of Cosines in the triangle formed by the position of the headquarters, the helicopter, and the UFO.

In kilometers:

$$\begin{aligned}r^2 &= 34 + 137.250 - 2 \cdot \sqrt{34} \cdot \sqrt{137.250} \cdot \cos\gamma \\ r^2 &= 216.250 \\ r &= 14.705 \text{ km} \\ &= 14{,}705 \text{ m.}\end{aligned}$$

From Figure 12.4 we see:

$$\begin{aligned}\tan\theta &= \frac{4500}{14{,}000} \\ \theta &= 17.819°.\end{aligned}$$

Therefore, the helicopter must fly 14,705 meters with an angle of 17.819° from the horizontal.

25. The vector $\vec{v} + \vec{w}$ is equivalent to putting the vectors $\overrightarrow{OA}$ and $\overrightarrow{AB}$ end-to-end as shown in Figure 12.5; the vector $\vec{w} + \vec{v}$ is equivalent to putting the vectors $\overrightarrow{OC}$ and $\overrightarrow{CB}$ end-to-end. Since they form a parallelogram, $\vec{v} + \vec{w}$ and $\vec{w} + \vec{v}$ are both equal to the vector $\overrightarrow{OB}$, we have $\vec{v} + \vec{w} = \vec{w} + \vec{v}$.

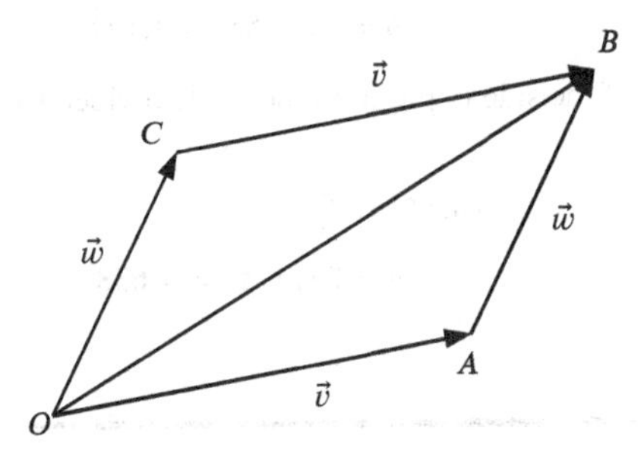

Figure 12.5

29. According to the definition of scalar multiplication, $1 \cdot \vec{v}$ has the same direction and magnitude as $\vec{v}$, so it is the same as $\vec{v}$.

Solutions for Section 12.2

Skill Refresher

S1. To find the distance between $(0, 0)$ and $(2, 3)$, we first draw a right triangle whose hypotenuse is the line segment from $(0, 0)$ to $(2, 3)$ (see Figure 12.6). Since the horizontal distance between the points is $2 - 0 = 2$ and the vertical distance is $3 - 0 = 3$, the side lengths of the right triangle are 2 and 3. Using the Pythagorean Theorem, we have:

$$x^2 = 2^2 + 3^2 = 13$$
$$x = \sqrt{13}.$$

Therefore, since the length of the hypotenuse is $\sqrt{13}$, the distance between $(0, 0)$ and $(2, 3)$ is $\sqrt{13}$.

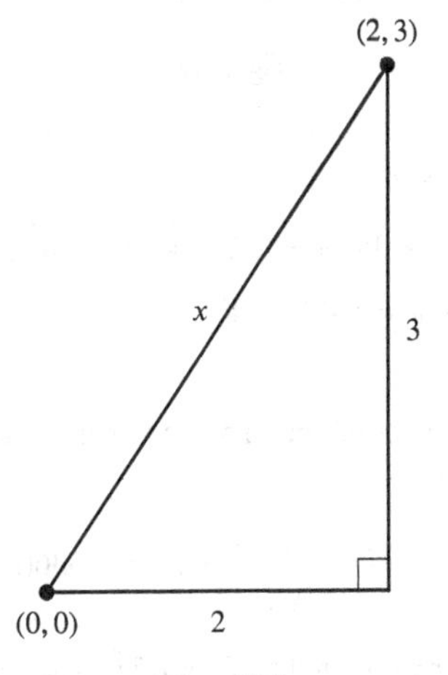

Figure 12.6

S5. From the triangle, we see that $\cos 55°$ equals the length of the side adjacent to the angle divided by the length of the hypotenuse. Therefore, we have:

$$\cos 55° = \frac{b}{2}$$
$$b = 2\cos 55° = 1.147.$$

Similarly, $\sin 55°$ equals the length of the side opposite to the angle divided by the length of the hypotenuse. Therefore, we have:

$$\sin 55° = \frac{a}{2}$$
$$a = 2\sin 55° = 1.638.$$

EXERCISES

1. The vector we want is the displacement from Q to P, which is given by

$$\overrightarrow{QP} = (1-4)\vec{i} + (2-6)\vec{j} = -3\vec{i} - 4\vec{j}.$$

5. $4\vec{i} + 2\vec{j} - 3\vec{i} + \vec{j} = \vec{i} + 3\vec{j}$.

9. $\|\vec{v}\| = \sqrt{1^2 + (-1)^2 + 3^2} = \sqrt{11} \approx 3.317$

PROBLEMS

13. The velocity of the ship in still water is $10\vec{j}$ knots and the velocity of the current is $-5\vec{i}$ since the current is east to west. The velocity of the ship is $-5\vec{i} + 10\vec{j}$ knots.

17. Figure 12.7 shows the vector $\vec{w}$ redrawn to show that it is perpendicular to the displacement vector $\overrightarrow{PQ}$, which lies along the dotted line. Thus, the angle is 90° or $\pi/2$.

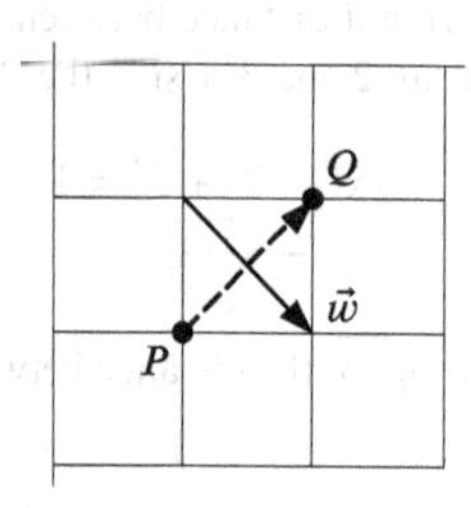

Figure 12.7

21. We need to calculate the length of each vector.

$$\|21\vec{i} + 35\vec{j}\| = \sqrt{21^2 + 35^2} = \sqrt{1666} \approx 40.8,$$
$$\|40\vec{i}\| = \sqrt{40^2} = 40.$$

So the first car is faster.

25. **(a)** The velocity, $\vec{v}$, is represented by a vector of length 5 in a northeasterly direction. The vector $\vec{i} + \vec{j}$ points northeast, but has length $\sqrt{1^2 + 1^2} = \sqrt{2}$. Thus,

$$\vec{v} = \frac{5}{\sqrt{2}}(\vec{i} + \vec{j}) = 3.536(\vec{i} + \vec{j})$$

(b) The current flows northward, so it is represented by $\vec{c} = 1.2\vec{j}$. The swimmer's velocity relative to the riverbed is

$$\vec{s} = \vec{c} + \vec{v} = 1.2\vec{j} + 3.536(\vec{i} + \vec{j}) = 3.536\vec{i} + 4.736\vec{j}.$$

29. We get displacement by subtracting the coordinates of the bottom of the tree, $(2,4,0)$, from the coordinates of the squirrel, $(2,4,1)$, giving:

$$\text{Displacement} = (2-2)\vec{i} + (4-4)\vec{j} + (1-0)\vec{k} = \vec{k}.$$

Solutions for Section 12.3

Skill Refresher

S1. First, we use Pythagorean Theorem to find a:

$$\begin{aligned} a^2 + 3^2 &= 4^2 \\ a^2 &= 4^2 - 3^2 = 7 \\ a &= \sqrt{7}. \end{aligned}$$

Since $\sin\theta$ is equal to the length of the side opposite to the angle θ divided by the length of the hypotenuse, we also have $\sin\theta = 3/4$. Therefore,

$$\theta = \sin^{-1}\left(\frac{3}{4}\right) = 48.59^\circ.$$

EXERCISES

1. $\vec{B} = 2\vec{M} = 2(1,1,2,3,5,8) = (2,2,4,6,10,16)$.

5. $\vec{K} = \dfrac{\vec{N}}{3} + \dfrac{2\vec{N}}{3} = \dfrac{3\vec{N}}{3} = \vec{N} = (5,6,7,8,9,10)$.

9. Since the components of $\vec{Q}$ represent millions of people, an increase of 120,000 people will increase each component by 0.12. Therefore,

$$\begin{aligned} \vec{S} &= \vec{Q} + (0.12, 0.12, 0.12, 0.12, 0.12, 0.12) \\ &= (3.57, 1.33, 6.55, 1.32, 1.05, 0.63) + (0.12, 0.12, 0.12, 0.12, 0.12, 0.12) \\ &= (3.69, 1.45, 6.67, 1.44, 1.17, 0.75). \end{aligned}$$

PROBLEMS

13. The total scores are out of 300 and are given by the total score vector $\vec{v} + 2\vec{w}$:

$$\begin{aligned} \vec{v} + 2\vec{w} &= (73, 80, 91, 65, 84) + 2(82, 79, 88, 70, 92) \\ &= (73, 80, 91, 65, 84) + (164, 158, 176, 140, 184) \\ &= (237, 238, 267, 205, 268). \end{aligned}$$

To get the scores as a percentage, we divide by 3, giving

$$\frac{1}{3}(237, 238, 267, 205, 268) \approx (79.000, 79.333, 89.000, 68.333, 89.333).$$

17. **(a)** See the sketch in Figure 12.8, where $\vec{v}$ represents the first part of the man's walk, and $\vec{w}$ represents the second part. Since the man first walks 5 miles, we know $\|\vec{v}\| = 5$. Since he walks 30° north of east, resolving gives

$$\vec{v} = 5\cos 30^\circ \vec{i} + 5\sin 30^\circ \vec{j} = 4.330\vec{i} + 2.500\vec{j}.$$

For the second leg of his journey, the man walks a distance x miles due east, so $\vec{w} = x\vec{i}$.

(b) The vector from finish to start is $-(\vec{v} + \vec{w}) = -(4.330 + x)\vec{i} - 2.500\vec{j}$. This vector is at an angle of 10° south of west. So, using the magnitudes of the sides in the triangle in Figure 12.9:

$$\begin{aligned}\frac{2.500}{4.330 + x} &= \tan(10^\circ) = 0.176\\ 2.500 &= 0.176(4.330 + x)\\ x &= \frac{2.500 - 0.176 \cdot 4.330}{0.176} = 9.848.\end{aligned}$$

This means that $x = 9.848$.

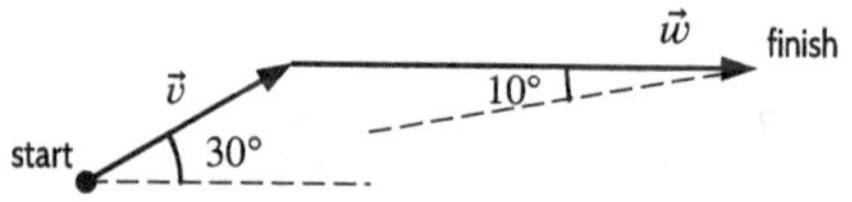

Figure 12.8

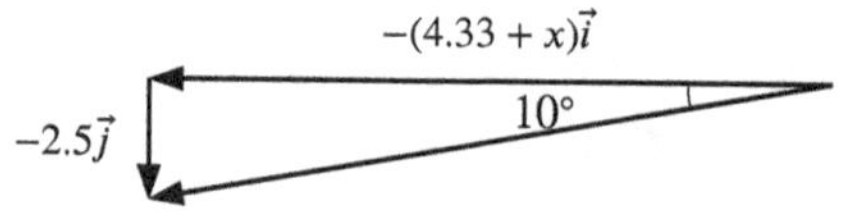

Figure 12.9

(c) The distance from the starting point is $\| -(4.330 + 9.848)\vec{i} - (2.500)\vec{j}\| = \sqrt{14.178^2 + 2.500^2} = 14.397$ miles.

21. In an actual video game, our rectangle would be replaced with a more sophisticated graphic (perhaps an airplane or an animated figure). But the principles involved in rotation about the origin are the same, and it will be easier to think about them using rectangles instead of fancy graphics.

We can represent the four corners of the rectangle (before rotation) using the position vectors $\vec{p}_a, \vec{p}_b, \vec{p}_c$, and $\vec{p}_d$. For instance, the components of $\vec{p}_a$ are $\vec{p}_a = 2\vec{i} + \vec{j}$.

After the rectangle has been rotated, its four corners are given by the position vectors $\vec{q}_a, \vec{q}_b, \vec{q}_c$, and $\vec{q}_d$. Notice that the lengths of these vectors have not changed; in other words,

$$\|\vec{p}_a\| = \|\vec{q}_a\|, \quad \|\vec{p}_b\| = \|\vec{q}_b\|, \quad \|\vec{p}_c\| = \|\vec{q}_c\|, \quad \text{and} \quad \|\vec{p}_d\| = \|\vec{q}_d\|.$$

This is because in a rotation the only thing that changes is orientation, not length.

When the rectangle is rotated through a 35° angle, the angle made by corner a increases by 35°. So do the angles made by the other three corners. Letting θ be the angle made by corner a, we have

$$\begin{aligned}\tan\theta &= \frac{1}{2}\\ \theta &= \arctan 0.5 = 26.565^\circ.\end{aligned}$$

This is the direction of the position vector $\vec{p}_a$. After rotation, the angle θ is given by

$$\theta = 26.565^\circ + 35^\circ = 61.565^\circ.$$

This is the direction of the new position vector $\vec{q}_a$. The length of $\vec{q}_a$ is the same as the length of $\vec{p}_a$ and is given by

$$\|\vec{q}_a\|^2 = \|\vec{p}_a\|^2 = 2^2 + 1^2 = 5,$$

and so $\|\vec{q}_a\| = \sqrt{5}$. Thus, the components of $\vec{q}_a$ are given by

$$\begin{aligned}\vec{q}_a &= (\sqrt{5}\cos 61.565^\circ)\vec{i} + (\sqrt{5}\sin 61.565^\circ)\vec{j}\\ &= 1.065\vec{i} + 1.966\vec{j}.\end{aligned}$$

This process can be repeated for the other three corners. You can see for yourself that the angles made with the origin by the corners a, b, and c, respectively, are 14.036°, 26.565°, and 45°. After rotation, these angles are 49.036°, 61.565°, and

80°. Similarly, the lengths of the position vectors for these three points (both before and after rotation) are $\sqrt{17}$, $\sqrt{20}$, and $\sqrt{8}$. Thus, the final positions of these three points are

$$\vec{q}_b = 2.703\vec{i} + 3.113\vec{j},$$
$$\vec{q}_c = 2.129\vec{i} + 3.933\vec{j},$$
$$\vec{q}_d = 0.491\vec{i} + 2.785\vec{j}.$$

Solutions for Section 12.4

Skill Refresher

S1. Since

$$3\cos\theta = 2$$
$$\cos\theta = \frac{2}{3},$$

we have

$$\theta = \cos^{-1}\left(\frac{2}{3}\right) = 48.19^\circ.$$

EXERCISES

1. $\vec{z}\cdot\vec{a} = (\vec{i} - 3\vec{j} - \vec{k})\cdot(2\vec{j} + \vec{k}) = 1\cdot 0 + (-3)2 + (-1)1 = 0 - 6 - 1 = -7.$

5. $\vec{a}\cdot\vec{b} = (2\vec{j} + \vec{k})\cdot(-3\vec{i} + 5\vec{j} + 4\vec{k}) = 0(-3) + 2\cdot 5 + 1\cdot 4 = 0 + 10 + 4 = 14.$

9. Since $\vec{a}\cdot\vec{b}$ is a scalar and $\vec{a}$ is a vector, the expression is a vector parallel to $\vec{a}$. We have

$$\vec{a}\cdot\vec{b} = (2\vec{j} + \vec{k})\cdot(-3\vec{i} + 5\vec{j} + 4\vec{k}) = 0(-3) + 2(5) + 1(4) = 14.$$

Thus,

$$(\vec{a}\cdot\vec{b})\cdot\vec{a} = 14\vec{a} = 14(2\vec{j} + \vec{k}) = 28\vec{j} + 14\vec{k}.$$

PROBLEMS

13. We use the dot product to find this angle.
We have $\vec{w} = \vec{i} - \vec{j}$ and $\vec{v} = \vec{i} + 2\vec{j}$ so

$$\vec{w}\cdot\vec{v} = (\vec{i} - \vec{j})\cdot(\vec{i} + 2\vec{j}) = -1,$$

therefore

$$\vec{w}\cdot\vec{v} = ||\vec{w}||\cdot||\vec{v}||\cos\theta.$$

Since $||\vec{w}|| = \sqrt{5}$ and $||\vec{v}|| = \sqrt{2}$, and $\vec{w}\cdot\vec{v} = -1$, we have

$$-1 = \sqrt{2}\sqrt{5}\cos\theta$$
$$\cos\theta = -\frac{1}{\sqrt{10}}$$
$$\theta = \arccos -\frac{1}{\sqrt{10}} = 108.435^\circ.$$

17. Using the dot product, the angle is given by

$$\cos\theta = \frac{(\vec{i}+\vec{j}+\vec{k})\cdot(\vec{i}-\vec{j}-\vec{k})}{\|\vec{i}+\vec{j}+\vec{k}\|\|\vec{i}-\vec{j}-\vec{k}\|} = \frac{1\cdot 1+1(-1)+1(-1)}{\sqrt{1^1+1^2+1^2}\sqrt{1^2+(-1)^2+(-1)^2}} = -\frac{1}{3}.$$

So, $\theta = \arccos(-\frac{1}{3}) \approx 1.911$ radians, or $\approx 109.471°$.

21. It is clear from the Figure 12.10 that only angle $\angle CAB$ could possibly be a right angle. Subtraction of x, y values for the points gives $\overrightarrow{AB} = 3\vec{i} - \vec{j}$ and $\vec{AC} = 1\vec{i} + 2\vec{j}$. Taking the dot product yields $\overrightarrow{AB}\cdot\overrightarrow{AC} = (3)(1) + (-1)(2) = 1$. Since this is nonzero, the angle cannot be a right angle.

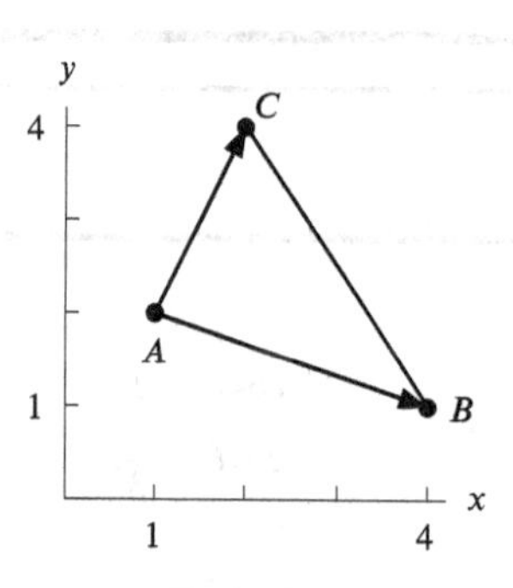

Figure 12.10

25. **(a)** We have the price vector $\vec{a} = (3, 2, 4)$. Let the consumption vector $\vec{c} = (c_b, c_e, c_m)$, then $3c_b + 2c_e + 4c_m = 40$ or $\vec{a}\cdot\vec{c} = 40$.

(b) Note $\vec{a}\cdot\vec{c}$ is the cost of consuming $\vec{c}$ groceries at Acme Store, so $\vec{b}\cdot\vec{c}$ is the cost of consuming $\vec{c}$ groceries at Beta Mart. Thus $\vec{b}\cdot\vec{c} - \vec{a}\cdot\vec{c} = (\vec{b}-\vec{a})\cdot\vec{c}$ is the difference in costs between Beta and Acme for the same $\vec{c}$ groceries.

For $\vec{b}-\vec{a}$ to be perpendicular to $\vec{c}$, we must have $(\vec{b}-\vec{a})\cdot\vec{c} = 0$. Since $\vec{b}-\vec{a} = (0.20, -0.20, 0.50)$, the vector $\vec{b}-\vec{a}$ is perpendicular to $\vec{c}$ if $0.20c_b - 0.20c_e + 0.50c_m = 0$. For example, this occurs when we consume the same number of loaves of bread as dozens of eggs, but no milk.

(c) Since $\vec{b}\cdot\vec{c}$ is the cost of groceries at Beta, you might think of $(1/1.1)\vec{b}\cdot\vec{c}$ as the "freshness-adjusted" cost at Beta. Then $(1/1.1)\vec{b}\cdot\vec{c} < \vec{a}\cdot\vec{c}$ means the "freshness-adjusted" cost is lower at Beta.

29. **(a)** We have

$$\begin{aligned}\vec{r}\cdot\vec{w} &= (x_1, y_1, x_2, y_2)\cdot(-1, 0, 1, 0)\\ &= -x_1 + x_2 = x_2 - x_1,\end{aligned}$$

so, since $x_2 > x_1$, this quantity represents the width w.

(b) We have

$$\begin{aligned}\vec{r}\cdot\vec{h} &= (x_1, y_1, x_2, y_2)\cdot(0, -1, 0, 1)\\ &= -y_1 + y_2 = y_2 - y_1,\end{aligned}$$

so, since $y_2 > y_1$, this quantity represents the height h.

(c) We have

$$\begin{aligned}2\vec{r}\cdot(\vec{w}+\vec{h}) &= 2\vec{r}\cdot\vec{w} + 2\vec{r}\cdot\vec{h}\\ &= 2w + 2h,\end{aligned}$$

where from part (a) w is the width and from part (b) h is the height. Thus, this quantity represents the perimeter p.

Solutions for Section 12.5

Skill Refresher

S1. Since the number of male students starts at 4000 and increases by 5%, the resulting number of male students is

$$4000 + 0.05(4000) = 4000 + 200 = 4200.$$

Therefore, since the number of female students attending the university remains unchanged at 6000, the resulting number of total students attending the university is

$$4200 + 6000 = 10{,}200.$$

EXERCISES

1. **(a)** We have

$$5\mathbf{R} = 5\begin{pmatrix} 3 & 7 \\ 2 & -1 \end{pmatrix} = \begin{pmatrix} 5\cdot 3 & 5\cdot 7 \\ 5\cdot 2 & 5\cdot -1 \end{pmatrix} = \begin{pmatrix} 15 & 35 \\ 10 & -5 \end{pmatrix}.$$

(b) We have

$$-2\mathbf{S} = -2\begin{pmatrix} 1 & -5 \\ 0 & 8 \end{pmatrix} = \begin{pmatrix} -2\cdot 1 & -2\cdot -5 \\ -2\cdot 0 & -2\cdot 8 \end{pmatrix} = \begin{pmatrix} -2 & 10 \\ 0 & -16 \end{pmatrix}.$$

(c) We have

$$\begin{aligned} \mathbf{R}+\mathbf{S} &= \begin{pmatrix} 3 & 7 \\ 2 & -1 \end{pmatrix} + \begin{pmatrix} 1 & -5 \\ 0 & 8 \end{pmatrix} \\ &= \begin{pmatrix} 3+1 & 7-5 \\ 2+0 & -1+8 \end{pmatrix} = \begin{pmatrix} 4 & 2 \\ 2 & 7 \end{pmatrix}. \end{aligned}$$

(d) Writing $\mathbf{S} - 3\mathbf{R} = \mathbf{S} + (-3)\mathbf{R}$, we first find $-3\mathbf{R}$:

$$-3\mathbf{R} = -3\begin{pmatrix} 3 & 7 \\ 2 & -1 \end{pmatrix} = \begin{pmatrix} -3\cdot 3 & -3\cdot 7 \\ -3\cdot 2 & -3\cdot -1 \end{pmatrix} = \begin{pmatrix} -9 & -21 \\ -6 & 3 \end{pmatrix}.$$

This gives

$$\begin{aligned} \mathbf{S}+(-3)\mathbf{R} &= \begin{pmatrix} 1 & -5 \\ 0 & 8 \end{pmatrix} + \begin{pmatrix} -9 & -21 \\ -6 & 3 \end{pmatrix} \\ &= \begin{pmatrix} 1-9 & -5-21 \\ 0-6 & 8+3 \end{pmatrix} = \begin{pmatrix} -8 & -26 \\ -6 & 11 \end{pmatrix}. \end{aligned}$$

(e) Writing $\mathbf{R} + 2\mathbf{R} + 2(\mathbf{R} - \mathbf{S}) = 5\mathbf{R} + (-2)\mathbf{S}$, we use our answers to parts (a) and (b):

$$\begin{aligned} 5\mathbf{R}+(-2)\mathbf{S} &= \begin{pmatrix} 15 & 35 \\ 10 & -5 \end{pmatrix} + \begin{pmatrix} -2 & 10 \\ 0 & -16 \end{pmatrix} \\ &= \begin{pmatrix} 15-2 & 35+10 \\ 10+0 & -5-16 \end{pmatrix} = \begin{pmatrix} 13 & 45 \\ 10 & -21 \end{pmatrix}. \end{aligned}$$

(f) We have

$$k\mathbf{S} = \begin{pmatrix} k\cdot 1 & k\cdot -5 \\ k\cdot 0 & k\cdot 8 \end{pmatrix} = \begin{pmatrix} k & -5k \\ 0 & 8k \end{pmatrix}.$$

5. **(a)** We have

$$\mathbf{A}\vec{u} = \begin{pmatrix} 2 & 5 & 7 \\ 4 & -6 & 3 \\ 16 & -5 & 0 \end{pmatrix}\begin{pmatrix} 3 \\ 2 \\ 5 \end{pmatrix}$$

$$= \begin{pmatrix} 2\cdot 3 + 5\cdot 2 + 7\cdot 5 \\ 4\cdot 3 - 6\cdot 2 + 3\cdot 5 \\ 16\cdot 3 - 5\cdot 2 + 0\cdot 5 \end{pmatrix} = \begin{pmatrix} 51 \\ 15 \\ 38 \end{pmatrix}.$$

(b) We have

$$\mathbf{B}\vec{v} = \begin{pmatrix} 8 & -6 & 0 \\ 5 & 3 & -2 \\ 3 & 7 & 12 \end{pmatrix}\begin{pmatrix} -1 \\ 0 \\ 3 \end{pmatrix}$$

$$= \begin{pmatrix} 8\cdot -1 - 6\cdot 0 + 0\cdot 3 \\ 5\cdot -1 + 3\cdot 0 - 2\cdot 3 \\ 3\cdot -1 + 7\cdot 0 + 12\cdot 3 \end{pmatrix} = \begin{pmatrix} -8 \\ -11 \\ 33 \end{pmatrix}.$$

(c) Letting $\vec{w} = \vec{u} + \vec{v} = (2, 2, 8)$, we have:

$$\mathbf{A}\vec{w} = \begin{pmatrix} 2 & 5 & 7 \\ 4 & -6 & 3 \\ 16 & -5 & 0 \end{pmatrix}\begin{pmatrix} 2 \\ 2 \\ 8 \end{pmatrix}$$

$$= \begin{pmatrix} 2\cdot 2 + 5\cdot 2 + 7\cdot 8 \\ 4\cdot 2 - 6\cdot 2 + 3\cdot 8 \\ 16\cdot 2 - 5\cdot 2 + 0\cdot 8 \end{pmatrix} = \begin{pmatrix} 70 \\ 20 \\ 22 \end{pmatrix}.$$

Another to work this problem would be to write $\mathbf{A}(\vec{u} + \vec{v})$ as $\mathbf{A}\vec{u} + \mathbf{A}\vec{v}$ and proceed accordingly.

(d) Letting $\mathbf{C} = \mathbf{A} + \mathbf{B}$, we have

$$\mathbf{C} = \begin{pmatrix} 2 & 5 & 7 \\ 4 & -6 & 3 \\ 16 & -5 & 0 \end{pmatrix} + \begin{pmatrix} 8 & -6 & 0 \\ 5 & 3 & -2 \\ 3 & 7 & 12 \end{pmatrix} = \begin{pmatrix} 10 & -1 & 7 \\ 9 & -3 & 1 \\ 19 & 2 & 12 \end{pmatrix}.$$

We can now write $(\mathbf{A} + \mathbf{B})\vec{v}$ as $\mathbf{C}\vec{v}$, and so:

$$\mathbf{C}\vec{v} = \begin{pmatrix} 10 & -1 & 7 \\ 9 & -3 & 1 \\ 19 & 2 & 12 \end{pmatrix}\begin{pmatrix} -1 \\ 0 \\ 3 \end{pmatrix}$$

$$= \begin{pmatrix} 10\cdot -1 - 1\cdot 0 + 7\cdot 3 \\ 9\cdot -1 - 3\cdot 0 + 1\cdot 3 \\ 19\cdot -1 + 2\cdot 0 + 12\cdot 3 \end{pmatrix} = \begin{pmatrix} 11 \\ -6 \\ 17 \end{pmatrix}.$$

(e) From part (a) we have $\mathbf{A}\vec{u} = (51, 15, 38)$, and from part (b) we have $\mathbf{B}\vec{v} = (-8, -11, 33)$. This gives

$$\begin{aligned} \mathbf{A}\vec{u} \cdot \mathbf{B}\vec{v} &= (51, 15, 38) \cdot (-8, -11, 33) \\ &= 51 \cdot -8 + 15 \cdot -11 + 38 \cdot 33 = 681. \end{aligned}$$

(f) We have $\vec{u}\cdot\vec{v} = 3\cdot -1 + 2\cdot 0 + 5\cdot 3 = 12$, and so

$$(\vec{u}\cdot\vec{v})\mathbf{A} = 12\mathbf{A} = 12\begin{pmatrix} 2 & 5 & 7 \\ 4 & -6 & 3 \\ 16 & -5 & 0 \end{pmatrix} = \begin{pmatrix} 24 & 60 & 84 \\ 48 & -72 & 36 \\ 192 & -60 & 0 \end{pmatrix}.$$

PROBLEMS

9. **(a)** We have

$$\begin{aligned} s_{\text{new}} &= s_{\text{old}} - \underbrace{0.10s_{\text{old}}}_{\text{10\% infected}} \\ &= 0.90s_{\text{old}} \\ i_{\text{new}} &= i_{\text{old}} + \underbrace{0.10s_{\text{old}}}_{\text{10\% infected}} - \underbrace{0.50i_{\text{old}}}_{\text{50\% recover}} + \underbrace{0.02r_{\text{old}}}_{\text{2\% reinfected}} \\ &= 0.10s_{\text{old}} + 0.50i_{\text{old}} + 0.02r_{\text{old}} \\ r_{\text{new}} &= r_{\text{old}} + \underbrace{0.50i_{\text{old}}}_{\text{50\% recover}} - \underbrace{0.02r_{\text{old}}}_{\text{2\% reinfected}} \\ &= 0.50i_{\text{old}} + 0.98r_{\text{old}}. \end{aligned}$$

Using matrix multiplication, we can rewrite these three equations as

$$\begin{pmatrix} s_{\text{new}} \\ i_{\text{new}} \\ r_{\text{new}} \end{pmatrix} = \begin{pmatrix} 0.90 & 0 & 0 \\ 0.10 & 0.50 & 0.02 \\ 0 & 0.50 & 0.98 \end{pmatrix}\begin{pmatrix} s_{\text{old}} \\ i_{\text{old}} \\ r_{\text{old}} \end{pmatrix},$$

and so $\vec{p}_{\text{new}} = \mathbf{T}\vec{p}_{\text{old}}$ where $\mathbf{T} = \begin{pmatrix} 0.90 & 0 & 0 \\ 0.10 & 0.50 & 0.02 \\ 0 & 0.50 & 0.98 \end{pmatrix}$.

(b) We have

$$\begin{aligned} \vec{p}_1 = \mathbf{T}\vec{p}_0 &= \begin{pmatrix} 0.90 & 0 & 0 \\ 0.10 & 0.50 & 0.02 \\ 0 & 0.50 & 0.98 \end{pmatrix}\begin{pmatrix} 2.0 \\ 0.0 \\ 0.0 \end{pmatrix} \\ &= \begin{pmatrix} 0.9(2) + 0(0) + 0(0) \\ 0.1(2) + 0.5(0) + 0.02(0) \\ 0(2) + 0.5(0) + 0.98(0) \end{pmatrix} = \begin{pmatrix} 1.8 \\ 0.2 \\ 0.0 \end{pmatrix} \\ \vec{p}_2 = \mathbf{T}\vec{p}_1 &= \begin{pmatrix} 0.90 & 0 & 0 \\ 0.10 & 0.50 & 0.02 \\ 0 & 0.50 & 0.98 \end{pmatrix}\begin{pmatrix} 1.8 \\ 0.2 \\ 0.0 \end{pmatrix} \\ &= \begin{pmatrix} 0.9(1.8) + 0(0.2) + 0(0) \\ 0.1(1.8) + 0.5(0.2) + 0.02(0) \\ 0(1.8) + 0.5(0.2) + 0.98(0) \end{pmatrix} = \begin{pmatrix} 1.62 \\ 0.28 \\ 0.10 \end{pmatrix} \\ \vec{p}_3 = \mathbf{T}\vec{p}_2 &= \begin{pmatrix} 0.90 & 0 & 0 \\ 0.10 & 0.50 & 0.02 \\ 0 & 0.50 & 0.98 \end{pmatrix}\begin{pmatrix} 1.62 \\ 0.28 \\ 0.10 \end{pmatrix} \end{aligned}$$

$$= \begin{pmatrix} 0.9(1.62) + 0(0.28) + 0(0.1) \\ 0.1(1.62) + 0.5(0.28) + 0.02(0.1) \\ 0(1.62) + 0.5(0.28) + 0.98(0.1) \end{pmatrix} = \begin{pmatrix} 1.458 \\ 0.304 \\ 0.238 \end{pmatrix}.$$

13. **(a)** We first find $\vec{v}$:

$$\vec{v} = \mathbf{A}\vec{u} = \begin{pmatrix} 2 & 1 \\ 3 & 2 \end{pmatrix}\begin{pmatrix} 3 \\ 5 \end{pmatrix} = \begin{pmatrix} 11 \\ 19 \end{pmatrix}.$$

Now, we show that $\vec{u} = \mathbf{A}^{-1}\vec{v}$:

$$\mathbf{A}^{-1}\vec{v} = \begin{pmatrix} 2 & -1 \\ -3 & 2 \end{pmatrix}\begin{pmatrix} 11 \\ 19 \end{pmatrix} = \begin{pmatrix} 3 \\ 5 \end{pmatrix} = \vec{u}.$$

(b) We first find $\vec{v}$:

$$\vec{v} = \mathbf{A}\vec{u} = \begin{pmatrix} 2 & 1 \\ 3 & 2 \end{pmatrix}\begin{pmatrix} -1 \\ 7 \end{pmatrix} = \begin{pmatrix} 5 \\ 11 \end{pmatrix}.$$

Now, we show that $\vec{u} = \mathbf{A}^{-1}\vec{v}$:

$$\mathbf{A}^{-1}\vec{v} = \begin{pmatrix} 2 & -1 \\ -3 & 2 \end{pmatrix}\begin{pmatrix} 5 \\ 11 \end{pmatrix} = \begin{pmatrix} -1 \\ 7 \end{pmatrix} = \vec{u}.$$

(c) We first find $\vec{v}$:

$$\vec{v} = \mathbf{A}\vec{u} = \begin{pmatrix} 2 & 1 \\ 3 & 2 \end{pmatrix}\begin{pmatrix} a \\ b \end{pmatrix} = \begin{pmatrix} 2a+b \\ 3a+2b \end{pmatrix}.$$

Now, we show that $\vec{u} = \mathbf{A}^{-1}\vec{v}$:

$$\begin{aligned} \mathbf{A}^{-1}\vec{v} &= \begin{pmatrix} 2 & -1 \\ -3 & 2 \end{pmatrix}\begin{pmatrix} 2a+b \\ 3a+2b \end{pmatrix} \\ &= \begin{pmatrix} 2(2a+b) - (3a+2b) \\ -3(2a+b) + 2(3a+2b) \end{pmatrix} \\ &= \begin{pmatrix} 4a+2b-3a-2b \\ -6a-3b+6a+4b \end{pmatrix} \\ &= \begin{pmatrix} a \\ b \end{pmatrix} = \vec{u}. \end{aligned}$$

17. **(a)** We have $\mathbf{C} = \begin{pmatrix} 3 & 5 \\ 2 & 4 \end{pmatrix}$ and so

$$\begin{aligned} \mathbf{C}\vec{u} &= \begin{pmatrix} 3 & 5 \\ 2 & 4 \end{pmatrix}\begin{pmatrix} a \\ b \end{pmatrix} \\ &= \begin{pmatrix} 3a+5b \\ 2a+4b \end{pmatrix} \\ &= \begin{pmatrix} 3a \\ 2a \end{pmatrix} + \begin{pmatrix} 5b \\ 4b \end{pmatrix} \\ &= a\begin{pmatrix} 3 \\ 2 \end{pmatrix} + b\begin{pmatrix} 5 \\ 4 \end{pmatrix} \\ &= a\vec{c}_1 + b\vec{c}_2. \end{aligned}$$

(b) Provided $\mathbf{C}^{-1}$ exists, we can write $\vec{u} = \mathbf{C}^{-1}v$. We can find $\mathbf{C}^{-1}$ using the approach from Problem 14: we have $a = 3$, $b = 5$, $c = 2$, $d = 4$, and so $D = ad - bc = 3(4) - 5(2) = 2$. We have

$$\mathbf{C}^{-1} = \frac{1}{D}\begin{pmatrix} d & -b \\ -c & a \end{pmatrix} = \frac{1}{2}\begin{pmatrix} 4 & -5 \\ -2 & 3 \end{pmatrix} = \begin{pmatrix} 2 & -2.5 \\ -1 & 1.5 \end{pmatrix}.$$

This means that

$$\begin{aligned} \vec{u} &= \mathbf{C}^{-1}\vec{v} \\ &= \begin{pmatrix} 2 & -2.5 \\ -1 & 1.5 \end{pmatrix}\begin{pmatrix} 2 \\ 5 \end{pmatrix} \\ &= \begin{pmatrix} 2(2) - 2.5(5) \\ -1(2) + 1.5(5) \end{pmatrix} \\ &= \begin{pmatrix} -8.5 \\ 5.5 \end{pmatrix}. \end{aligned}$$

(c) From parts (a) and (b), we see that

$$\vec{v} = \begin{pmatrix} 2 \\ 5 \end{pmatrix} = \mathbf{C}\vec{u} = \begin{pmatrix} 3 & 2 \\ 5 & 4 \end{pmatrix}\begin{pmatrix} -8.5 \\ 5.5 \end{pmatrix}.$$

Referring to Problem 16, we see that

$$\begin{aligned} \vec{v} = \begin{pmatrix} 2 \\ 5 \end{pmatrix} &= \Bigg(\underbrace{\boxed{\begin{matrix} 3 \\ 2 \end{matrix}}}_{\vec{c_1}} \; \underbrace{\boxed{\begin{matrix} 5 \\ 4 \end{matrix}}}_{\vec{c_2}} \Bigg) \underbrace{\begin{pmatrix} -8.5 \\ 5.5 \end{pmatrix}}_{\vec{u}} \\ &= -8.5 \underbrace{(3, 2)}_{\vec{c_1}} + 5.5 \underbrace{(5, 4)}_{\vec{c_2}} \\ &= -8.5\vec{c_1} + 5.5\vec{c_2}, \end{aligned}$$

which gives $\vec{v}$ as a combination of $\vec{c_1}$ and $\vec{c_2}$.

STRENGTHEN YOUR UNDERSTANDING

1. False. The length $||0.5\vec{i} + 0.5\vec{j}|| = \sqrt{(0.5)^2 + (0.5)^2} = \sqrt{0.5} \neq 1$.

5. False. The dot product $(2\vec{i} + \vec{j}) \cdot (2\vec{i} - \vec{j}) = 3$ is not zero.

9. True. If $\vec{u} = (u_1, u_2)$ and $\vec{v} = (v_1, v_2)$, then $\vec{u} + \vec{v}$ and $\vec{v} + \vec{u}$ both equal $(u_1 + v_1, u_2 + v_2)$.

13. False. Distance is not a vector. The vector $\vec{i} + \vec{j}$ is the displacement vector from P to Q. The distance between the points is the length of the displacement vector: $||\vec{i} + \vec{j}|| = \sqrt{2}$.

17. True. Both vectors have length $\sqrt{13}$.

21. True. In the subscript, the first number gives the row and the second number gives the column.

CHAPTER THIRTEEN

Solutions for Section 13.1

EXERCISES

1. Not arithmetic. The differences are 5, 4, 3.

5. Arithmetic, with $a = 6$, $d = 3$, so $a_n = 6 + (n-1)3 = 3 + 3n$.

9. Not geometric. The ratios of successive terms are $2, 2, \frac{3}{2}$.

13. Not geometric, since the ratios of successive terms are $1/4, 1/4, 1/2$.

17. Geometric, since the ratios of successive terms are all $1/1.2$. Thus, $a = 1$ and $r = 1/1.2$, so $a_n = 1(1/1.2)^{n-1} = 1/(1.2)^{n-1}$.

PROBLEMS

21. $a_1 = \cos(\pi) = -1$, $a_2 = \cos(2\pi) = 1$, $a_3 = \cos(3\pi) = -1$, $a_4 = \cos(4\pi) = 1$. This is a geometric sequence.

25. Since the first term is 5 and the difference is 10, the arithmetic sequence is given by

$$a_n = 5 + (n-1)10.$$

For $a_n = 5 + (n-1) \cdot 10 > 1000$, we must have

$$\begin{aligned} 10(n-1) &> 995 \\ n-1 &> 99.5 \\ n &> 100.5. \end{aligned}$$

The terms of the sequence exceed 1000 when $n \geq 101$.

29. Since $a_6 - a_3 = 3d = 9 - 5.7 = 3.3$, we have $d = 1.1$, so $a_1 = a_3 - 2d = 5.7 - 2 \cdot 1.1 = 3.5$. Thus

$$\begin{aligned} a_5 &= 3.5 + (5-1)1.1 = 7.9 \\ a_{50} &= 3.5 + (50-1)1.1 = 57.4 \\ a_n &= 3.5 + (n-1)1.1 = 2.4 + 1.1n. \end{aligned}$$

33. Since $a_2 = ar = 6$ and $a_4 = ar^3 = 54$, we have

$$\frac{a_4}{a_2} = \frac{ar^3}{ar} = r^2 = \frac{54}{6} = 9,$$

so

$$r = 3.$$

We have $a = 6/r = 6/3 = 2$. Thus

$$\begin{aligned} a_6 &= ar^5 = 2 \cdot (3)^5 = 486 \\ a_n &= ar^{n-1} = 2 \cdot 3^{n-1}. \end{aligned}$$

37. **(a)** The growth factor of the population is

$$r = \frac{19.042}{18.934} = 1.0057.$$

Thus, one and two years after 2012, the population is

$$a_1 = 19.042(1.0057) = 19.151$$
$$a_2 = 19.042(1.0057)^2 = 19.260.$$

(b) Using the growth factor in part (a), n years after 2012, the population is

$$a_n = 19.042(1.0057)^n.$$

(c) The doubling time is the value of n for which

$$\begin{aligned} a_n &= 2 \cdot 19.042 \\ 19.042(1.0057)^n &= 2 \cdot 19.042 \\ (1.0057)^n &= 2 \\ n \ln(1.0057) &= \ln 2 \\ n &= \frac{\ln 2}{\ln(1.0057)} = 121.951. \end{aligned}$$

Thus the doubling time is about 121.951 years

41. Arithmetic, because points lie on a line. The sequence is decreasing, so $d < 0$.

45. Since $a_1 = 3$ and $a_n = 2a_{n-1} + 1$, we have $a_2 = 2a_1 + 1 = 7$, $a_3 = 2a_2 + 1 = 15$, $a_4 = 2a_3 + 1 = 31$. Writing out the terms without simplification to try and guess the pattern, we have

$$\begin{aligned} a_1 &= 3 \\ a_2 &= 2 \cdot 3 + 1 \\ a_3 &= 2(2 \cdot 3 + 1) + 1 = 2^2 \cdot 3 + 2 + 1 \\ a_4 &= 2(2^2 \cdot 3 + 2 + 1) + 1 = 2^3 \cdot 3 + 2^2 + 2 + 1. \end{aligned}$$

Thus

$$a_n = 2^{n-1} \cdot 3 + 2^{n-2} + 2^{n-3} + \cdots + 2 + 1.$$

49. **(a)** You send the letter to 4 friends; at that stage your name is on the bottom of the list. The 4 friends send it to 4 each, so 4^2 people have the letter when you are third on the list. These 4^2 people send it to 4 each, so 4^3 people have the letter when you are second on the list. By similar reasoning, 4^4 people have the letter when you are at the top of the list. Thus, you should receive $\$4^4 = \$256$. (The catch is, of course, that someone usually breaks the chain.)

(b) By the same reasoning as in part (a), we see that $d_n = 4^n$.

Solutions for Section 13.2

EXERCISES

1. This series is not arithmetic, as each term is twice the previous one.

5. $\sum_{i=-1}^{5} i^2 = (-1)^2 + 0^2 + 1^2 + 2^2 + 3^2 + 4^2 + 5^2.$

9. $\sum_{j=2}^{10} (-1)^j = (-1)^2 + (-1)^3 + (-1)^4 + \cdots + (-1)^{10}.$

13. The pattern is 1/2, 2/2, 3/2, …, 8/2. Thus, a possible solution is

$$\sum_{n=1}^{8} \frac{1}{2}n.$$

17. One way to work this problem is to complete the first few values of a_n, and then continue the pattern across the row. Then, the values of S_n can be found by summing the values of a_n from left to right. See Table 13.1. We see that $a_1 = 3$ and $d = a_2 - a_1 = 7 - 3 = 4$.

Table 13.1

n	1	2	3	4	5	6	7	8
a_n	3	7	11	15	19	23	27	31
S_n	3	10	21	36	55	78	105	136

21. We use the formula $S_n = 1 + 2 + 3 + \cdots + n = \frac{1}{2}n(n+1)$ with $n = 1000$:

$$S_{1000} = \frac{1}{2} \cdot 1000 \cdot 1001 = 500{,}500.$$

25. $\sum_{n=0}^{10}(8 - 4n) = 8 + 4 + 0 + (-4) + \cdots$. This is an arithmetic series with 11 terms and $d = -4$.

$$S_{11} = \frac{1}{2} \cdot 11(2 \cdot 8 + 10(-4)) = -132.$$

29. This is an arithmetic series with $a_1 = -3.01$, $n = 35$, and $d = -0.01$. Thus

$$S_{35} = \frac{1}{2} \cdot 35(2(-3.01) + 34(-0.01)) = -111.3.$$

PROBLEMS

33. (a) (i) From the table, we have $S_4 = 248.7$, the population of the US in millions 4 decades after 1950, that is, in 1990. Similarly, $S_5 = 281.4$, the population in millions in 2000, and $S_6 = 308.7$, the population in 2010.

(ii) We have $a_2 = S_2 - S_1 = 203.3 - 179.3 = 24$; that is, the increase in the US population in millions in the 1960s. Similarly, $a_5 = S_5 - S_4 = 281.4 - 248.7 = 32.7$, the population increase in millions during the 1990s. In the same way, $a_6 = S_6 - S_5 = 308.7 - 281.4 = 27.3$, the population increase during the 2000s.

(iii) Using the answer to (ii), we have $a_6/10 = 27.3/10 = 2.73$, the average yearly population growth during the 2000s.

(b) We have

S_n = US population, in millions, n decades after 1950.
$a_n = S_n - S_{n-1}$ = growth in US population in millions, during the nth decade after 1950.
$a_n/10$ = Average yearly growth, in millions, during the nth decade after 1950.

37. Expanding both sums, we see

$$\begin{aligned}\sum_{i=1}^{15} i^3 - \sum_{j=3}^{15} j^3 &= (1^3 + 2^3 + 3^3 + 4^3 + \cdots + 15^3) - (3^3 + 4^3 + \cdots + 15^3)\\ &= 1^3 + 2^3 = 9.\end{aligned}$$

41. We have $a_1 = 16$ and $d = 32$. At the end of n seconds, the object has fallen

$$S_n = \frac{1}{2}n(2 \cdot 16 + (n-1)32) = \frac{1}{2}n(32 + 32n - 32) = 16n^2.$$

The height of the object at the end of n seconds is

$$h = 1000 - S_n = 1000 - 16n^2.$$

When the object hits the ground, $h = 0$, so

$$0 = 1000 - 16n^2$$

$$n^2 = \frac{1000}{16} \quad \text{so} \quad n = \pm\sqrt{\frac{1000}{16}} = \pm 7.906 \text{ sec.}$$

Since n is positive, it takes 7.906 seconds, that is, nearly 8 seconds, for the object to hit the ground.

45. **(a)** Since there are 9 terms, we can group them into 4 pairs each totaling 66. The middle or fifth term, $a_5 = 33$, remains unpaired. This means that

$$\text{Sum of series} = 4(66) + 33 = 297.$$

This is the same answer we got by adding directly.

(b) We can use our formula derived for the sum of an even number of terms to add the first 8 terms. We have $a_1 = 5$, $n = 8$, and $d = 7$. This gives

$$\text{Sum of first 8 terms} = \frac{1}{2}(8)(2(5) + (8-1)(7)) = 236.$$

Adding the ninth term gives

$$\text{Sum of series} = 236 + 61 = 297.$$

This is the same answer we got by adding directly.

(c) Using the method from part (a), we add the first and last terms and obtain $a_1 + a_n$. Notice that $a_n = a_1 + (n-1)d$ and so this expression can be rewritten as $a_1 + a_1 + (n-1)d = 2a_1 + (n-1)d$. We then add the second and next to last terms and obtain $a_2 + a_{n-1}$. We know that $a_2 = a_1 + d$ and that $a_{n-1} = a_1 + (n-2)d$, and so this expression can be rewritten as $a_1 + d + a_1(n-2)d = 2a_1 + (n-1)d$. Continuing in this manner, we see that the sum of each pair is $2a_1 + (n-1)d$. The total number of such terms is given by $\frac{1}{2}(n-1)$. For instance, if there are 9 terms, the total number of pairs is $\frac{1}{2}(9-1) = 4$. Thus, the subtotal of these pairs is given by

$$\text{Subtotal of pairs} = \frac{1}{2}(n-1)(2a_1 + (n-1)d).$$

As you can check for yourself, the unpaired (middle) term is given by

$$\text{Unpaired (middle) term} = a_1 + \frac{1}{2}(n-1)d.$$

For instance, in the case of the series given in the question, the unpaired term is given by

$$5 + \frac{1}{2}(9-1)(7) = 5 + 4(7) = 33.$$

Therefore, the sum of the arithmetic series is given by

$$\begin{aligned}\text{Sum} &= \text{Subtotal of pairs} + \text{Unpaired term} \\ &= \underbrace{\frac{1}{2}(n-1)(2a_1 + (n-1)d)}_{\text{Subtotal of pairs}} + \underbrace{a_1 + \frac{1}{2}(n-1)d}_{\text{Unpaired term}}.\end{aligned}$$

One way to simplify this expression is to first factor out 1/2:

$$\text{Sum} = \frac{1}{2}\left[(n-1)(2a_1 + (n-1)d) + 2a_1 + (n-1)d\right].$$

The bracketed part of this expression involves $(n-1)$ terms equaling $2a_1 + (n-1)d$, plus one more such term, for a total number of n such terms. We have

$$\text{Sum} = \frac{1}{2}n(2a_1 + (n-1)d),$$

which is the same as the formula derived in the text.

Using the method from part (b), we see that since n is odd, $(n-1)$ is even, so we can use the formula derived for an even number of terms to add the first $(n-1)$ terms. Substituting $(n-1)$ for n in our formula, we have

$$\text{Sum of first } (n-1) \text{ terms} = \frac{1}{2}(n-1)(2a_1 + (n-2)d).$$

The total sum is given by

$$\text{Sum} = \text{Sum of first } (n-1) \text{ terms} + n^{\text{th}} \text{ term}.$$

Since the nth term is given by $a_1 + (n-1)d$, we have

$$\text{Sum} = \underbrace{\frac{1}{2}(n-1)(2a_1 + (n-2)d)}_{\text{Sum of first } (n-1) \text{ terms}} + \underbrace{a_1 + (n-1)d}_{n^{\text{th}} \text{ term}}.$$

To simplify this expression, we first factor out 1/2, as before:

$$\text{Sum} = \frac{1}{2}\left[(n-1)(2a_1 + (n-2)d) + 2a_1 + 2(n-1)d\right].$$

We can rewrite $2a_1 + 2(n-1)d$ as $2a_1 + 2nd - 2d$, and then as $2a_1 + nd - 2d + nd$, and finally as $2a_1 + (n-2)d + nd$. This gives

$$\text{Sum} = \frac{1}{2}\left[(n-1)(2a_1 + (n-2)d) + 2a_1 + (n-2)d + nd\right].$$

We have $n-1$ terms each equaling $2a_1 + (n-2)d$ plus one more such term plus a term equaling nd. This gives a total of n terms equaling $2a_1 + (n-2)d$ plus the nd term:

$$\begin{aligned}
\text{Sum} &= \frac{1}{2}\left[n(2a_1 + (n-2)d) + nd\right] \\
&= \frac{1}{2}n\left[2a_1 + (n-2)d + d\right] \qquad \text{factoring out } n \\
&= \frac{1}{2}n(2a_1 + (n-1)d),
\end{aligned}$$

which is the same answer as before.

Solutions for Section 13.3

EXERCISES

1. Since

$$\sum_{j=5}^{18} = 3 \cdot 2^5 + 3 \cdot 2^6 + 3 \cdot 2^7 + \cdots + 3 \cdot 2^{18},$$

there are $18 - 4 = 14$ terms in the series. The first term is $a = 3 \cdot 2^5$ and the ratio is $r = 2$, so

$$\text{Sum} = \frac{3 \cdot 2^5(1 - 2^{14})}{1-2} = 1{,}572{,}768.$$

5. This is a geometric series with a first term is $a = 1/125$ and a ratio of $r = 5$. To determine the number of terms in the series, use the formula, $a_n = ar^{n-1}$. Calculating successive terms of the series $(1/125)5^{n-1}$, we get

$$1/125, 1/25, 1/5, 1, 5, 25, 125, 625.$$

Thus we sum the first eight terms of the series:

$$S_8 = \frac{(1/125)(1-5^8)}{1-5} = \frac{97656}{125} = 781.248.$$

9. Yes, $a = 1$, ratio $= -1/2$.

13. These numbers are all power of 3, with signs alternating from positive to negative. We need to change the signs on an alternating basis. By raising -1 to various powers, we can create the pattern shown. One possible answer is: $\sum_{n=1}^{6}(-1)^{n+1}(3^n)$.

17. $\sum_{n=0}^{5} 3/(2^n) = 3 + 3/2 + 3/4 + 3/8 + 3/16 + 3/32 = 3(1 - 1/2^6)/(1 - 1/2)) = 189/32$

PROBLEMS

21. **(a)** Let a_n be worldwide oil consumption n years after 2015, so that the consumption in 2016 was 35.3 billion barrels. Then, $a_1 = 35.3$, $a_2 = 35.3(1.016)$, and $a_n = 35.3(1.016)^{n-1}$. Thus, between 2016 and 2036,

$$\text{Total oil consumption} = \sum_{n=1}^{20} 35.3(1.016)^{n-1} \text{ billion barrels.}$$

(b) Using the formula for the sum of a finite geometric series, we have

$$\text{Total oil consumption} = \frac{35.3\left(1 - (1.016)^{20}\right)}{1 - 1.016} = 824.35 \text{ billion barrels.}$$

25. The answer to Problem 24 is given by

$$B_{20} = \frac{1000(1 - (1.03)^{20})}{1 - 1.03} = 26{,}870.37.$$

(a) Replacing \$1000 by \$2000 doubles the answer, giving \$53,740.75.
(b) Doubling the interest rate to 6% by replacing 1.03 by 1.06 less than doubles the answer, giving \$36,785.60.
(c) Doubling the number of deposits to 40 by replacing $(1.03)^{20}$ by $(1.03)^{40}$ more than doubles the answer, giving \$75,401.26.

Solutions for Section 13.4

EXERCISES

1. Yes, $a = 1$, ratio $= -x$.

5. Yes, $a = e^x$, ratio $= e^x$.

9. Sum $= \dfrac{y^2}{1 - y}$, $|y| < 1$.

13.

$$\sum_{i=4}^{\infty}\left(\frac{1}{3}\right)^i = \left(\frac{1}{3}\right)^4 + \left(\frac{1}{3}\right)^5 + \cdots = \left(\frac{1}{3}\right)^4\left(1 + \frac{1}{3} + \left(\frac{1}{3}\right)^2 + \cdots\right) = \frac{(\frac{1}{3})^4}{1 - \frac{1}{3}} = \frac{1}{54}.$$

17. Since

$$\sum_{i=1}^{\infty} x^{2i} = x^2 + x^4 + x^6 \cdots,$$

the first term is $a = x^2$ and the ratio is $r = x^2$. Since $|x| < 1$,

$$\text{Sum} = \sum_{i=1}^{\infty} x^{2i} = \frac{a}{1 - r} = \frac{x^2}{1 - x^2}.$$

PROBLEMS

21. $0.122222\ldots = 0.1 + \frac{2}{100} + \frac{2}{1000} + \frac{2}{10000} + \frac{2}{100000} + \cdots$. Thus,

$$S = 0.1 + \frac{\frac{2}{100}}{1 - \frac{1}{10}} = \frac{1}{10} + \frac{2}{90} = \frac{11}{90}.$$

25. (a)

$$\begin{aligned}
P_1 &= 0 \\
P_2 &= 250(0.04) \\
P_3 &= 250(0.04) + 250(0.04)^2 \\
P_4 &= 250(0.04) + 250(0.04)^2 + 250(0.04)^3 \\
&\vdots \\
P_n &= 250(0.04) + 250(0.04)^2 + \cdots + 250(0.04)^{n-1}
\end{aligned}$$

(b) Factoring our formula for P_n, we see that it involves a geometric series of $n - 2$ terms:

$$P_n = 250(0.04)\underbrace{\left[1 + 0.04 + (0.04)^2 + \cdots + (0.04)^{n-2}\right]}_{n-2 \text{ terms}}.$$

The sum of this series is given by

$$1 + 0.04 + (0.04)^2 + \cdots + (0.04)^{n-2} = \frac{1 - (0.04)^{n-1}}{1 - 0.04}.$$

Thus,

$$\begin{aligned}
P_n &= 250(0.04)\left(\frac{1 - (0.04)^{n-1}}{1 - 0.04}\right) \\
&= 10\left(\frac{1 - (0.04)^{n-1}}{1 - 0.04}\right).
\end{aligned}$$

(c) In the long run, that is, as $n \to \infty$, we know that $(0.04)^{n-1} \to 0$, and so

$$P_n = 10\left(\frac{1 - (0.04)^{n-1}}{1 - 0.04}\right) \to 10\left(\frac{1 - 0}{1 - 0.04}\right) = 10.417.$$

Thus, P_n gets closer to 10.417 and Q_n gets closer to 260.42. We'd expect these limits to differ because one is right before taking a tablet and one is right after. We'd expect the difference between them to be exactly 250 mg, the amount of ampicillin in one tablet.

29.

$$\begin{aligned}
\text{Present value of first coupon} &= \frac{100}{1.02} \\
\text{Present value of second coupon} &= \frac{100}{(1.02)^2}, \text{etc.}
\end{aligned}$$

$$\begin{aligned}
\text{Total present value} &= \underbrace{\frac{100}{1.02} + \frac{100}{(1.02)^2} + \cdots + \frac{100}{(1.02)^{10}}}_{\text{coupons}} + \underbrace{\frac{10{,}000}{(1.02)^{10}}}_{\text{principal}} \\
&= \frac{100}{1.02}\left(1 + \frac{1}{1.02} + \cdots + \frac{1}{(1.02)^9}\right) + \frac{10{,}000}{(1.02)^{10}} \\
&= \frac{100}{1.02}\left(\frac{1 - \left(\frac{1}{1.02}\right)^{10}}{1 - \frac{1}{1.02}}\right) + \frac{10{,}000}{(1.02)^{10}} \\
&= 898.259 + 8203.483 \\
&= \$9101.74.
\end{aligned}$$

STRENGTHEN YOUR UNDERSTANDING

1. True. $a_1 = (1)^2 + 1 = 2$.

5. True. The differences between successive terms are all 1.

9. True. The first partial sum is just the first term of the sequence.

13. True. The sum is n terms of 3. That is, $3 + 3 + \cdots + 3 = 3n$.

17. False. The terms of the series can be negative so partial sums can decrease.

21. True. If $a = 1$ and $r = -\frac{1}{2}$, it can be written $\sum_{i=0}^{5}(-\frac{1}{2})^i$.

25. False. If payments are made at the end of each year, after 20 years, the balance at 1% is about \$44,000, while at 2% it would be about \$48,600. If payments are made at the start of each year, the corresponding figures are \$44,500 and \$49,600.

29. False. The series does not converge since the odd terms (Q_1, Q_3, etc.) are all -1.

33. False. An arithmetic series with $d \neq 0$ diverges.

CHAPTER FOURTEEN

Solutions for Section 14.1

EXERCISES

1. We use a parameter t so that when $t = 0$ we have $x = 1$ and $y = 3$. One possible parameterization is $x = 1 + 2t$, $y = 3 + t$, $0 \le t \le 1$.

5. True. Eliminating t gives $x = 3(y - 3)$. Thus we have the straight line $y = 3 + \frac{x}{3}$. We must also check the end points $t = 0$ and $t = 1$. When $t = 0$ we have $x = 0$ and $y = 3$, and when $t = 1$ we have $x = 3$ and $y = 3 + 1 = 4$.

9. The graph of the parametric equations is in Figure 14.1.

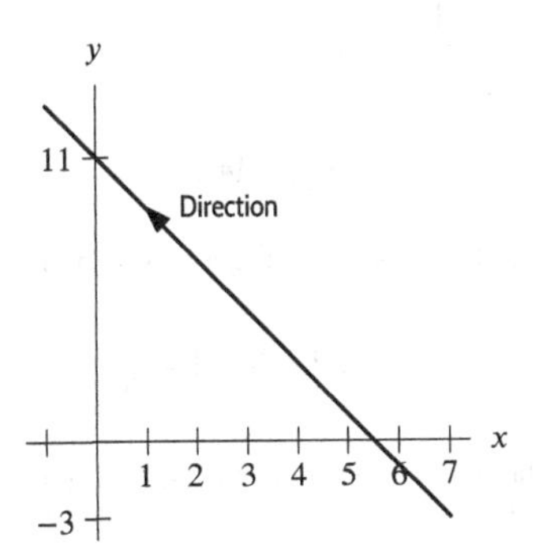

Figure 14.1

It is given that $x = 5 - 2t$, thus $t = (5 - x)/2$. Substitute this into the second equation:

$$y = 1 + 4t$$
$$y = 1 + 4\frac{(5 - x)}{2}$$
$$y = 11 - 2x.$$

13. The graph of the parametric equations is in Figure 14.2.

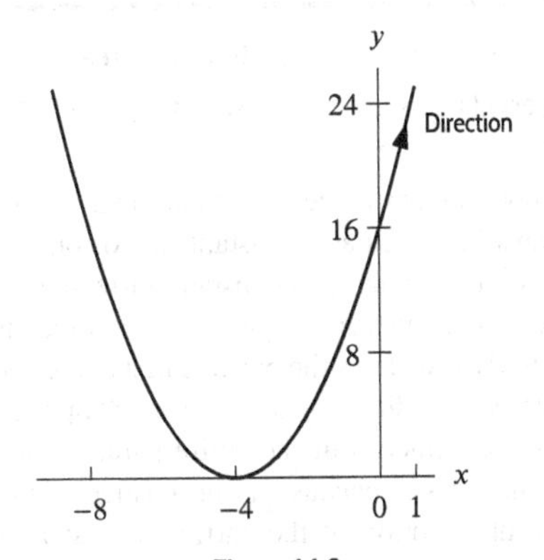

Figure 14.2

Since $x = t - 3$, we have $t = x + 3$. Substitute this into the second equation:

$$\begin{aligned} y &= t^2 + 2t + 1 \\ y &= (x+3)^2 + 2(x+3) + 1 \\ &= x^2 + 8x + 16 = (x+4)^2. \end{aligned}$$

17. The graph of the parametric equations is in Figure 14.3.

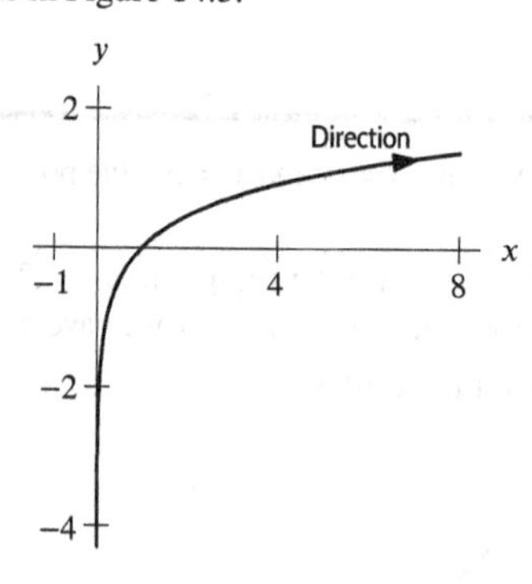

Figure 14.3

Since $x = t^3$, we take the natural log of both sides and get $\ln x = 3\ln t$ or $\ln t = 1/3 \ln x$. We are given that $y = 2\ln t$; thus,

$$y = 2\left(\frac{1}{3}\ln x\right) = \frac{2}{3}\ln x.$$

21. This is like Example 5 on page 14-8 of the text, except that the x-coordinate goes all the way to 2 and back. So the particle traces out the rectangle shown in Figure 14.4.

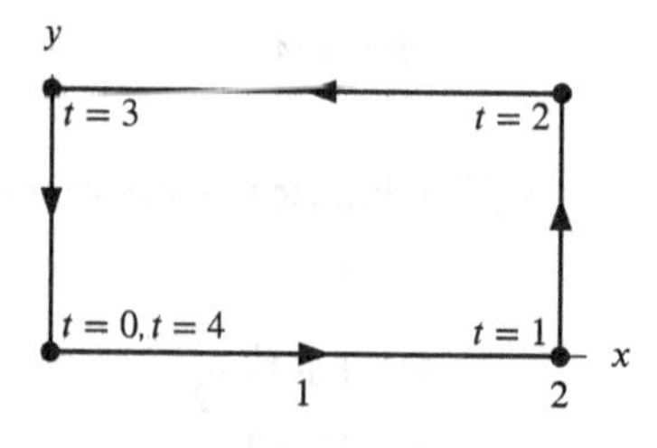

Figure 14.4

PROBLEMS

25. The particle moves clockwise: For $0 \le t \le \frac{\pi}{2}$, starting at $(1, 0)$ as t increases, we have $x = \cos t$ decreasing and $y = -\sin t$ decreasing. Similarly, for the time intervals $\frac{\pi}{2} \le t \le \pi, \pi \le t \le \frac{3\pi}{2}$, and $\frac{3\pi}{2} \le t \le 2\pi$, we see that the particle moves clockwise. The same is true for all $-\infty < t < +\infty$.

29. In all three cases, $y = x^2$, so that the motion takes place on the parabola $y = x^2$.

In case (a), the x-coordinate always increases at a constant rate of one unit distance per unit time, so the equations describe a particle moving to the right on the parabola at constant horizontal speed.

In case (b), the x-coordinate is never negative, so the particle is confined to the right half of the parabola. As t moves from $-\infty$ to $+\infty$, $x = t^2$ goes from ∞ to 0 to ∞. Thus the particle first comes down the right half of the parabola, reaching the origin $(0, 0)$ at time $t = 0$, where it reverses direction and goes back up the right half of the parabola.

In case (c), as in case (a), the particle traces out the entire parabola $y = x^2$ from left to right. The difference is that the horizontal speed is not constant. This is because a unit change in t causes larger and larger changes in $x = t^3$ as t approaches $-\infty$ or ∞. The horizontal motion of the particle is faster when it is farther from the origin.

33. For $0 \leq t \leq 2\pi$, the graph is in Figure 14.5.

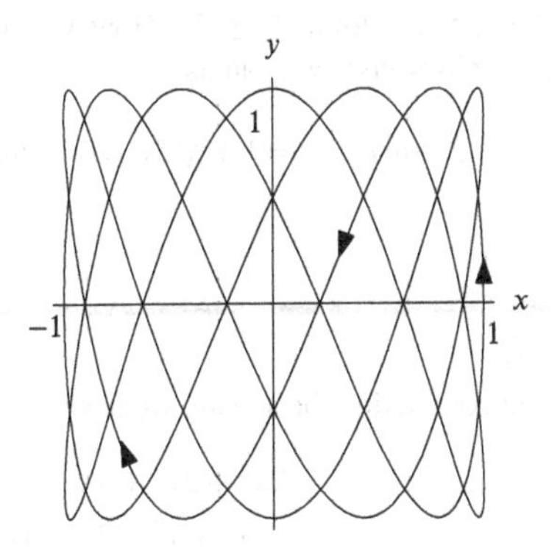

Figure 14.5

37. **(a)** Since the x-coordinate and the y-coordinate are always the same (they both equal t) , the bug follows the path $y = x$.

(b) The bug starts at $(1, 0)$ because $\cos 0 = 1$ and $\sin 0 = 0$. Since the x-coordinate is $\cos x$, and the y-coordinate is $\sin x$, the bug follows the path of a unit circle, traveling counterclockwise. It reaches the starting point of $(1, 0)$ when $t = 2\pi$, because $\sin t$ and $\cos t$ are periodic with period 2π.

(c) Now the x-coordinate varies from 1 to -1, while the y-coordinate varies from 2 to -2; otherwise, this is much like part (b) above. If we plot several points, the path looks like an ellipse, which is a circle stretched out in one direction.

41. The particle moves back and forth between -1 and 1. See Figure 14.6.

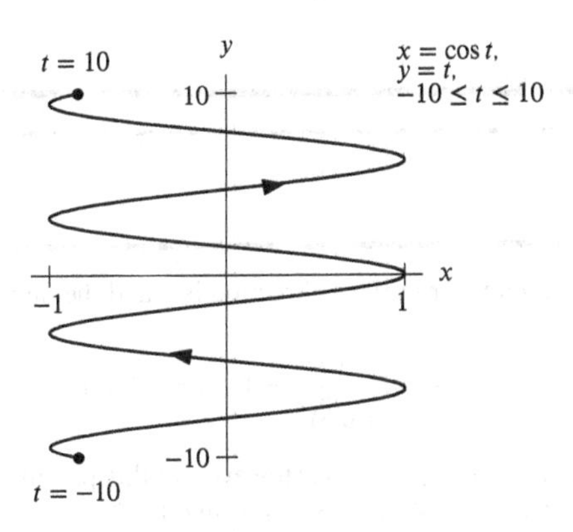

Figure 14.6

Solutions for Section 14.2

EXERCISES

1. Explicit. For each x we can write down the corresponding values of y.

5. Implicit. This is a quadratic in y so solving for y by completing the square, we obtain $y = 1 + \sqrt{2 + x}$ and $y = 1 - \sqrt{2 + x}$.

9. Multiplying by $(y - 4)^2$ gives

$$4 - (x - 4)^2 - (y - 4)^2 = 0$$
$$(x - 4)^2 + (y - 4)^2 = 4.$$

Thus, the center is $(4, 4)$, and the radius is 2.

13. We can use $x = 5\sin t$, $y = -5\cos t$ for $0 \le t \le 2\pi$.

17. We consider $x = -2 + \sqrt{5}\sin t$, $y = 1 + \sqrt{5}\cos t$ for $0 \le t \le 2\pi$. However, as t increases from 0, the value of x increases, so this starts at the correct point but goes clockwise. We can use

$$x = -2 - \sqrt{5}\sin t, \quad y = 1 + \sqrt{5}\cos t, \quad \text{for } 0 \le t \le 2\pi.$$

PROBLEMS

21. **(a)** Center is $(2, -4)$ and radius is $\sqrt{20}$.

(b) Rewriting the original equation and completing the square, we have

$$\begin{aligned} 2x^2 + 2y^2 + 4x - 8y &= 12 \\ x^2 + y^2 + 2x - 4y &= 6 \\ (x^2 + 2x + 1) + (y^2 - 4y + 4) - 5 &= 6 \\ (x+1)^2 + (y-2)^2 &= 11. \end{aligned}$$

So the center is $(-1, 2)$, and the radius is $\sqrt{11}$.

25. Since $y - 3 = \sin t$ and $x = 4\sin^2 t$, this parameterization traces out the parabola $x = 4(y-3)^2$ for $2 \le y \le 4$.

29. Explicit: $y = \sqrt{4 - x^2}$
Implicit: $y^2 = 4 - x^2$ or $x^2 + y^2 = 4$, $y > 0$
Parametric: $x = 4\cos t$, $y = 4\sin t$, with $0 \le t \le \pi$.

Solutions for Section 14.3

EXERCISES

1. **(a)** The center is at the origin. The diameter in the x-direction is 4 and the diameter in the y-direction is $2\sqrt{5}$.

(b) The equation of the ellipse is

$$\frac{x^2}{2^2} + \frac{y^2}{(\sqrt{5})^2} = 1 \quad \text{or} \quad \frac{x^2}{4} + \frac{y^2}{5} = 1.$$

5. We consider $x = -2\cos t$, $y = 5\sin t$, for $0 \le t \le 2\pi$. However, in this parameterization, y increases as t increases from 0, so it traces clockwise. Thus, we take $x = -2\cos t$, $y = -5\sin t$, for $0 \le t \le 2\pi$.

9. The fact that the parameter is called s, not t, makes no difference. The minus sign means that the ellipse is traced out in the opposite direction. The graph of the ellipse in the xy-plane is the same as the ellipse in the example, and it is traced out once as s increases from 0 to 2π.

PROBLEMS

13. Completing the square on $x^2 + 4x$ and $y^2 + 10y$:

$$\begin{aligned} \frac{1}{9}(x^2 + 4x) + \frac{1}{25}(y^2 + 10y) &= -\frac{4}{9} \\ \frac{1}{9}((x+2)^2 - 4) + \frac{1}{25}((y+5)^2 - 25) &= -\frac{4}{9} \\ \frac{1}{9}(x+2)^2 - \frac{4}{9} + \frac{1}{25}(y+5)^2 - 1 + \frac{13}{4} &= -\frac{4}{9} \\ \frac{(x+2)^2}{9} + \frac{(y+5)^2}{25} &= 1. \end{aligned}$$

The center is $(-2, -5)$, and $a = 3$, $b = 5$.

17. Factoring out the 9 from $9x^2 - 54x = 9(x^2 - 6x)$ and the 4 from $4y^2 - 16y = 4(y^2 - 4)$ and completing the square on $x^2 - 6x$ and $y^2 - 4y$:

$$\begin{aligned} 9(x^2 - 6x) + 4(y^2 - 4y) - 61 &= 0 \\ 9((x-3)^2 - 9) + 4((y-2)^2 - 4) + 61 &= 0 \\ 9(x-3)^2 - 81 + 4(y-2)^2 - 16 + 61 &= 0 \\ 9(x-3)^2 + 4(y-2)^2 - 36 &= 0. \end{aligned}$$

Moving the -36 to the right and dividing by 36 to get 1 on the right side,

$$\begin{aligned} \frac{9(x-3)^2}{36} + \frac{4(y-2)^2}{36} &= \frac{36}{36} \\ \frac{(x-3)^2}{4} + \frac{(y-2)^2}{9} &= 1. \end{aligned}$$

The center is $(3, 2)$, and $a = 2$, $b = 3$.

21. We have

$$\begin{aligned} Ax^2 - Bx + y^2 &= r_0^2 \\ A\left(x^2 - \frac{B}{A}x\right) + y^2 &= r_0^2 \\ A\left(x^2 - \frac{B}{A}x + \left(\frac{B}{2A}\right)^2\right) + y^2 &= r_0^2 + \frac{B^2}{4A} && \text{completing the square} \\ A\left(x - \frac{B}{2A}\right)^2 + y^2 &= \frac{4Ar_0^2 + B^2}{4A}. \end{aligned}$$

Dividing both sides by $(4Ar_0^2 + B^2)/(4A)$, we obtain

$$\frac{(x - B/2A)^2}{a^2} + \frac{y^2}{b^2} = 1,$$

where $a^2 = (4Ar_0^2 + B^2)/(4A^2)$ and $b^2 = (4Ar_0^2 + B^2)/(4A)$.

Solutions for Section 14.4

EXERCISES

1. **(a)** The vertices are at $(0, 7)$ and $(0, -7)$. The center is at the origin.

(b) The asymptotes have slopes $7/2$ and $-7/2$. The equations of the asymptotes are

$$y = \frac{7}{2}x \quad \text{and} \quad y = -\frac{7}{2}x.$$

(c) The equation of the hyperbola is

$$\frac{y^2}{7^2} - \frac{x^2}{2^2} = 1 \quad \text{or} \quad \frac{y^2}{49} - \frac{x^2}{4} = 1.$$

5. The hyperbola is centered at the origin, and $a = 2$, $b = 7$. We can use $x = 2\tan t$, $y = 7\sec t = 7/\cos t$.
If $0 < t < \pi/2$, then $x > 0$, $y > 0$, so we have Quadrant I.
If $\pi/2 < t < \pi$, then $x < 0$, $y < 0$, so we have Quadrant III.
If $\pi < t < 3\pi/2$, then $x > 0$, $y < 0$, so we have Quadrant IV.
If $3\pi/2 < t < 2\pi$, then $x < 0$, $y > 0$, so we have Quadrant II.
So the upper half is given by $0 \le t < \pi/2$ together with $3\pi/2 < t < 2\pi$.

PROBLEMS

9. Factoring out -1 from $-y^2+4y=-(y^2-4y)$ and completing the square on x^2-2x and y^2-4y gives

$$\begin{aligned}\frac{1}{4}(x^2-2x)-(y^2-4y)&=\frac{19}{4}\\ \frac{1}{4}((x-1)^2-1)-((y-2)^2-4)&=\frac{19}{4}\\ \frac{(x-1)^2}{4}-\frac{1}{4}-(y-2)^2+4&=\frac{19}{4}\\ \frac{(x-1)^2}{4}-(y-2)^2&=1.\end{aligned}$$

The center is $(1,2)$, the hyperbola opens right-left, and $a=2$, $b=1$.

13. Factoring out 4 from $4x^2-8x=4(x^2-2x)$ and 36 from $36y^2-36y=36(y^2-y)$ and completing the square on x^2-2x and y^2-y gives

$$\begin{aligned}4(x^2-2x)&=36(y^2-y)-31\\ 4((x-1)^2-1)&=36\left(\left(y-\frac{1}{2}\right)^2-\frac{1}{4}\right)-31\\ 4(x-1)^2-4&=36\left(y-\frac{1}{2}\right)^2-9-31\\ 4(x-1)^2&=36\left(y-\frac{1}{2}\right)^2-36.\end{aligned}$$

Moving $36(y-\frac{1}{2})^2$ to the left and dividing by -36 to get 1 on the right:

$$\begin{aligned}\frac{4(x-1)^2}{-36}-\frac{36\left(y-\frac{1}{2}\right)^2}{-36}&=-\frac{36}{-36}\\ -\frac{(x-1)^2}{9}+\left(y-\frac{1}{2}\right)^2&=1\\ \left(y-\frac{1}{2}\right)^2-\frac{(x-1)^2}{9}&=1.\end{aligned}$$

The center is $(1,\frac{1}{2})$, the hyperbola opens up-down, and $a=3$, $b=1$.

17. The center is $(-5,2)$, vertices are $(-5+\sqrt{6},2)$ and $(-5-\sqrt{6},2)$. The asymptotes are $y=\pm\frac{2}{\sqrt{6}}(x+5)+2$. Figure 14.7 shows the hyperbola.

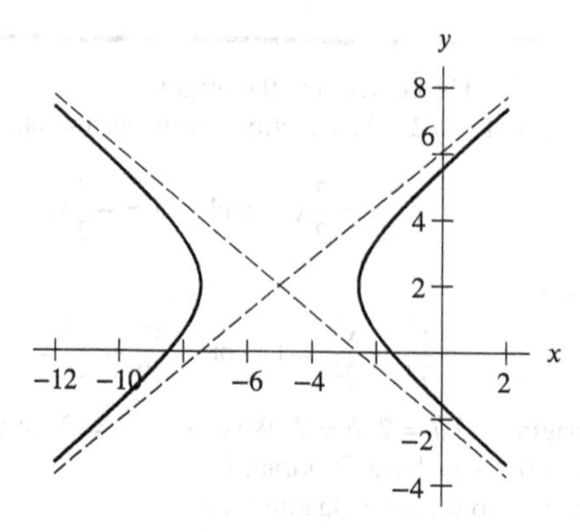

Figure 14.7

Solutions for Section 14.5

EXERCISES

1. This is the standard form of an ellipse. Since $a > b$, the major axis is the x-axis.

5. Subtract the x^2 term from both sides to get the standard form of a hyperbola. Since the y^2 term is positive, the vertices are on the y-axis.

9. The two focal points lie on the major axis, and are equidistant from the center. Therefore, the major axis is vertical and the other focal point is at $(0, -2)$.

13. In the form

$$\frac{x^2}{a^2} - \frac{y^2}{b^2} = 1$$

we see that $a^2 = b^2 = 1$. The focal points lie on the same axis as the positive squared term. Therefore the focal points are at $(\pm c, 0)$, where $c = \sqrt{a^2 + b^2} = \sqrt{1+1} = \sqrt{2}$. The two focal points are $(\sqrt{2}, 0)$ and $(-\sqrt{2}, 0)$.

PROBLEMS

17. The reflective properties of a parabola are somewhat helpful, because the transmission can be focused to go in only one direction. However, the location of the transmission will be perfectly clear to anyone noticing the transmission. A series of elliptical mirrors would be much more complicated and expensive to build, and by following the signal, someone could figure out where the transmission was coming from without your being aware, because surveying a series of mirrors is complicated. A hyperbolic reflection allows you to survey a small mirror near your camp while at the same time making it appear that your camp is somewhere else (at the other focus of the hyperbola). So you should follow the mess sergeant's advice.

21. From the figure, we see the vertex has coordinates $(1, 0)$ and the equation must be of the form

$$x = ay^2 + 1$$

The distance, c from the vertex to the focus is 2. Using $c = 1/(4a)$, we solve for a in $2 = 1/(4a)$ and find $a = 1/8$. The equation is $x = (1/8)y^2 + 1$. The directrix is the line $x = -1$, since it is 2 units to the left of the vertex.

25. Rewrite the formula as

$$(x^2 + 4x) + 2(y^2 - 6y) = 3.$$

Complete the square of both the x and y expressions,

$$(x^2 + 4x + 4) + 2(y^2 - 6y + 9) = 3 + 4 + 2(9),$$

and simplify to

$$(x+2)^2 + 2(y-3)^2 = 25$$

or

$$\frac{(x+2)^2}{25} + \frac{(y-3)^2}{(25/2)} = 1$$

This is an ellipse with center at $(-2, 3)$, and $a = 5$ and $b = \sqrt{25/2}$. Since $a > b$, the distance c, from the center to a focus point is found from

$$c = \sqrt{a^2 - b^2} = \sqrt{25 - 25/2} = \sqrt{25/2} = 5/\sqrt{2}.$$

With $c = 5/\sqrt{2}$ the focal points are at $(-2 + 5/\sqrt{2}, 3)$ and $(-2 - 5/\sqrt{2}, 3)$.

29. First rewrite the equation in the form,

$$\frac{x^2}{8} - \frac{y^2}{16} = 1.$$

Because the x^2 term is positive and the y^2 term is negative this hyperbola has vertices on a horizontal line through its center. The center is at the origin. The length from the center to the vertex is the distance $a = 2\sqrt{2}$ which is read from the equation since $a^2 = 8$. Thus the vertices are at the points $(2\sqrt{2}, 0)$ and $(-2\sqrt{2}, 0)$. Since $b^2 = 16$ and $a^2 = 8$, we find $c^2 = a^2 + b^2 = 24$, and $c = \sqrt{24}$. The focal points are $(\sqrt{24}, 0)$ and $(-\sqrt{24}, 0)$.

33. After passing through the second focal point it is reflected back to the original focal point.

37. For a cross-section of the dish centered at the origin, the equation of the parabola is $y = ax^2$. We know that the rim of the dish is the point $(12, 4)$, thus $4 = a(12)^2$ and solving for a we find $a = 1/36$. Using the equation $c = 1/(4a)$, we find

$$c = \frac{1}{4(1/36)} = 9.$$

The end of the arm is placed 9 inches above the center of the dish to be the focus of the incoming signal.

41. **(a)** The major axis has length $88 + 5250 = 5338$ million km.

(b) Since $2a = 5338$, we find $a = 2669$. Let $(\pm c, 0)$ be the two focal points. Since $a = 2669 = c + 88$, we find $c = 2581$. To find b we use the formula $c^2 = a^2 - b^2$ and find $b^2 = 2669^2 - 2581^2 = 462{,}000$, so $b = 680$.

Shifting the axis so the sun is at the origin, we have

$$\frac{(x - 2581)^2}{2669^2} + \frac{y^2}{680^2} = 1.$$

(c) Use the parametric equations of an ellipse centered at (h, k),

$$x = h + a\cos t, \quad y = k + b\sin t, \quad 0 \le t \le 2\pi$$

to obtain

$$x = 2581 + 2669\cos t, \quad y = 680\sin t, \quad 0 \le t \le 2\pi.$$

45. Consider the set of points (x, y) so that the difference of the distances to two focal points $(\pm c, 0)$ is constant. Note that the x-intercepts $(\pm a, 0)$ satisfy this condition.

Using the distance formula, we see that the distance from (x, y) to $(c, 0)$ is $\sqrt{(x-c)^2 + y^2}$ and the distance from (x, y) to $(-c, 0)$ is $\sqrt{(x+c)^2 + y^2}$. The distance from $(a, 0)$ to $(-c, 0)$ is $c + a$ and the distance from $(a, 0)$ to $(c, 0)$ is $c - a$ since $c > a$. We have:

$$\begin{aligned}
\text{Difference of distances from } (x, y) \text{ to focal points} &= \text{Difference of distances from } (a, 0) \text{ to focal points} \\
\sqrt{(x-c)^2 + y^2} - \sqrt{(x+c)^2 + y^2} &= (c + a) - (c - a) \\
\sqrt{(x-c)^2 + y^2} - \sqrt{(x+c)^2 + y^2} &= 2a \\
\sqrt{(x-c)^2 + y^2} &= 2a + \sqrt{(x+c)^2 + y^2}.
\end{aligned}$$

We square both sides and simplify:

$$\begin{aligned}
(x-c)^2 + y^2 &= 4a^2 + 4a\sqrt{(x+c)^2 + y^2} + ((x+c)^2 + y^2) \\
x^2 - 2cx + c^2 + y^2 &= 4a^2 + 4a\sqrt{(x+c)^2 + y^2} + x^2 + 2cx + c^2 + y^2.
\end{aligned}$$

We cancel x^2, c^2, and y^2 from both sides, and subtract $2cx$ from both sides to obtain:

$$-4cx = 4a^2 + 4a\sqrt{(x+c)^2 + y^2}.$$

We divide through by 4:

$$-cx = a^2 + a\sqrt{(x+c)^2 + y^2},$$

and then isolate the square root:

$$-a\sqrt{(x+c)^2 + y^2} = a^2 + cx.$$

We square both sides again to obtain:

$$\begin{aligned}
a^2((x+c)^2 + y^2) &= a^4 + 2cxa^2 + c^2x^2 \\
a^2(x^2 + 2cx + c^2 + y^2) &= a^4 + 2cxa^2 + c^2x^2 \\
a^2x^2 + 2cxa^2 + a^2c^2 + a^2y^2 &= a^4 + 2cxa^2 + c^2x^2 \\
a^2x^2 + a^2c^2 + a^2y^2 &= a^4 + c^2x^2.
\end{aligned}$$

Since $c > a$, there is a positive number b such that $c^2 = a^2 + b^2$:

$$\begin{aligned} a^2x^2 + a^2(a^2+b^2) + a^2y^2 &= a^4 + (a^2+b^2)x^2 \\ a^2x^2 + a^4 + a^2b^2 + a^2y^2 &= a^4 + a^2x^2 + b^2x^2 \\ a^2b^2 + a^2y^2 &= b^2x^2 \\ b^2x^2 - a^2y^2 &= a^2b^2. \end{aligned}$$

Dividing through by a^2b^2, we arrive at the equation for a hyperbola given in Section 14.3:

$$\begin{aligned} \frac{b^2x^2}{a^2b^2} - \frac{a^2y^2}{a^2b^2} &= \frac{a^2b^2}{a^2b^2} \\ \frac{x^2}{a^2} - \frac{y^2}{b^2} &= 1. \end{aligned}$$

This is the equation for a hyperbola.

Solutions for Section 14.6

EXERCISES

1. We use $x = \sinh t$, $y = \cosh t$, for $-\infty < t < \infty$.

5. We use $x = 1 + 2\sinh t$, $y = -1 - 3\cosh t$, for $-\infty < t < \infty$.

9. Divide by 36 to rewrite the equation as

$$\frac{(y+3)^2}{4} - \frac{(x+1)^2}{9} = 1, \quad y > -3,$$

and use $x = -1 + 3\sinh t$, $y = -3 + 2\cosh t$, for $-\infty < t < \infty$.

13. The graph of $\sinh x$ in the text suggests that

$$\begin{aligned} &\text{As } x \to \infty, && \sinh x \to \tfrac{1}{2}e^x \\ &\text{As } x \to -\infty, && \sinh x \to -\tfrac{1}{2}e^{-x}. \end{aligned}$$

Using the facts that

$$\begin{aligned} &\text{As } x \to \infty, && e^{-x} \to 0, \\ &\text{As } x \to -\infty, && e^{x} \to 0, \end{aligned}$$

we can predict the same results algebraically:

$$\begin{aligned} &\text{As } x \to \infty, && \sinh x = \tfrac{e^x - e^{-x}}{2} \to \tfrac{1}{2}e^x \\ &\text{As } x \to -\infty, && \sinh x = \tfrac{e^x - e^{-x}}{2} \to -\tfrac{1}{2}e^{-x}. \end{aligned}$$

PROBLEMS

17. Factoring out 2 from $2x^2 - 12x = 2(x^2 - 6x)$ and 4 from $4y^2 + 4y = 4(y^2 + y)$ and completing the square on $x^2 - 6x$ and $y^2 + y$ gives

$$\begin{aligned} 25 + 2(x^2 - 6x) &= 4(y^2 + y), \quad y > -\frac{1}{2} \\ 25 + 2((x-3)^2 - 9) &= 4\left(\left(y + \frac{1}{2}\right)^2 - \frac{1}{4}\right), \quad y > -\frac{1}{2} \\ 25 + 2(x-3)^2 - 18 &= 4\left(y + \frac{1}{2}\right)^2 - 1, \quad y > -\frac{1}{2} \\ 2(x-3)^2 &= 4\left(y + \frac{1}{2}\right)^2 - 8, \quad y > -\frac{1}{2}. \end{aligned}$$

Then, moving $4(y+\frac{1}{2})^2$ to the left and dividing by -8 to get 1 on the right,

$$\frac{2(x-3)^2}{-8} - \frac{4\left(y+\frac{1}{2}\right)^2}{-8} = 1, \quad y > -\frac{1}{2}$$

$$\frac{\left(y+\frac{1}{2}\right)^2}{2} - \frac{(x-3)^2}{4} = 1, \quad y > -\frac{1}{2},$$

so we use $x = 3 + 2\sinh t$, $y = -\frac{1}{2} + \sqrt{2}\cosh t$ for $-\infty < t < \infty$.

21. Yes. First, we observe that

$$\cosh 2x = \frac{e^{2x} + e^{-2x}}{2}.$$

Now, using the fact that $e^x \cdot e^{-x} = 1$, we calculate

$$\begin{aligned}\cosh^2 x &= \left(\frac{e^x + e^{-x}}{2}\right)^2 \\ &= \frac{(e^x)^2 + 2e^x \cdot e^{-x} + (e^{-x})^2}{4} \\ &= \frac{e^{2x} + 2 + e^{-2x}}{4}.\end{aligned}$$

Similarly, we have

$$\begin{aligned}\sinh^2 x &= \left(\frac{e^x - e^{-x}}{2}\right)^2 \\ &= \frac{(e^x)^2 - 2e^x \cdot e^{-x} + (e^{-x})^2}{4} \\ &= \frac{e^{2x} - 2 + e^{-2x}}{4}.\end{aligned}$$

Thus, to obtain $\cosh 2x$, we need to add (rather than subtract) $\cosh^2 x$ and $\sinh^2 x$, giving

$$\begin{aligned}\cosh^2 x + \sinh^2 x &= \frac{e^{2x} + 2 + e^{-2x} + e^{2x} - 2 + e^{-2x}}{4} \\ &= \frac{2e^{2x} + 2e^{-2x}}{4} \\ &= \frac{e^{2x} + e^{-2x}}{2} \\ &= \cosh 2x.\end{aligned}$$

Thus, we see that the identity relating $\cosh 2x$ to $\cosh x$ and $\sinh x$ is

$$\cosh 2x = \cosh^2 x + \sinh^2 x.$$

25. We know that $\sinh(iz) = i\sin z$, where z is real. Substituting $z = ix$, where x is real so z is imaginary, we have

$$\begin{aligned}\sinh(iz) &= i\sin z \\ \sinh(i \cdot ix) &= i\sin(ix) \qquad \text{substituting } z = ix \\ \sinh(-x) &= i\sin(ix).\end{aligned}$$

But $\sinh(-x) = -\sinh(x)$; thus we have

$$-\sinh x = i\sin(ix).$$

Multiplying both sides by i gives

$$-i\sinh x = -1\sin(ix).$$

Thus,

$$i\sinh x = \sin(ix).$$

STRENGTHEN YOUR UNDERSTANDING

1. True. The path is on the line $3x - 2y = 0$.

5. False. Starting at $t = 0$ the object is at $(0, 1)$. For $0 \le t \le \pi$, $x = \sin(t/2)$ increases and $y = \cos(t/2)$ decreases, thus the motion is clockwise.

9. False. The standard form of the circle is $(x+4)^2 + (y+5)^2 = 141$. The center is $(-4, -5)$.

13. False. There are many parameterizations; $x = \cos t$, $y = \sin t$ and $x = \sin t$, $y = \cos t$ are two of them.

17. True. Since

$$\frac{x^2}{4} + y^2 = \frac{(2\cos t)^2}{4} + (\sin t)^2 = \cos^2 t + \sin^2 t = 1,$$

these equations parameterize the ellipse $(x^2/4) + y^2 = 1$.

21. False. It is defined as $\sinh x = \dfrac{e^x - e^{-x}}{2}$.

25. True. We have $\cosh(-x) = (e^{-x} + e^{-(-x)})/2 = (e^x + e^{-x})/2 = \cosh x$, so the hyperbolic cosine is an even function.

29. False. We have

$$\sinh \pi = \frac{e^{\pi}}{2} - \frac{e^{-\pi}}{2} = 11.549 \ne 0.$$

We do have $\sin \pi = 0$.

33. False. The two asymptotes are entirely in the region between the two branches of the hyperbola. The two focal points are outside this region. To move from an asymptote to a focal point you must cross the hyperbola.